T0199301

"A vital and vibrant contribution to the emerging area of sound studies! Held together with the thread of how we play with sound in art, web, games and other interactive media, *Foundations in Sound Design for Interactive Media* combines theory and practice to explore the state of the art in interactive sound and music. With authors drawing on areas of psychology, philosophy, film studies, media theories, computer science and sound studies, the book invites the reader to explore sound as an increasingly important element in our interactive media. As we become increasingly immersed in our new interactive technological world, this book will serve as a guide for future researchers and practitioners interested in the contribution sound makes to our everyday experiences. Thought provoking and informative!"

—Karen Collins, Associate Professor and Canada Research
Chair in Interactive Audio, University of Waterloo

"A volume full of state-of-the-art information on sound design in the interactive media era. Excellently balanced between engineering and creative thinking, it is a must-read reference for the modern sound designer and artist."

—Dr. Andreas Floros, Professor, Audio Engineering
and Technology, Ionian University, Greece

"A fundamental volume that covers all relevant production techniques, exemplified by specific works—ranging from cinematic aesthetics to game sound design, from spatialized sound, online sound, sound effects, to earcons, from soundscape design to sound installations. A truly indispensable manual to teach and to learn the design of interactive sound."

—Holger Schulze, Author of *The Sonic Persona* (2018)
and *Sound Works* (2019)

"This is an important collection for anyone thinking about or engaging with the field of interactive media, from sound and music in video games to installations and locative media. Each chapter provides a useful summary of current thought and goes beyond this to raise intriguing new questions. It's going straight onto my 'essential' reading list."

—Richard Stevens, Course Director, Leeds Beckett University
and Co-Author of *Game Audio Implementation* (Routledge)

Foundations in Sound Design for Interactive Media

This volume provides a comprehensive introduction to foundational topics in sound design for interactive media, such as gaming and virtual reality; compositional techniques; new interfaces; sound spatialization; sonic cues and semiotics; performance and installations; music on the web; augmented reality applications; and sound-producing software design.

The reader will gain a broad understanding of the key concepts and practices that define sound design for its use in computational media and design. The chapters are written by international authors from diverse backgrounds who provide multidisciplinary perspectives on sound in its interactive forms.

The volume is designed as a textbook for students and teachers, as a handbook for researchers in sound, design and media, and as a survey of key trends and ideas for practitioners interested in exploring the boundaries of their profession.

Michael Filimowicz, PhD, is Senior Lecturer in the School of Interactive Arts and Technology (SIAT) at Simon Fraser University and co-editor of *The Soundtrack* journal. He develops new forms of general-purpose multimodal and audiovisual display technology, exploring novel product lines across different application contexts including gaming, immersive exhibitions, control rooms, telepresence and simulation-based training. He has published across disciplines in journals such as *Organised Sound, Arts and Humanities in Higher Education, Leonardo, Sound Effects, Parsons Journal for Information Mapping* and *Semiotica*. His art has been exhibited internationally at venues such as SIGGRAPH, Re-New, Design Shanghai, ARTECH, Les Instants Vidéo, IDEAS, Kinsey Institute and Art Currents, and published in monographs such as *Spotlight: 20 Years of the Biel/Bienne Festival of Photography*, *Reframing Photography* and *Infinite Instances*. His personal website is http://filimowi.cz.

Sound Design

The Sound Design series takes a comprehensive and multidisciplinary view of the field of sound design across linear, interactive and embedded media and design contexts. Today's sound designers might work in film and video, installation and performance, auditory displays and interface design, electroacoustic composition and software applications, and beyond. These forms and practices continuously cross-pollinate and produce an ever-changing array of technologies and techniques for audiences and users, which the series aims to represent and foster.

Series Editor
Michael Filimowicz

Titles in the Series

Foundations in Sound Design for Linear Media

A Multidisciplinary Approach

Foundations in Sound Design for Interactive Media

A Multidisciplinary Approach

Foundations in Sound Design for Embedded Media

A Multidisciplinary Approach

For more information about this series, please visit: www.routledge.com/Sound-Design/book-series/SDS

Foundations in Sound Design for Interactive Media

A Multidisciplinary Approach

Edited by Michael Filimowicz

Routledge
Taylor & Francis Group

NEW YORK AND LONDON

First published 2020
by Routledge
52 Vanderbilt Avenue, New York, NY 10017

and by Routledge
2 Park Square, Milton Park, Abingdon, Oxon, OX14 4RN

Routledge is an imprint of the Taylor & Francis Group, an informa business

© 2020 Taylor & Francis

Library of Congress Cataloging-in-Publication Data
Names: Filimowicz, Michael, editor.
Title: Foundations in sound design for interactive media :
   a multidisciplinary approach / [edited by] Michael Filimowicz.
Description: New York, NY : Routledge, 2019. | Series: Sound design series;
   volume 2 | Includes bibliographical references and index.
Identifiers: LCCN 2019004476 (print) | LCCN 2019018540 (ebook) |
   ISBN 9781315106342 (Master e-book) | ISBN 9781351603867 (pdf) |
   ISBN 9781351603843 (Mobi) | ISBN 9781351603850 (ePub3) |
   ISBN 9781138093935 (hbk : alk. paper) | ISBN 9781138093942
   (pbk : alk. paper) | ISBN 9781315106342 (ebk)
Subjects: LCSH: Sound in virtual reality. | Sound—Recording
   and reproducing—Digital techniques.
Classification: LCC TK7881.94 (ebook) | LCC TK7881.94.F68 2019 (print) |
   DDC 006.5—dc23
LC record available at https://lccn.loc.gov/2019004476

ISBN: 978-1-138-09393-5 (hbk)
ISBN: 978-1-138-09394-2 (pbk)
ISBN: 978-1-315-10634-2 (ebk)

Typeset in Times New Roman
by Apex CoVantage, LLC

Contents

Contributors

Enda Bates is a composer, musician, producer and academic based in Dublin, Ireland. He is an Assistant Professor and deputy director of the Music and Media Technologies program in Trinity College Dublin. In 2010 he completed a PhD entitled The Composition and Performance of Spatial Music. Research interests include spatial music, spatial audio for VR, AR and 360 Video, the aesthetics of electroacoustic music, and the development of music software and hardware such as the Augmented Electric Guitar. He is an active composer of acoustic and electroacoustic contemporary music and scores and other material are available from his page at the Irish Contemporary Music Centre. He is also active as a performer, both of his own music and with groups such as the Irish folk-rock band the Spook of the Thirteenth Lock.

Brian Bridges is a composer and lecturer based in Derry~Londonderry, Northern Ireland, where he lectures in music technology at Ulster University and is Research Director for music and associated subjects. He is the current president of ISSTA (the Irish Sound, Science and Technology Association) and serves on the editorial board of *Interference: A Journal of Audio Culture*. Much of his work is inspired by connections between perceptual processes, creative practices and technologies, and his creative output includes sound-based installations, audiovisual pieces and electroacoustic and acoustic composition, including microtonal and spatial music. He is a member of the Spatial Music Collective and is represented by the Irish Contemporary Music Centre.

Kristin Carlson is an Assistant Professor in the Arts Technology Program at Illinois State University, exploring the role that computation plays in embodied creative processes. She has a history of working in choreography, computational creativity, media performance, interactive art and design tools.

Greg Corness is an Assistant Professor in the Interactive Art and Media Department at Columbia College Chicago, working with embodied interaction in media environments. His research focuses on generative music, interdisciplinary improvisation, distributed cognition in performance and methodologies for researching experience in performance. He has developed several generative sound systems as well as computer-vision and tangible interfaces for use in interactive performance and installation works.

Stuart Cunningham is a Senior Lecturer in Computer Science and Information Systems at Manchester Metropolitan University, and has been working in the UK University sector for over 15 years in a variety of academic, research, and managerial roles. Stuart is Visiting Professor at Wrexham Glyndŵr University, where he was formerly Head of the Department of Creative Industries (2010–2014). He holds a PhD in similarity-based audio compression from the University of Wales, having previously completed a BSc and MSc at the University of Paisley. His research interests include: audio compression; affective technologies; and sonic interaction. He leads the Affective Audio multidisciplinary research team. Dr. Cunningham is a Fellow of the British Computer Society, Chartered IT Professional, Member of the Institution of Engineering and Technology, and Fellow of the Higher Education Academy.

Georg Essl is a Research Professor in the Department of Mathematical Sciences at the University of Wisconsin–Milwaukee. He received his PhD from Princeton University in 2002, working with Perry Cook on real-time sound synthesis method for solid objects. He completed his undergraduate degree in Telematics at Graz University of Technology in 1996. He has been on the faculty of the Universities of Michigan and Florida, and has worked at MIT Media Lab Europe and the Deutsche Telekom Laboratories at the Technical University of Berlin. He is a member of IEEE, ASA, ACM, AMS and ICMA, and serves on the advisory board of the New Interfaces for Musical Expression conference. His research interests are computer music, mobile HCI and mobile music making, human-computer interfaces, real-time physical simulation of audio, tactile feedback and foundations of numerical methods for interactive applications.

Tom A. Garner is a Postdoctoral Fellow and Course Leader in Virtual and Augmented Reality at the University of Portsmouth. Tom has been working in sound design research for just over nine years. He has a PhD in Psychology (auditory perception in virtual environments) from the University of Aalborg and has published two books on the topic of virtual sound: *Sonic Virtuality* and *Echoes of Other Worlds*.

Hans-Peter Gasselseder (Mag. rer. nat., Dipl.-Psy.) has studied and taught Psychology, Communication Science and Musicology/Dance Science at the University of Salzburg (Austria) and at Aalborg University (Denmark). Working as a tutor, research assistant and lecturer in Salzburg, Aalborg (Communication and Psychology) and St. Pölten (Media Production and Psychology), Gasselseder's published work focuses on the cognitive psychology of immersive experience and situational context, but also includes contributions to media production, such as the implementation and effects of nonlinear music and sound in film and video games as well as stage-of-the-art ambisonics and six-degrees-of-freedom recording methods for virtual reality. Further research topics at the crossroads of psychology and human-computer interaction include agency theory, sound perception, situational trait models as well as applications of augmented reality for performer interaction. Currently he is lecturing at the University of Applied Sciences St. Pölten and is preparing his PhD thesis "Re-Dramatizing the Game Orchestra" at Aalborg University.

Rolf Inge Godøy is Professor of Music Theory at the Department of Musicology, University of Oslo. His main interest is in phenomenological approaches to music theory, meaning taking our subjective experiences of music as the point of departure for music theory. This work has been expanded to include research on music-related body motion in performance and listening, using various conceptual and technological tools to explore the relationships between sound and body motion in the experience of music.

Mark Grimshaw-Aagaard is the Obel Professor of Music at Aalborg University, Denmark. He has published widely across subjects as diverse as sound, biofeedback in computer games, virtuality, the Uncanny Valley, and IT systems and also writes free, open source software for virtual research environments (WIKINDX). Mark is series editor for the Palgrave Macmillan series *Studies in Sound*, and his books include the anthologies *Game Sound Technology & Player Interaction* (IGI Global 2011) and *The Oxford Handbook of Virtuality* (Oxford University Press 2014) and, with co-author Tom A. Garner, a monograph entitled *Sonic Virtuality* (Oxford University Press 2015). Among recent works, a two-volume co-edited anthology, *The Oxford Handbook of Sound & Imagination*, was published in 2019 by Oxford University Press as was the co-authored *The Recording, Mixing, & Mastering Reference*.

Maria Kallionpää obtained her PhD in composition in 2015 (University of Oxford) and has graduated from the Royal Academy of Music (2009) and Universität für Musik und Darstellende Kunst Wien (2010). She won the first prize of the OUPHIL composition competition in 2013. Kallionpää works currently as a postdoctoral fellow at the university of Aalborg (funded by Kone Foundation and the Finnish Ministry of Education and Culture) and as an artist in residence of the Mixed Reality Laboratory of the Nottingham University. Her artistic research project focuses on designing a music engine that uses gamification as a composition technique. Furthermore, as a winner of the Fabbrica Young Artist Development Program of Opera di Roma, Kallionpää was commissioned to compose an opera, which premiered at Teatro Nazionale, Rome, in 2017. In collaboration with her colleague Markku Klami, Kallionpää has also composed the first full length puppet opera produced in the Nordic Countries (premiered in 2018). Kallionpää was a laureate of Académie de France à Rome in 2016.

Sang Won Lee is an Assistant Professor in the Department of Computer Science at Virginia Tech. His work lies at the intersection of human-computer interaction and computer music, exploring how to computationally mediate musical collaboration and enable novel musical expression. More broadly, his research aims to bring the collaborative, live nature of music making to computational systems by developing interactive systems that facilitate real-time collaboration in music making and broader creative domains.

David McGookin is an Assistant Professor at Aalto University, Finland. He received his PhD at Glasgow University, studying how concurrent presentation of earcons affects identification and how this might be improved. He has since studied the application of sound mappings in a variety of domains, including supporting information access to visually impaired individuals and ambient auditory display of social media. He has over 50 publications in leading HCI and Auditory Display venues.

Adam Melvin is a Composer and Lecturer in Popular and Contemporary Music at Ulster University, Derry~Londonderry, Northern Ireland. A great deal of both his compositional and research practice is concerned with interrogating the relationship between music, sound, site and the visual arts, particularly moving images. He has received numerous international performances and broadcasts of his music; his research has been published in *The Soundtrack, Short Film Studies* (Intellect) and in *The Palgrave Handbook of Sound Design and Music in Screen Media*. He is a member of the Spatial Music Collective and is represented by the Irish Contemporary Music Centre.

David Moffat is a Postdoctoral Researcher within the Audio Engineering Group of the Centre for Digital Music at Queen Mary University London. His research focuses on intelligent and assistive mixing and audio production tools through the implementation of semantic tools and machine learning. His PhD, from Queen Mary University London, focused on sound effect synthesis and evaluation. He has since developed new methods and approaches for sound synthesis, from this understanding of the field. He received an MSc in Digital Music Processing from Queen Mary University of London in 2014, and graduated from Edinburgh University with a BSc in Artificial Intelligence and Computer Science in 2011.

Dafna Naphtali is a singer/instrumentalist, electronic-musician and educator. She composes and performs experimental, interactive electroacoustic music, for 20+ years, drawing on a wide-ranging musical background in jazz, classical, rock and near-eastern music and using her custom Max/MSP programming. She's performed in the US, Canada, Europe, India, Russia and the Middle East, with current projects including "Audio Chandelier," multichannel audio piece presented in US, Berlin and Montreal (IX Symposium 2017 @Satosphére); "Robotica" (music robots and voice) and continuing; "Walkie Talkie Dream Angles," an Audio Augmented-Reality soundwalk (using U-GRUVE AR platform) to create personalized interactive composition written for NY's Washington Square Park; and a new walk "Walkie Talkie Dream Garden" that premiered in June 2018 for the waterfront area of Williamsburg, Brooklyn.

Philippe Pasquier is Associate Professor in Interactive Arts and Technology and an adjunct professor in Cognitive Science at Simon Fraser University. He is both a scientist and a multidisciplinary artist. His contributions range from theoretical research in artificial intelligence, multi-agent systems and machine learning to applied artistic research and practice in digital art, computer music and generative art. Philippe is the chair and investigator of the AAAI international workshops on Musical Metacreation (MUME), the MUME concerts series, the international workshops on Movement and Computation (MOCO), and he was director of the Vancouver edition of the International Symposium on Electronic Arts (ISEA2015). He has co-authored over 120 peer-reviewed contributions, and presented in forums ranging from the most rigorous scientific venues to the most subjective artistic ones.

Joshua D. Reiss is a Professor of Audio Engineering with the Centre for Digital Music at Queen Mary University of London. He has published more than 200 scientific papers (including over 50 in premier journals and five best paper awards), and co-authored the textbook *Audio*

Effects: Theory, Implementation and Application. His research has been featured in dozens of original articles and interviews on TV, radio and in the press. He is a former Governor of the Audio Engineering Society (AES), chair of their Publications Policy Committee, and co-chair of the Technical Committee on High-resolution Audio. He co-founded the highly successful spin-out company, LandR, and recently co-founded a second start-up, FXive. He has investigated psychoacoustics, sound synthesis, multichannel signal processing, intelligent music production and digital audio effects. His primary focus of research, which ties together many of the above topics, is on the use of state-of-the-art signal processing techniques for professional sound engineering. He maintains a popular blog, YouTube channel and twitter feed for scientific education and dissemination of research activities.

Richard Rodkin is the founder of Memetic Arts, Inc., a New York City-based digital media publishing company focused on creative technology applications for the web and mobile devices. Richard is also the creator of U-GRUVE AR, which launched during New York Creative Tech Week 2016. A lifelong musician, composer and creative technologist, Richard first began developing U-GRUVE in the early 1990s, while working in the then-emerging virtual reality industry. Richard holds a Bachelor of Music in Theory and Composition from the University of Cincinnati College— Conservatory of Music.

Sheldon Schiffer has designed sound for film and interactive media throughout his 30-year career as a media artist and academic. As a graduate of UCLA's School of Theater, Film and Television, he studied with award-winning sound editor Patrick Drummond. In mid-career, Schiffer embarked on the journey of developing games and sound for interactive media. He is currently a Professor of Digital Media Production at the Creative Media Industries Institute of Georgia State University in Atlanta, Georgia, and is the Creative Director of Mysterious Pictures.

Rod Selfridge is Technical Lead for sound synthesis and procedural audio at FXive, a Queen Mary University of London start-up company. His PhD was focused on creating real-time synthesis models of aeroacoustic sounds, derived from fundamental fluid dynamics principles. His PhD research has produced two journal articles and six peer-reviewed conference papers, including two that received best paper awards. Rod is also a Post-Doctoral researcher at the University of Edinburgh restoring historical buildings, including the acoustics, within a virtual reality environment, enabling historic musical performances to be recreated. In 1995, Rod obtained a First Class Honors Degree in Communication Engineering from Edinburgh

Napier University and in 2011 completed a Masters with Distinction in Digital Music Processing at Queen Mary University of London.

Prophecy Sun is a PhD Candidate at the School of Interactive Arts and Technology at Simon Fraser University. Her interdisciplinary performance practice threads together both conscious and unconscious choreographies, sound and environment to create exploratory works that invoke deep body memory. Her research is supported by the Joseph-Armand Bombardier Canada Graduate Scholarship, the Social Sciences and Humanities Research Council of Canada and the British Columbia Arts Council.

Benjamin Taylor is a composer and creative coder who specializes in Web Audio, networked performance practices and computational approaches to music composition. Ben has presented his research internationally at conferences and festivals, including the Pixilerations New Media Festival, New Interfaces for Musical Expression, the Web Audio Conference, Leaders in Software and Art, the International Conference on Live Coding and others. He received an MFA in Electronic Music and Recording Media from Mills College, and has studied with members of the League of Automatic Music Composers and The Hub, Brian Harnetty, and Pauline Oliveros. Prior to Mills, he received a BA in Music with honors from Kenyon College. In 2016, he earned a PhD in Experimental Music and Digital Media from LSU while doing research at the LSU Center for Computation and Technology.

Miles Thorogood is an artist/engineer at the University of British Columbia with research centered around the practice and theory in media arts for developing engaging interactive experiences. He engages with quantitative and qualitative methods toward cutting-edge research in the development of computational systems for community engaged artistic creation. Miles' research contributions have produced new knowledge in the fields of soundscape studies, affective computing and cognitive science focusing on sound design practice. This research seeks to identify formal models of creativity as it is by investigating aspects of human perception and design process in order to encode creative structures for computer assisted technologies in art making environments. Building on this research, he has leveraged methods to design multimedia systems that combine audio, video and electronics that explore the human and community experience. As a service from the output of this research the work has been featured as interactive museum exhibits, installations and performances. The interactive installation and performance works frame the research in creative practice that brings meaningful contexts of experience and environment to the foreground using algorithmic processes combining art making, audio and visual media, databases, artificial intelligence and physical and net-work computing.

Jonathan Weinel is a London-based artist, writer and researcher whose main expertise is in electronic music and audiovisual media. He is the author of *Inner Sound: Altered States of Consciousness in Electronic Music and Audio-Visual Media* (Oxford University Press, 2018). In 2012 Jon completed his AHRC-funded PhD in Music at Keele University regarding the use of altered states of consciousness as a basis for compos-ing electroacoustic music. His work operates within the nexus of sound, psychedelic culture and immersive computer technologies. His electronic music and audio-visual compositions have been presented at a variety of international festivals. Jon lectures at Coventry University, London South Bank University and Aalborg University, Denmark, where he is currently a Visiting Research Fellow. He is a Full Professional Member of the Brit-ish Computer Society (MBCS), a member of the Computer Arts Society (CAS) specialist interest group and a co-chair of the EVA London (Elec-tronic Visualisation and the Arts) conference.

Series Preface
Foundations in Sound Design: A Multidisciplinary Approach

Edited by Michael Filimowicz

This series organizes topics in sound design that combine multi-disciplinary perspectives across linear, interactive and embedded technologies. Such an approach is needed as today the practices of sound design are diversifying beyond what could adequately be captured by any single author or discipline. Today's sound designers need to be prepared just as much for games as for films, for programming in coding languages as for mastery of proprietary industry software, and for prototyping web applications or new industrial designs as for traditional occupations in film, television, music and audio.

The volumes are designed to be more future proof than most volumes on media technologies, by focusing on high-level concepts that can be easily put into practice. The first three volumes in the series are sequenced as follows:

Volume 1: Linear Media covers traditional topics such as audiovisual preproduction, production and postproduction but adds other important aspects of linear media as well, such as electronica music production and basic music theory for sound designers, as well as artistic compositional practices such as soundscape design and electroacoustic music.

Volume 2: Interactive Media expands the cinematic soundtrack developed in Volume 1 by developing interaction approaches through consideration of gaming technologies, music programming, installations, spatial audio, real-time synthesis, performances and web-based interfaces and databases including mobile and locative media.

Volume 3: Embedded Media brings much needed coverage to emerging areas such as auditory display, data sonification, the role of sound in the internet of things, wearables and multimodal interaction by integrating physical computing technologies and product development in contexts ranging from toys to automobiles.

This approach to the foundations of sound design does justice to the ever growing uses, content variations, audiences and professional roles that sound design is brought to bear on in the contemporary context. Each volume in itself constitutes an introductory text to its respective area of sound design: linear, interactive and embedded media. Taken together, they comprise a comprehensive introduction to the many forms, technologies and practices of sound design.

These first three volumes set up the possibility for other books that can expand upon the foundational topics for deeper explorations of specialized topics. Prospective authors are encouraged to send ideas for other volumes that can be added to the initial three-volume set, either single- and co-authored monographs or edited volumes, by submitting a proposal to the series editor, Michael Filimowicz (michael@filimowi.cz). Finally, feedback from readers on the content is always welcome.

Michael Filimowicz

Volume Introduction

Michael Filimowicz

Foundations in Sound Design for Interactive Media introduces key concepts for the production of interactive media forms including games, installations, performance, apps, interfaces, software and websites. Such media draw on computational systems features that allow for a high degree of interactivity and generativity in media control and display. Intersecting with rich cinematic, ludic, performative and musical traditions, newer resources from fields such as user experience and interface design have richly informed new practices for sound interactions. The chapters in this volume cover gaming and virtual reality; compositional techniques; new interfaces; sound spatialization; sonic cues and semiotics; performance and installations; music on the web; augmented reality applications; and sound-producing software design.

Chapter 1—"Cinematic Sound Design for Players" by Tom A. Garner—draws from relevant theory concerning film, audio signal processing, auditory perception and musical composition to connect game sound aesthetics to cinematic styles and practices, exploring the notion of what makes games "cinematic" in their sound design.

Chapter 2—"Aleatory Acoustics, Mechanics and Aesthetics: Game Sound Design With Audio Controllers and their Interfaces" by Sheldon Schiffer—focuses on the middleware between the game engine and audio hardware, the software system that provides more sophisticated techniques and artistic practices to optimize a game sound's combinatorial possibilities

Chapter 3—"Designing Game Audio Based on Avatar-Centered Subjectivity" by Jonathan Weinel and Stuart Cunningham—takes up the issues around recreating first-person perspectives and conveying a sense of altered states of consciousness. Prototypes are discussed which explore the positioning of the avatar in rendering the sonic environment, which has wide application in games and simulations.

Chapter 4—"Presence and Biofeedback in First-Person Perspective Computer Games: The Potential of Sound" by Mark Grimshaw-Aagaard—presents an overview of the current state of thinking around sound, presence and immersion. It explores the potentials for incorporating bio-feedback technologies to enhance the feelings of immersion and presence for users engaged in a virtual environment.

Chapter 5—"Composed to Experience: The Cognitive Psychology of Interactive Music for Video Games" by Hans-Peter Gasselseder—brings a perspective informed by cognitive psychology to the design of interactive music systems. Narrative, characters and game play can be made to feel more convincing and immersive by integrating experiential dimensions in the interaction design.

Chapter 6—"Interfaces for Sound Installation" by Greg Corness, Kristin Carlson and Prophecy Sun—explores interface design for sound installations. Interfaces for how we engage with sound can be as simple as hearing with our ears and feeling vibrations in our bodies, but also can extend to devices, tangible designs and motion that can interact with and control sound.

Chapter 7—"Sound Spatialization" by Enda Bates, Brian Bridges and Adam Melvin—discusses perceptual cues for spatial hearing, including details on how different types of sound materials affect ease of localization. It further discusses sound spatialization technologies, including stereophonic, Ambisonic and binaural approaches, room acoustics, and how these influence installation art practices.

Chapter 8—"Thinking Sound-Motion Objects" by Rolf Inge Godøy—analyzes "sound objects" which are event-based and of short durations, and which we perceive as closely linked to bodily sensations, especially of movement. Recent scientific findings are integrated into the discussion of how these produce multimodal shape images, providing us with solid and tangible images of otherwise ephemeral sensations.

Chapter 9— "Climb!—A Composition Case Study: Actualizing and Replicating Virtual Spaces in Classical Music Composition and Performance" by Maria Kallionpää and Hans-Peter Gasselseder—is a case study of one of the authors' performance works. The role of space, either physical or virtual, is highlighted as a source for technical manipulations and aesthetic impact.

Chapter 10—"Interactive Music on the Web" by Sang Won Lee, Benjamin Taylor and Georg Essl—introduces the Web Audio platform for encoding interactive audio within the browser. Based on JavaScript, Web Audio enables easy access, online collaboration and platform independence, and has by now developed abundant online resources to support sound designers interested in web-based audio.

Chapter 11—"Auditory Cue Design" by David McGookin—outlines the main semiotic approaches to the design of sounds for auditory displays. Such cues quickly and efficiently convey important system information, such as errors, warnings and user feedback. Approaches based on the study of sign systems have been developed, such as earcons, spearcons, musicons and auditory icons, and are widely in use today.

Chapter 12—"Soundscape Generation Systems" by Miles Thorogood—describes a soundscape generation system, which uses algorithms, information retrieval and machine learning to produce artificial soundscapes for a variety of media, including games, animation and virtual reality. Such systems automate many aspects of tasks associated with soundscape design.

Chapter 13—"Sound Effect Synthesis" by David Moffat, Rod Selfridge and Joshua D. Reiss—reviews the wide range of techniques associated with sound effects synthesis. The contextual reasons and importance of synthesizing sound effects is presented. Procedural audio is explored in detail, drawing attention to why this is an active development area for games and virtual reality environments.

Chapter 14—"Audio Augmented Reality For Interactive Soundwalks, Sound Art and Music Delivery" by Dafna Naphtali and Richard Rodkin—details approaches to augmented reality and locative media projects, called Audio Augmented Reality (AAR). The authors discuss approaches for building an AAR system as well as the considerations that composers must work through to capitalize on the technology.

Finally, Chapter 15—"Soundscape Online Databases State of the Art and Challenges" by Miles Thorogood and Philippe Pasquier—summarizes the characteristics of online audio databases that catalog sound files by taxonomy, ontology representation and collaborative tagging. A general outline of database environments is examined, focusing on the user-contributed model, expertly managed databases and sound maps.

Acknowledgment

The chapter summaries here have in places drawn from the authors' chapter abstracts, the full versions of which can be found in Routledge's online reference for the volume.

Cinematic Sound Design for Players

Tom A. Garner

1.1. Introduction

By definition, the term "cinematic" describes something possessing a quality that is "of the cinema." Consequently, if the nature of cinema changes, so too should that which aims to evoke a cinematic aesthetic. Contemporary video games are most certainly a prime example of an artistic form with cinematic ambitions. There are numerous games that illustrate moments in which video games have appeared rather enamored with film, such as in the development of *Fable* (Molyneux et al. 2004). For this title, the studio enlisted Danny Elfman to produce the main theme, then passed the remainder of the scoring duties to Russell Shaw, instructing Shaw to directly imitate the cinematic template set by Elfman (Gibbons 2017).

  While something of an obvious statement, it is important to observe that video games are not films. As we shall see throughout this chapter, the fact that films and games are substantially different creatures remains a persistent issue when attempting to adapt theory from the former to the latter. As Whalen (2007) observes, the primary distinction between film and video games is interactivity. This presents an immediate difficulty with regard to nonlinearity of games' narrative structures. While a film follows a predetermined and singular timeline, a game must offer multiple-choice, freedom of exploration and the ability to stop, start and replay particular sections of game play at the player's discretion. As a result, the creative process of game sound design cannot possibly utilize the same methods used for film. So how can game sound designers evoke a cinematic aesthetic with such dramatic practical differences?

1.2. Defining Cinematic Sound

To reach a convincing definition of cinematic sound it makes sense to employ a bit of divide and conquer. This section therefore begins with a very brief look at sound, specifically at a couple of issues with terminology. Next is a quick overview of some of the more generalized definitions of cinematic. Here the term's alternative and developing meanings in film, television and video games are outlined. Finally we bring the two halves together to introduce the meaning of cinematic sound and consider its relevance to contemporary video games.

1.2.1. Defining "Sound"

As is often the case in the more creative disciplines, flexible terminology tends to be favored over that which is precise and consistent. When professionals discuss sound, they may be referring to speech, musical timbres, aesthetics or just music in general. They may also be describing the most stubbornly elusive of auditory elements, one so difficult to pin down it is often referred to as "non-music, non-speech sound." Other possible monikers include "environmental sound," "foley," "ambient sound" and "sound effects." For better clarity, this chapter brings together these various elements under "audio" as an umbrella term. The extensive range of auditory content types is acknowledged in section 2 where the intention is to provide a broad overview of how the notion of cinematic quality relates to the various aspects of sound design. While there is certainly much overlap between the various types of cinematic sound, this chapter does exclude speech. The main reason for this is that speech, while admittedly containing acoustic, timbral features, is commonly more associated with language than sound. This also frees up some space to explore musical and audio aspects of cinematic sound in greater depth.

1.2.2. What Do We Mean by "Cinematic"?

As you might expect, one of the broadest and most widely accepted meanings of "cinematic" is as a quality derived from cinema (Ebrahimian 2004). The term "quality" can of course encompass seemingly any aspect of an artifact. In film, this could extend to elements such as narrative, acting (even qualities of the actors themselves), dialogue, cinematography, environment/setting, visual effects and of course, musical scoring and sound design. Facilitated by rapidly increasing budgets, a great deal of contemporary television and video games have revealed their cinematic aspirations. A prominent example of the former would undoubtedly be the fantasy serial drama, *Game of Thrones* (Benioff and Weiss 2011–present). A review by Mills (2013) identifies several broader cinematic qualities relating to television that include: experimentation, greater use of special effects, cutting-edge computer-generated imagery, originality, richer audiovisual textures and more lingering scenes that encourage an audience to reflect and consider the meaning of the work. Referring back to *Game of Thrones*, Mills' interpretation of cinematic television certainly rings true. From its almost photorealistic environment rendering to its pushing of boundaries, the series continues to epitomize the meaning of cinematic in television (Alexander 2017; Epstein 2016).

Within the video game industry, the term "cinematic" has been acknowledged to the extent that it has essentially become a genre (or at least, type of game) in its own right (Abdou 2015). Examples of "cinematic games" include *Red Dead Redemption* (Martin et al. 2010), *Alan Wake* (Ranki et al. 2010) and *L.A. Noire* (McNamara et al. 2011). Such games are often described in similar terms to Mills' (2013) cinematic qualities of television. The opening scene of *Alan Wake* for example, is a lingering aerial shot, while *L.A. Noire's* development experimented with Motion Scan technology as a means of rendering highly detailed character facial animations (see Chandler 2017). However, as Newman (2013) observes, "cinematic" has also become synonymous with "expensive," and games development often plays a delicate balancing act between budget, player-expectation and cinematic quality. As the known budget increases, so does the player's expectation. However, meeting such expectation by increasing the cinematic quality puts increased strain on the budget. In terms of musical scoring for instance, "[once] the consumer demand is high enough, developers of big-budget games will have to use live orchestras to remain competitive" (Newman 2013, p. 157).

In terms of video games' history, the definition of cinematics has certainly changed over the years. Throughout the fifth generation of video games consoles (roughly 1993 to 2001, beginning with the lesser-known Fujitsu *FM Towns Marty* and including the much more well-known *Sony PlayStation*, *Sega Saturn* and *Nintendo 64* consoles), cinematics typically referred to noninteractive "cutscenes." These cutscenes would characteristically be scripted and linear, primarily serving as a means of churning out exposition and progressing the narrative. During the 1990s, games utilized cinematics for various functions. As Salen and Zimmerman (2004) state, such functions include: rhythm and pacing (a means of continuing the game's narrative while also facilitating brief moments of relief from the game play), player reward, "game play catapult" (moving the player into a new situation quickly without breaking narrative continuity), scene/mood setting, and as a way of presenting the consequences of a player's actions/choices. Making use of the fifth generation's uptake of optical disks, cinematics were usually full motion videos (such as *Command & Conquer* [Sperry et al. 1995]) or pre-rendered computer-generated graphics (*Final Fantasy VII* [Kitase et al. 1997]). Push forward to the sixth console generation (Sony *PlayStation 2*, Microsoft *Xbox* and Nintendo *Game Cube*) and pioneering games such as *Half-Life* (Lombardi et al. 1998) began to respond to criticisms that were being raised against noninteractive cutscene cinematics. The primary problem of noninteractive cinematics was instigated by communities of more experienced players who did not require cutscenes in order to feel situated and comfortable within the game world and, instead, perceived such cinematics as both distractors and delayers of game play (Salen and Zimmerman 2004). The response to this complaint appeared in pioneering games such as *Half-life*. In this particular game, scripted dialogue and character animations provided the cinematic functions but player-control was retained and the cinematic itself was integrated into the diegetic and interactive game world.

As we shall observe later in this chapter, the above tension between interactivity and cinematic quality is one of the most significant issues for video game sound designers to consider.

1.2.3.   *Cinematic Sound in Film and Video Games*

From the above we can observe that a cinematic aesthetic is synonymous with rich atmosphere, narrative exposition, experimentation, boundary pushing, high production costs and, overall, a finely crafted and controlled presentation. These points can be broadly carried over to inform cinematic sound design. For many games industry professionals, cinematic sound design typically refers to measures of fidelity. Such measures include high sample rates and resolution depths, implementation of a wide range of digital signal processing tools, and careful attention to postproduction. Cinematic video game sound is also expected to be realistic, vivid and dynamic in terms of spatialization, timbre and loudness (Newman 2013). For composer Bryce Jacobs (2017), cinematics relates heavily to the use of surround sound, echoing a commonplace assertion that cinematic sound is derived from the mechanical reproduction of sound within a theater that encircles the audience by way of multiple loudspeakers. Jacobs also refers to cinematics with regard to musical arrangement, asserting that the use of large ensembles (particularly those employing orchestral and/or choral elements) are akin to a cinematic score. Ultimately, cinematic game sound is generally expected to reflect the craft and financial investment of cinema (Bridgett 2007). We shall return to these points in greater detail in the next section of this chapter.

1.3.   Approaches to a Cinematic Aesthetic in Video Game Sound

In this section, we will be exploring what are generally perceived to be the most current approaches to designing cinematic sound for video games. The discussion is structured into classifications of game sound as either creative (within which we look into musical scoring and implementation) or technical (audio processing effects). This is not to say that the development and use of audio processing is not creative, nor that scoring requires no technical skill. This distinction is only drawn as a way of helping compartmentalize the numerous aspects of game sound that must be considered when seeking to evoke a cinematic aesthetic.

1.3.1.   Creative Approaches: Music Production and Implementation

1.3.1.1.   Scores and Soundtracks

The compositional aspect of a musical score encompasses what we might call "classical" aspects of music production, including melodic/harmonic construction, texture, structure, rhythm, timbre, orchestration and dynamics. Precisely how these numerous elements can be utilized to evoke a cinematic aesthetic arguably encompasses a vast collection of techniques that are further compounded by matters of method and genre. As a result, a full review of contemporary compositional techniques and their relevance to cinematic game music is beyond the scope of this chapter. That said, several approaches appear more prominent than others as effective routes to cinematic sound.

A useful overview of the technical requirements of a cinematic score is presented by Davis (2010). These requirements include physical, structural and psychological functions. Physical function refers to the use of a score to reinforce or punctuate the physical environments and actions displayed on the screen. This could include the use of regional instrumentation to signify a particular place, or melodic conventions that typify a particular musical era as a means of establishing the time period in which the narrative is set. Structural functions, by contrast, relate primarily to issues of continuity and overarching identity. For Davis, cinematic scoring must support the fluidity and consistency of transitions between shots, enabling the visuals to jump to different angles and focus upon different subjects with an auditory continuation that gives the audience the experience of a single, unified scene. Regarding structural identity, Davis asserts that a cinematic score will often utilize leitmotifs (brief musical themes representative of a specific character, place or action) to add sensory weight to particular aspects of the narrative. Through attaching such leitmotifs to a character for example, a film can reintroduce the idea of that character at a later point in time without actually having to make the character appear on screen, or it can vary the tone of the music to apply an emotional overlay to our concept of the character. Film composer Michael Giacchino (Patches 2013) gives an example of this practice in describing his score for the opening scene in *Up* (Docter 2009). Giacchino describes his use of a brief, constantly repeating melodic theme intended to embed an idea in the audience's minds, then retaining that melody while slowing the tempo and shifting the orchestration from a larger ensemble to a solo piano. The effect of this technique is a poignant tonal shift that applies a saddening event to a character we feel we know, despite them only being on screen very briefly and without uttering a word of dialogue. Giacchino's process is a good illustration of a cinematic score that exploits Davis' (2010) structural requirement (the use of a repeated motif) but also satisfies an emotional function (sadness at a character's passing away). This technique is also brilliantly exploited in the horror-shooter *Left 4 Dead* (Booth et al. 2008), in which each major enemy type is accompanied by a unique leitmotif whenever it spawns in the map. This gives the player an indication of that enemy's presence well in advance of them appearing on the screen as a particularly effective means of raising tension.

Psychological function, Davis' (2010) third requirement of a cinematic score, is arguably the most important requirement for a cinematic aesthetic. It describes the power of music to be emotionally evocative, which is of course also a fundamental function of cinema in general. As Davis explains, the score is required to complement the emotional direction of the visuals, but not necessarily to directly reinforce them. For instance, a film's narrative may be particularly dark and serious in tone to the extent that an equally solemn score would be too intense or aggressively one-dimensional. In such an instance, a good cinematic score would seek to counterbalance the tone (adding nuance and tonal layering) without undermining it. This technique may also be extended to enhance the narrative's emotional color by encouraging an ironic interpretation from the music's tone. Video games already display a keen awareness of this cinematic approach in games such as the narrative-driven science fiction shooter *Bioshock* (Levine et al. 2007), in which the relentlessly cheery 1930s soundtrack serves only to highlight the many blood-stained corpses that litter the sunken city of Rapture.

Sticking with *Bioshock* and the use of a soundtrack as opposed to musical score, Gibbons (2011) observes that a carefully constructed set of popular songs can be subtly woven into both the narrative and the actions of the player. As well as creating a jarring dichotomy by presenting fundamentally optimistic and upbeat songs within an otherwise bleak and decaying world, *Bioshock's* soundtrack fulfills two further functions. Firstly, it effectively establishes a sense of the narrative's time period. Secondly, several of the songs are used to subtly raise key narrative points that contextualize some of the player's actions[1]. As is something of a recurring statement within this chapter, cinematics and interactivity are, in certain ways, almost diametrically opposed and the great challenge is to effectively satisfy both features. Games that strike a balance between cinematics and interactivity, can do so in part by utilizing both an original score and a soundtrack of popular songs (Gibbons 2011). In *Bioshock*, this is achieved by presenting the score extra-diegetically (i.e. the orchestra and musicians exist outside the virtual/narrative world) while the soundtrack appears diegetically by being played on old gramophones and loudspeaker sources that exist within Rapture. Many of these soundtrack sources can be interacted with by the player (speakers can be shot, immediately halting the music, while gramophones can be switched on and off). This technique is particularly powerful as the soundtrack itself can effectively evoke an atmospheric and cinematic quality, while the ability to interact with the music within the game world directly supports interactivity.

In further practical terms, expanded background textures are repeatedly related to the richness commonly associated with a cinematic aesthetic. This approach describes the use of multiple, evolving musical phrases woven around the main melody. The intention is for these strands to be simple and unobtrusive, to avoid obscuring the melody, and, instead, to reduce its exposure with subtle decoration, adding rhythmic contrast and thickening the texture (Andersson 2017; Midi Film Scoring 2015). Use of an orchestra's string section to provide sustained, sweeping harmonies also features prominently in approaches to cinematic scoring (Midi Film Scoring 2015), possibly due to its association within Western tradition with cinematic quality resulting from the sheer frequency of its use in that context.

As we shall discuss in greater detail later in this chapter, cinematic qualities of a musical score are conflated with matters of genre, cultural conventions and trends. As Abramovitch (2015) observes, in 2010, "braaams" (not a typo but rather the colloquial term used to describe a sudden, low-register blast from the horn section) "ate Hollywood." Although a point of contention, braaams first featured prominently in the trailers for *District 9* (Blomkamp 2009) and *Inception* (Nolan 2010) and since then have appeared profusely in both trailers and films up to and including the present day. Fearing accusations of being derivative, film executives are keen to avoid overt reuse of previously successful cinematic techniques. As a result, techniques such as this continue to evolve, with trailer companies crafting signature braaams with alternative instruments or non-musical sources (Abramovitch 2015).

The specifics of cinematic scoring are best approached as a constantly changing collection of techniques. However, irrespective of trend, the underlying requirements for physical, structural and (most importantly) psychological function remains constant. For the video game composer/ sound designer, the priority is to be guided by these functions but also to look at the current trends in cinema to inform the specifics of their score. That said, using a different creative form is not necessarily a justification for producing derivative work, and composers should ideally be analyzing trends in cinema, not merely to emulate them, but to build upon them and develop ideas forward. Doing so will arguably help the completed game not only effectively evoke a cinematic aesthetic, but also contribute to the increasing recognition of video game music as an independent artistic form.

1.3.1.2. Implementation

One of the most prominent problems in the creation of cinematic video game music arrives at the implementation phase. After all that hard craft (creating and refining, then creating some more) to realize the desired physical and psychological requirements of the score, composers finds themselves tasked with retaining all they have achieved while also fully supporting interactivity and all the complexity, nonlinearity and unpredictability that comes with it. The response to this issue comes in the form of implementation, a collection of techniques for integrating musical content into video games.

Alternative methods of implementation can be readily positioned along a continuum (see Figure 1.1), based on the extent to which either the

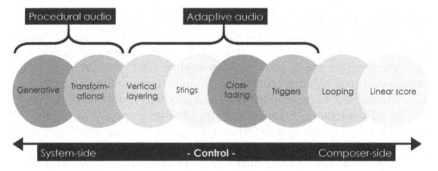

Figure 1.1 Continuum Separating Different Approaches to Game Sound by the Extent to Which Their Presentation Is Directed by the Game Itself.

designer/composer or the system has control. Towards the composer-led side of this continuum, we would position music and other sounds that are presented to us in a way that is largely similar to a film. Composer-side implementation typically presents a musical score that is unresponsive to the player's behaviors or the game's action. In its most extreme form, the music is essentially a fixed soundtrack that the game itself is built around. An interesting example of this is *TRIHAYWBFRFYH*,[2] a first-person "walking-simulator" in which the action is fixed to precisely 20 minutes and is designed around the soundtrack, not the other way around. In most cases, traditional game music loops indefinitely until the level (or potentially even the entire game) ends. While this approach would seem to reflect cinematic practice, it is often only utilized in simple, casual games, a good example of which would be *Tetris* (Pajitnov and Pokhilko 1984). Moving slightly to the left but still very much on the composer-led side, basic implementation techniques such as triggers and cross-fades appear. These are basic forms of adaptive game sound and they enable the score to better reflect the specifics of a particular game scene but often at the cost of a rather obvious and clunky transition. One step further and we may find the use of stings (short, sharp musical phrases sometimes referred to as 'stingers' or 'bumpers') as a means of partially masking the transitions between musical sections. Collectively, these techniques belong to the overarching group of "adaptive music," which also includes the next step, "vertical layering." Vertical layering thus brings us to a crucial point at which the way the musical score is crafted relates directly to the process of implementation. As Philips (2015) explains, vertical layering is a compositional approach in which "music is not captured in a single audio recording [but instead] several audio recordings play in sync with one another." Each layer typically features less "horizontal" musical development and is often a short, loopable phrase. However, its design is intended to complement the other layers, ideally in various combinations so that, when brought together, an exponentially larger amount of content is created—specially, content that responds smoothly to numerous game states or player actions. Philips cites the game for which she produced the music, *Little Big Planet 3* (Livesey et al. 2014), as an example of the vertical layering approach. This exemplifies the flexibility of a vertically layered score through its use within the game as an interactive music system (a game mechanic in which the player controls the game's music actively and with intent) and highlights this as a compositional approach that is significantly different to traditional film scoring and exclusive to video games.

Positioned more towards the system-led side of the continuum are procedural techniques, the use of algorithms to create game content independently of a human developer (Shaker, Togelius and Nelson 2016). As Wooller and colleagues (2005) point out, procedural techniques can be broadly separated into two categories, "transformational" and "generative." A transformational approach can use algorithms to procedurally adapt sound and music that is otherwise preproduced. This could describe the vertical layering method discussed above but also extends to real-time algorithmically controlled sound processing, two common examples of which are footsteps (that typically use only a couple of original samples that are replayed more frequently when the player runs or change timbre when the player walks upon a different surface material) and explosions (a small collection of samples are combined with their attack, decay, sustain and release,[3] all controlled by the system based on several variables in the game state). By contrast, generative algorithms typically denote sounds and music that are constructed by the system and based only upon foundational rules established by the designer. Within generative production however, there is still a wide range of options, based on the amount of control the designer may wish to exert. Control can be varied by either extending or limiting the system's range of options. On the one side, a generative system may be able to select any note length, pitch, key, instrument, or tempo (etc.) either at random or based upon variables in the game. The effect of this freedom for the system is almost certainly a sound that listeners perceive as random, disjointed and lacking in musicality. Therefore, sound designers typically opt to rein in this freedom by limiting the options and blending the generative aspect with some preproduced musical content.

For the video game sound designer aiming to deliver a cinematic aesthetic, the continuum of implementation and procedural generation is extremely important. As stated earlier in this chapter, games are not films, and interactivity presents significant challenges to those hoping to recreate a "Hollywood sound." The dichotomy between the interactive nature of video games and the linear, fixed nature of cinematic scoring often raises issues when attempting to marry the two. Such issues are to be ignored at the composer's peril. As Gibbons (2017) explains, one of the most significant problems is with repetition, particularly within game genres such as RPGs (role-playing games) that utilize recurring game play (often called "grind") as a core mechanic while encouraging extended sessions of game play. In such circumstances, repetition of a fixed score can quickly become apparent, then very irritating.

Producing music towards the right of the continuum enables the score to evoke cinematic quality, but at the sacrifice of interactivity and responsiveness to the game's state and action. However, while the left side can provide much more fluid and adaptive musical content, it becomes increasingly difficult to convey the desired cinematic musicality and tonality (most current generative techniques can only use synthesized instruments as extensive sample libraries would be far too resource intensive). As a final note however, it is important to recognize that technology is developing quickly and, as a result, the optimal approach to cinematic sound is expected to change. For example, recent research has revealed that the algorithms behind generative composition are developing in sophistication to the extent that they can display "emotional intelligence," enabling sound designers/composers to specify a desired mood or emotional state to which the system can respond by automatically creating musical content that reliably expresses and evokes that emotion (Williams et al. 2017). Such developments, combined with ever-increasing computing resources, suggest that while video game music must presently sit to the right of the implementation continuum in order to evoke cinematic quality, this is not going to be the case for much longer.

1.3.2. Technical Approaches: Audio Processing Effects

As discussed earlier in this chapter, by "audio" we are referring in shorthand to sonic elements that are typically categorized outside of music and speech. This includes ambient (background/environmental) sounds, action/event sounds and also extra-diegetic sounds that support game play but are not part of the narrative world (e.g. user interface/head-up display sounds). As with music, audio elements of a video game are of equal relevance when considering the cinematic aesthetic. There is however a key difference between audio and music in a cinematic context. While music prioritizes a qualitative and emotion-focused experience, audio appears to favor more objective technical standards pertinent to perceived sonic accuracy, detail and realism.

1.3.2.1. Positional Audio

In terms of more objective cinematic standards, one of the most significant is arguably the potential of the content (be it sound or visuals) to deliver immersion. Jerald (2015) provides us with a good example of immersion's more contemporary definition: the extent to which content physically surrounds the listener/viewer and its resolution. By this definition, a curved screen is visually more immersive than a flat screen as it encompasses more degrees of our field of view. Likewise, a multichannel speaker system is sonically more immersive than a stereo system as it projects sound to the side of (and typically behind) the listener rather than just from the front. Immersion as it relates to visual resolution is something most of us will be highly familiar with as it has become marketing's technical specification of choice for numerous vision electronics. The underlying concept here is that a higher resolution is more immersive because it reduces unwanted graphical noise (blur, pixelation, aliasing, screen door effect, etc.)

that reminds the viewer that what they are looking at is not "real." This also applies to audio content, with higher resolution signals more able to convey a "natural" sound quality that is more immersive by way of greater perceived realism.

The nature of sound provides it with an inherent advantage over traditional visual displays (possibly excluding virtual reality) in terms of immersion. As Grimshaw (2008) observes, the visual display of a video game (and this is also equally true of the cinema screen) is a two-dimensional representation of a virtual three-dimensional world. While the way we process visual data enables us to perceive depth upon a screen, each pixel (or the projected image) remains fixed to a flat plane. Sound by comparison is able to interact within our physical space, giving us a genuine experience of immersive, three-dimensional content rather than a mere perceptual illusion.

Surround sound technology was born of the cinema. With their origins in nineteenth-century inventions such as the Théâtrophone and Electrophone (which were used to transmit live performances from London Theaters into people's homes by way of telephony), multichannel audio systems began to appear in cinemas in the 1940s, with arrays of loudspeakers used to effectively bounce sound objects around the auditorium (Miller 2004). Following numerous post-war developments in surround technology, *Star Wars* (Lucas 1977), featuring Dolby Stereo, solidified the mainstream acceptance of contemporary surround sound technology in cinemas. Since then, and continuing to present day, arrays of loudspeakers adorn the walls of every cinema screen, facilitating surround sound as a quintessential aspect of the cinematic experience.

Concepts and technologies of surround sound are certainly not new to video games, and have featured in games titles dating back to 16-bit console games, such as *Jurassic Park* (Beard et al. 1993) on the Super Nintendo, which featured surround sound by way of an early Dolby Pro Logic processing system. When we consider current (at the time of writing, of course) eighth-generation consoles, video games' connection to surround sound has only deepened. For the contemporary game sound designer, surround sound (commonly referred to as "spatialization") methods primarily consist of Ambisonics and head-related transfer function, at which we now take an introductory look.

Ambisonics originated in the 1970s, pioneered by the work of Peter Fellgett (1975) and Michael Gerzon (1975). Put simply, Ambisonics uses four, closely positioned microphones to record the sounds within an environment. The first of these microphones is omnidirectional and collects sound wave information similarly to a normal recording method. The other three microphones however, utilize a figure-eight field that records sound positioned only in front and behind the microphone capsule, capturing sounds along a straight line. Each microphone is positioned using Euclidean geometry (X, Y and Z axis) to build up a complete three-dimensional capture of the environment. Following recording, some particularly clever processing is used to "decode" the audio information so that it can be played back on stereo headphones and manipulated in real time, responding to the position and orientation of the listener's head. When done correctly, the effect of Ambisonics is a capture of a real soundscape that can be "walked around" within in virtual space. What also sets Ambisonics apart from traditional cinematic surround sound is the addition of height information, enabling designers to clearly position sound objects above and below listeners, rather than simply around them.

Currently a more popular alternative to spatialization, head-related transfer function (HRTF) begins, like Ambisonics, with the audio recording of a physical environment. Unlike Ambisonics however, HRTF recording does not capture sounds in the traditional sense. Instead, it captures head-related impulse responses (HRIRs), sound wave information from which we can emulate the acoustic characteristics of the head and ears. The process is *relatively* straightforward (see Gardner and Martin 1995 for a more detailed explanation). A microphone and loudspeaker are positioned within an anechoic chamber so that only sound waves that reach the microphone directly are captured, avoiding environmental coloration of the sound. The microphone capsule is typically embedded within the ear canal of a prosthetic (dummy) head. The loudspeaker emits a steady sine wave that gradually increases and decreases in frequency. The sound waves emitted from the loudspeaker project towards the microphone, but before reaching the capsule, the waves interact with the prosthetic head, pinnae and ear canal. This process is repeated, with the loudspeaker aimed at the microphone from various angles across all three dimensions. This provides two streams of audio: firstly, the original sine wave at various frequencies and, secondly, the "processed" sine wave that has interacted with the prosthetic head from multiple positions. Comparing the two streams, and again involving some very clever math, enables us to determine the HRIRs then transform them into an HRTF algorithm. The HRTF algorithm can then be applied to any other monophonic recorded sounds, with the effect of recreating the acoustic properties of the head.

HRTF processing is a particularly powerful effect in virtual reality, where the user's head is tracked in three dimensions, enabling the virtual soundscape to respond to its precise orientation and creating the sensation that virtual sound objects are clearly positioned within three-dimensional space. It can also be used to recreate so-called nearfield sound. By placing the loudspeaker very close to the microphone, the collected impulse responses can be used to make a virtual sound appear as though it is extremely close to our ear. This enables the sound designer to present highly intense and emotionally evocative audio that is, at the same time, very quiet and intimate for the listener.

While HRTF's do have great potential for delivering positional audio, they are not without current limitations. For example, the "front-back problem," describing a listener's difficulty in distinguishing whether the source of a sound is in front of or behind them, is a particularly common issue in both binaural/HRTF and ambisonic systems (see Zahorik et al. 2006). Another issue is that of generalized HRTF processing (i.e. algorithms created using a single dummy head for all listeners, despite each listener possessing significant differences in the shape and size of their ears and head) limiting the accuracy of positional audio by essentially inviting the listener to hear sound through someone else's ears. Issues such as these are presenting themselves as barriers to the more widespread uptake of positional sound technology, but research is ongoing and solutions to such problems may be on the horizon. For those interested in the cutting edge within this field, one particularly interesting research development is in the use of frontal projection headphones as a means of overcoming the problem of generalized HRTF by projecting sound waves to the ear from the front rather than side. This allows the sound waves to interact with more points on the pinnae, essentially creating a (albeit still limited) form of individualized HRTF (see Sunder et al. 2013).

1.3.2.2. Environmental Modeling

With positional audio, immersion is arguably the most significant cinematic quality. Environmental modeling however, is important to delivering immersion in video game sound, but additionally has cinematic value by way of its potential to increase perceived realism. As with cinematic scoring, the interactive nonlinearity of video games dictates that sound designers cannot utilize the same techniques as they would in film production when seeking to create an acoustically detailed and realistic soundscape. The more passive experience of viewing a film means that the audience is relatively fixed in physical space, remaining in their seats with no control over the camera position. By contrast, a video game, particularly those using a first-person perspective (and even more so in virtual reality), requires the player to actively move (their avatar) around the virtual world and interact in multiple ways with the many elements that inhabit it. Unlike a film, it is often of great value to the player to be able to discern from the audio the relative position of an object and whether an object is in front of or behind an obstacle. While positional audio supports the former, environmental modeling addresses the latter.

In video game audio, environmental modeling chiefly refers to acoustic processes of occlusion, obstruction, extraction and convolution (see

Stevens and Raybould 2013). The first three of these processes refer to variations upon a single theme, namely the effect an intermediary object, positioned between the listener and the sound source, has upon what they actually hear. Occlusion describes an intermediary object that completely obscures the sound wave and its surface reflections from the listener, but can still be faintly heard as it allows certain low-frequency components of the sound wave to pass through. A good example of this is when we hear a car radio while standing outside the vehicle. The high frequency components are absorbed by the car's various materials while lower frequencies can escape to reach our ears. The end result is a muted or muffled sound. By comparison, obstruction is when the intermediary object obscures the direct sound wave (i.e. the straight line between source and listener) but not its reflections. This reduces the clarity of the sound and often makes it difficult to localize. In our car radio example this would be analogous to the far side door being open but the body of the vehicle still stops direct sound from reaching the listener. Extraction is essentially the opposite of obstruction, describing situations where the direct sound wave reaches the listener, but the environment absorbs the reflections to create a flat, "deadened" sound.

Convolution is broadly separate from the previous three aspects of environmental modeling and, in some ways, draws similarities to HRTF. Convolution refers to the process of filtering a sound recording with an impulse response pattern. As with HRTF, a loudspeaker (again emitting a sine wave of varying frequency) and microphone are positioned closely together. In this instance however, the microphone is not encased within a prosthetic head, nor is the recording carried out in an anechoic chamber. Instead the microphone and loudspeaker are placed within the particular acoustic space that the sound designer wishes to model. This could be any location from a cathedral to an underground tunnel. Here, the sine wave is filtered by the environment, with some frequencies absorbed and others reflected into the microphone. The difference between the original and filtered sine wave recordings can then be used to generate an impulse response pattern, which can then be applied to any other sound recording to virtually recreate the acoustics of the relevant physical environment (see Nair 2014). Such effects present the player with an interactive sonic environment that feels more comparable to reality, enabling the content of the game to feel more immersive, and thereby more meaningful, emotionally evocative and ultimately more cinematic.

1.4. The Changing Face of Cinematic Sound

This final section takes a step towards the future with a short review of recent developments in film theory's treatment of the cinematic aesthetic.

As noted in the introduction, the fundamental meaning of cinematic is that which evokes something of the cinema. Therefore changes in the nature of film (and our relationship with it) will have implications for the design of video games that wish to evoke cinematic qualities.

1.4.1. Bring Back the Melody

Thompson (2016) sets us a challenge: "Hum something from *Star Wars* to yourself. Easy right? Now try to think of the music from a Marvel Movie. Any of them." While I feel obliged at this point to play the juvenile contrarian and state that the main theme from Marvel's *The Avengers* (Whedon 2012) does come to mind with relative ease, Thompson's point is still one worth considering. Other recent articles have attempted to explain this apparent lack of identity in contemporary cinematic music by way of the "temp track phenomenon," a notion that the overlaying of a film's visuals with preexisting music, is the underlying problem. As Liptak (2016) explains, using a temporary track to generate ideas for the score inherently limits compositional diversity, as it encourages the director to form an immediate association between their visuals and the temporary music, the end result being a request of the composer to essentially recreate the aesthetic and atmosphere of the temporary score, but with (barely) original material. Irrespective of how this problem has arisen, it is arguably becoming increasingly significant. In recent years, cinematic sound has favored a modernistic approach, prioritizing atmosphere and affective tone over a catchy melody. However, the rise in criticisms against such priorities indicates that a resurgence of interest in more Classical and Romantic era compositions that give precedence to catchy melodies could be on the horizon.

1.4.2. The Integrated Soundtrack

In an article that concisely reviews van Leeuwen's (1999) notions on sound semiotics, Gabriel (2017) discusses the concept of the "integrated soundtrack," an assertion that "it is important to explore the common ground between speech, music and sound rather than separating each into categories." In practical terms this can describe various activities, from exploring the musicality of everyday sounds, to the use of musical phrases that reflect the tonality of speech. Contemporary games already evidence existing practice of this approach to cinematic sound design, for example, with the charming point-and-click adventure game *Samorost 3* (Dvorský et al. 2016). In the game, the player-character primarily communicates with various creatures with the use of a straight bugle. The bugle acts bidirectionally, with the player-character able to place the mouthpiece to their ear and direct the bell towards the creature to listen. This action prompts a flurry of speech-like music that reflects the circumstance and emotional state of the creature. Throughout game play, *Samorost 3* also evidences consideration of sonic common ground as both environmental and action-feedback sounds display complementary timbres and rhythms similar to that of the musical score. It should also be noted that, the integrated soundtrack concept has been argued to originate from video games, primarily due to the limited sound technology of early games that required designers to use a small number of tones for multiple purposes including sound effects and music (Kassabian 2003).

1.4.3. Embodied Cognition and Hapticity

In addition to the integrated soundtrack, one particularly prominent concept being applied to film theory is "embodied cognition." As a relatively recent psychological framework, embodied cognition has become more widely known thanks to the work of Margaret Wilson (2002), who presents several key assertions as the theory's foundation. For Wilson, our cognitive processing (essentially how we think, feel and our perception of the world) is deeply integrated with both our bodies and the world around us. One of Wilson's assertions is that cognition is situated, meaning that our immediate physical, geographical and situational circumstances exert a powerful influence over our present thoughts. Embodied cognition is a bottom-up perspective that asserts even our most intricate cognitive processes (from our creative expression to our interpretation of complex emotions) are formed of many lower-level components interacting with one another.

As Ward (2015) explains, an embodied approach to cinematic sound is one in which "the primary function of sound design is to elicit affective imagery, which, in turn, shapes cognition and consciousness." For Ward, the designer's priority is not to express complex, high-level concepts, but to evoke a series of more basic, fundamental emotional responses that encourage the audience to mentally construct higher meaning interactively. Approaching film studies from such a phenomenological perspective has raised the notion of cinema as a means of evoking illusory feelings of touch, specifically haptic and tactile sensations. Mera (2016) explores this under the title of "hapticity": "a mode of perception and expression through which the body is enacted" (p. 4). As Mera states, hapticity encapsulates numerous sensations, namely, tension, balance, energy, inertia, languor (tiredness), velocity and rhythm (see Barker 2009). Within his discussion on hapticity, Mera (2016) notes that his primary intention is to integrate feeling into the experience of film sound: "I do not hear solely with my ears, I hear with my whole body" (p. 5). Mera's (2016) account of hapticity continues, asserting that when a cinema audience feels, they are doing so by way of their physical bodies. Donaldson (2017) reinforces this notion, suggesting that sound is "a physical and tactile phenomenon [that] impacts the bodies of the audience" (p. 34). Here sound is presented in film as a powerful means of both expressing and evoking "virtual" sensory experiences of texture, pressure and even (to a certain extent, of course) pain. Underriner (2017) presents a practical means of exploiting hapticity in describing the "audio reality effect," referring to when a sound evokes a memory in the listener, who then subconsciously superimposes that past experience over their present experience of the sound. Underriner's example of this is when a designer presents the sound of a shower, with running water hitting a bathtub, with the intention of stimulating the audience's memory and evoke re-experiences of warm water on their skin.

Whilst the specific sensations of hapticity are arguably not traditional features of haptics (which typically refers to the resistance we experience when touching an object) and more descriptive of proprioceptive feelings, hapticity as a concept points to contemporary cinematic sound "thinking smaller," concentrating on the lower-level building blocks of our perception in an effort to exert more control over our experience and intensify our emotional responses to their content.

1.4.4. Hyperreality

A further concept being revisited in contemporary film sound is that of hyperreality. As Underriner (2017) observes, hyperreality is "the idea that it is no longer possible to distinguish between the real and the imaginary in life in general" (p. 20). Underriner describes the effect of mimesis (the various artistic representations of reality) as destabilizing our understanding of reality and what is (and converse, isn't) realistic. deBeer (2012) explains hyperreal sound as that which is "generally over the top with extreme detail and extensive EQ" (p. 91). Firearm shots are classic examples of hyperreal sound, particularly the firing of a silenced pistol (in reality, much louder, less of a "pew" and more of a "thunk") and a shotgun (much flatter and less reverberant). The cinematic experience of hyperreal sound has caused our expectations for sonic realism to skew away from that which is realistic, towards that which is more emotionally evocative. The relevance to video game sound here is that cinematic sound is not necessarily only a requirement if the designer wishes to evoke a cinematic quality, but may actually be required simply for the player to accept the sound as real, even if ironically, it isn't.

1.4.5.   The Transnational Effect

Lastly, a quick word on the so-called "transnational effect" (see Beck 2013). Throughout this chapter, the terms "cinematic" and "Hollywood" have been relatively interchangeable. For many years this has been understandable due to the dominance of American cinema. However, in a contemporary and more globally connected world, cinematic now has relevance outside of Hollywood. The transnational effect is now well observed within the film industry. It describes the alternative definitions of "cinematic" between cultures. For instance, while contemporary Hollywood may associate pounding percussions and foghorn-like brass swells (braaams) with notions of a cinematic aesthetic, Japanese cinema is more likely produce cinematic scoring by way of woodwinds and plucked strings (Lin 2017). In relating this back to video game sound, it is more than fair to suggest that the production of video games is potentially more broadly spread across the globe than even film. Therefore it is of great importance that game sound designers consider the national expectations (or tropes) for cinematic sound based upon their intentions to market the game to different countries around the world.

1.5.   Chapter Summary

The cinematic aesthetic is arguably both multifaceted and evolving. Its meaning can vary between cultures and it has proven itself susceptible to trends. When considered within the context of sound, the cinematic aesthetic raises various design approaches depending upon whether we are considering audio or music. For the former, cinematic appears to be synonymous with more objective and technical qualities such as fully three-dimensional spatialization, intricately detailed soundscapes and acoustically accurate environmental modeling effects. For the latter, large-scale orchestral and choral ensembles, high dynamic and timbral ranges, and atmospheric tonalities are prescriptive of a cinematic aesthetic. The nature of cinematic is also highly responsive to developments in film theory, which draws from wider philosophical (e.g. phenomenology) and psychological (e.g. embodied cognition) concepts. The end game, however, arguably remains constant, with developing techniques and theories ultimately endeavoring to evoke more intense and meaningful affective experiences for the audience. In video games, the crucial aspect of interactivity dictates the use of entirely bespoke approaches to cinematic sound design, as both audio and music are required to evoke the controlled craft of cinematic sound while simultaneously remaining flexible to the changes in game state and the actions of the player. For video game sound designers, the solution lies somewhere on the continuum of sound implementation. Here, they may utilize a combination of various techniques along this spectrum, from system-led procedural generations and transformations, to designer-led preproduced content. The great challenge for the designer is ultimately to know how and where to make the necessary compromises between cinematics and interactivity based on careful consideration of the particular requirements of the game.

Notes

1. Gibbons himself provides a great example of this in *Bioshock*: when the player dispatches a key antagonist, the song "It Had to Be You" begins to play, hinting at the later revelation that many of the player's actions are preordained.
2. Full name: *The Rapture Is Here and You Will Be Forcibly Removed From Your Home*. See: https://gamejolt.com/games/trihaywbfrfyh/20342 (accessed 01.05.2018).
3. These terms refer to the loudness curve that is applied to each sample. Attack is the rate at which the sound reaches its initial loudness peak and decay is the rate at which the loudness then dips. Sustain denotes the amount of time the sound continues at a steady intensity before decay, which describes the rate at which the sound fades to nothing.

References

Abdou, D., 2015. Best Cinematic Games: The Most Beautiful Out There. *TGN*. Viewed Dec 7, 2017 http://central.tgn.tv/video/best-cinematic-games-the-most-beautiful-out-there/

Abramovitch, S., 2015. 'Braaams' for Beginners: How a Horn Sound Ate Hollywood. *The Hollywood Reporter*. Viewed Dec 19, 2017 www.hollywoodreporter.com/news/braaams-beginners-how-a-horn-793220

Alexander, J., 2017. Game of Thrones Showrunners Ignite Debate Over Whether It's a TV Show or Movie. *Polygon* www.polygon.com/tv/2017/3/13/14911318/game-of-thrones-tv-movie

Andersson, A., 2017. Making Your Music More Cinematic. *Evenant*. Viewed Dec 8, 2017 www.evenant.com/music/making-your-music-more-cinematic/

Barker, J.M., 2009. *The Tactile Eye: Touch and the Cinematic Experience*. Los Angeles: University of California Press.

Beard, J.H., Kerry, C. et al., 1993. *Jurassic Park*. Manchester: Ocean Software.

Beck, J., 2013. Acoustic Auteurs and Transnational Cinema. In: Vernallis, C., Herzog, A. and Richardson, J. (Eds.), *The Oxford Handbook of Sound and Image in Digital Media*. New York: Oxford University Press.

Benioff, D. and Weiss, D.B., 2011-Present. *Game of Thrones*. Warner Bros. Television. USA.

Blomkamp, N., 2009. *District 9*. TriStar Pictures. USA/New Zealand/South Africa.

Booth, M., Faliszek, C., Morasky, M. et al., 2008. *Left 4 Dead*. Valve Corporation. USA.

Bridgett, R., 2007. Post-Production Sound: A New Production Model for Interactive Media. *The Soundtrack*, 1, no. 1, pp. 29–39.

Chandler, N., 2017. How MotionScan Technology Works. *How Stuff Works*. Viewed Dec 21, 2017 https://electronics.howstuffworks.com/motionscan-technology.htm

Davis, R., 2010. *Complete Guide to Film Scoring: The Art and Business of Writing Music for Movies and TV*. Boston: Hal Leonard Corporation.

deBeer, G., 2012. *Pro Tools 10 for Game Audio*. Boston: CENGAGE Learning.

Docter, P., 2009. *Up*. Walt Disney Studios. USA.

Donaldson, L.F., 2017. Feeling and Filmmaking: The Design and Affect of Film Sound. *The New Soundtrack*, 7, no. 1, pp. 31–46.

Dvorský, J., Dvořák, T., Kratochvílová, B. et al., 2016. *Samorost 3*. Amanita Design. Czech Republic.

Ebrahimian, B.A., 2004. *The Cinematic Theater*. Lanham: Scarecrow Press.

Epstein, A., 2016. Game of Thrones Is the Most Cinematic TV Show Ever Made. *Quartz*. Viewed Dec 8, 2017 https://qz.com/712430/game-of-thrones-is-the-most-cinematic-tv-show-ever-made/

Fellgett, P., 1975. Ambisonics. Part One: General System Description. *Studio Sound*, 17, no. 8, pp. 20–22.

Gabriel, G., 2017. A Sound Semiotic Investigation of How Subjective Experiences Are Signified in Ex Machina. In: Zhoa, S., Djonov, E., Björkvall, A. and Boeriis, M. (Eds.), *Advancing Multimodal and Critical Discourse Studies: Interdisciplinary Research Inspired by Theo Van Leeuwen's Social Semiotics*. New York: Routledge.

Gardner, W.G. and Martin, K.D., 1995. HRTF Measurements of a KEMAR. *The Journal of the Acoustical Society of America*, 97, no. 6, pp. 3907–3908.

Gerzon, M.A., 1975. Ambisonics. Part Two: Studio Techniques. *Studio Sound*, 17, no. 8, pp. 24–26.

Gibbons, W., 2011. Wrap Your Troubles in Dreams: Popular Music, Narrative, and Dystopia in Bioshock. *Game Studies*, 11, no. 3.

Gibbons, W., 2017. Music, Genre and Nationality in the Postmillennial Fantasy Role-Playing Game. In: Mera, M., Sadoff, R. and Winters, B. (Eds.), *The Routledge Companion to Screen Music and Sound*. New York: Routledge, Taylor & Francis Group.

Grimshaw, M., 2008. Sound and Immersion in the First-Person Shooter. *International Journal of Intelligent Games and Simulation*, 5, no. 1, pp. 119–124.

Jacobs, B., 2017. Interview with Bryce Jacobs. *The Sound Architect*. Viewed Jul 12, 2017 www.thesoundarchitect.co.uk/brycejacobs/

Jerald, J., 2015. *The VR Book: Human-Centered Design for Virtual Reality*. Williston: Morgan & Claypool.

Kassabian, A., 2003. The Sound of a New Film Form. In: Inglis, I. (Ed.), *Popular Music and Film*. London: Wallflower.

Kitase, Y., Sakaguchi, H., Narita, K. et al., 1997. *Final Fantasy VII*. Squaresoft. Japan.

Levine, K., Sinclair, S., Schyman, G. et al., 2007. *Bioshock*. 2K Games. Australia.

Liptak, A., 2016. How Hollywood's Temp Scores Are Hurting Your Favorite Action Movies. *The Verge*. Viewed June 12, 2017 www.theverge.com/2016/9/12/12893622/hollywood-temp-scores-every-frame-a-painting-film

Lin, V., 2017. Elliot Leung's 'epic scores': We Talk with a Cinematic Composer About the Music of Movies. *Young Post*. Viewed Jul 12, 2017 http://yp.scmp.com/entertainment/music/article/105399/elliot-leung's-'epic-scores'-we-talk-cinematic-composer-about

Livesey, S., O'Brien, J., Leonard, M. et al., 2014. *Little Big Planet 3*. Sumo Digital. UK.

Lombardi, D. et al., 1998. *Half-Life*. Valve. USA.

Lucas, G., 1977. *Star Wars: A New Hope*. Twentieth Century Fox. USA.

Martin, S., Needleman, J., Kunkler, D. et al., 2010. *Red Dead Redemption*. Rockstar Games. USA.

McNamara, B., Hirani, N., Needleman, J. et al., 2011. *L.A. Noire*. Rockstar Games. USA.

Mera, M., 2016. Materialising Film Music. In: Cooke, M. and Ford, F. (Eds.), *The Cambridge Companion to Film Music*. Cambridge, UK: Cambridge University Press.

Midi Film Scoring., 2015, 3 Music Composition Techniques for Supporting a Melody. *Midi Film Scoring*. Viewed Dec 18, 2017 www.midifilmscoring.com/music-composition-techniques-supporting-a-melody/

Miller, M., 2004. The History of Surround Sound. *Inform IT*. Viewed Dec 18, 2017 www.informit.com/articles/article.aspx?p=337317

Mills, B., 2013. What Does It Mean to Call Television 'Cinematic'? *Television Aesthetics and Style*, pp. 57–66.

Molyneux, P. et al., 2004. *Fable*. Big Blue Box Studios. USA.

Nair, V., 2014. Audio and VR. *Designing Sound*. Viewed Dec 19, 2017 http://designingsound.org/2014/05/audio-and-vr/

Newman, R., 2013. *Cinematic Game Secrets for Creative Directors and Producers: Inspired Techniques from Industry Legends*. Burlington, MA: Taylor & Francis.

Nolan, C., 2010. *Inception*. Warner Bros. Pictures. USA.

Pajitnov, A. and Pokhilko, V., 1984. *Tetris*. Various developers. Russia.

Patches, M., 2013. How to Compose a Killer Film Score, by Michael Giacchino. *Vulture*. Viewed Dec 18, 2017 www.vulture.com/2013/10/michael-giacchino-how-to-score-a-movie.html

Philips, W., 2015. Arrangement for Vertical Layers. *Winifred Philips*. Viewed Dec 21, 2017 https://winifredphillips.wordpress.com/2015/09/29/arrangement-for-vertical-layers-pt-1-a-game-composers-guide/

Ranki, J., Kasurinen, M., Tervo, O. et al., 2010. *Alan Wake*. Remedy/Microsoft Game Studios. USA.

Salen, K. and Zimmerman, E., 2004. *Rules of Play: Game Design Fundamentals*. Cambridge, USA: The MIT Press.

Shaker, N., Togelius, J. and Nelson, M., 2016. *Procedural Content Generation in Games*. Switzerland: Springer International Publishing.

Sperry, B., Yeo, E., Bostic, J. et al., 1995. *Command & Conquer*. Westwood Studios. USA.

Stevens, R. and Raybould, D., 2013. *The Game Audio Tutorial: A Practical Guide to Creating and Implementing Sound and Music for Interactive Games*. Burlington, MA: Taylor & Francis.

Sunder, K., Tan, E.L. and Gan, W.S., 2013. Individualization of Binaural Synthesis Using Frontal Projection Headphones. *Journal of the Audio Engineering Society*, 61, no. 12, pp. 989–1000.

Thompson, A., 2016. Why You Can't Remember What Modern Movies Sound Like. *Popular Mechanics*. Viewed Dec 5, 2017 www.popularmechanics.com/culture/movies/a22907/movie-soundtracks-forgettable/

Underriner, C.F., 2017. The Sound-Poetry of the Instability of Reality: The Audio Reality Effect and Mimesis. *Organised Sound*, 22, no. 1, pp. 20–31.

van Leeuwen, T., 1999. *Speech, Music, Sound*. London: Palgrave Macmillan.

Ward, M., 2015. Art in Noise: An Embodied Simulation Account of Cinematic Sound Design. In: Coëgnarts, M. and Kravanja, P. (Eds.), *Embodied Cognition and Cinema*. Belgium: Leuven University Press, pp. 155–186.

Whalen, Z., 2007. Case Study: Film music vs. Video-Game Music: The Case of Silent Hill. In: Sexton, J. (Ed.), *Music, Sound and Multimedia: From the Live to the Virtual*. Edinburgh, UK: Scholarship Online, pp. 68–81.

Whedon, J., 2012. *The Avengers*. Marvel/Walt Disney Studios. USA.

Williams, D., Kirke, A., Miranda, E., Daly, I., Hwang, F., Weaver, J. and Nasuto, S., 2017. Affective Calibration of Musical Feature Sets in an Emotionally Intelligent Music Composition System. *ACM Transactions on Applied Perception (TAP)*, 14, no. 3, article no. 17.

Wilson, M., 2002. Six Views of Embodied Cognition. *Psychonomic Bulletin & Review*, 9, no. 4, pp. 625–636.

Wooller, R., Brown, A.R., Miranda, E., Berry, R. and Diederich, J., 2005. A Framework for Comparison of Processes in Algorithmic Music Systems. *Generative Arts Practice*, pp. 109–124. Creativity and Cognition Studios Press. Australia.

Zahorik, P., Bangayan, P., Sundareswaran, V., Wang, K. and Tam, C., 2006. Perceptual Recalibration in Human Sound Localization: Learning to Remediate Front-Back Reversals. *The Journal of the Acoustical Society of America*, 120, no. 1, pp. 343–359.

Aleatory Acoustics, Mechanics and Aesthetics

Game Sound Design With Audio Controllers and Their Interfaces

Sheldon Schiffer

Imagine you are playing a rather simple game, one where you control an avatar through a maze of rooms as you go in search of some valuable objects, like glittering doorstops made of solid gold. Each room you search is a little different in a postapocalyptic world of abandoned offices and factories that once made valued swag for corporate conferences around the world. The surfaces you walk on are made of stone in one room, dirt in another, steel grating in a third. A debris field of objects made of wood, stone and broken concrete litters a fourth room. As you walk through these rooms searching for gold wedges, you hear the arrival of some menacing robots who, as it turns out, are looking for swag looters, just like you. So, as you scamper through the maze of rooms, you sometimes run fast, jump over things, squat low to hide and shuffle very quietly. You slide on tiptoes, sometimes barefoot to avoid being heard. If you fail to elude the patrolling robots, you know from the introductory cutscene to this game that you will be tased until you relinquish your pilfered iridescent loot, heels and fists thrashing on the hard surfaces on which you may fall.

In this goofy fictional game, the quality of footsteps changes in a variety of ways. Your avatar will change speed and your feet will collide with different material, affecting the sound of your footstep's volume, pitch, direction and reverberation throughout the game. Your avatar will move through the game world in a finite set of step styles determined by the animations assigned to your avatar's state and the game play situation. These footstep variations create such a huge set of possible combinations of sounds, no programmer or game designer would be wise to provide a recording of every footstep to match a given sequence in the game. Recording, editing, layering, labeling, randomizing and grouping together these individual sounds, with all the possible necessary filtering processes described so far, is a complex task in game development best completed with a software that solves the problem algorithmically during the real time of game play.

Most commercially available game engines, such as Unity3D and Unreal, often include features that allow developers to implement sound directly with some combination of GUI (graphical user interface) and an engine-specific API (application program interface) with its set of programming language protocols (Horowitz and Looney 2014a, p. 46; Unity3d 2018; Unreal Engine 2018). But more often developers of complex games use a separate audio engine, such as Audiokinetic's *Wwyse*, *FMOD* or Tazman Audio's *Fabric*. These audio engines can more efficiently organize a robust set of possible combinations of pre-edited audio files, and process the sounds for a range of parameter values for the many game and avatar states. Audio engines that effectively "plug into" a game engine, are often called *middleware* (Collins 2008, p. 82), though some industry observers have called such software a *SmartAudio Engine* (Collins 2008, pp. 103–104). During the game production process, an audio engine of this kind is used in the "middle," between the recording or editing of sounds for a game, and the general game programming interface, usually referred to as the game engine. Middleware facilitates programming of the playback of a game's audio assets within the game engine (Horowitz and Looney 2014b, p. 66). While the term "middleware" describes more a software's position in the workflow of game audio development, this chapter shall refer to two aspects of this tool separately. First, we will examine middleware for its programming function as a *controller* that allows the sound designer to develop logical and aesthetic rules that govern *when* a sound plays, *what* combination of sounds play, and *how* a sound plays using assigned digital audio processes. Second, we will consider the *interface* that allows sounds to be grouped together hierarchically so that game events can select the designated sound with appropriate processes applied in real time. While the common understanding of the phrase "interface" is usually associated with a graphic system of manipulating settings and preferences, application developers also use the concept of interface as a symbolic system (with some software object) within one application that allows exchange of data with another external software system (with some other software object). An interface for game audio middleware is its mechanism that exchanges data with the game engine (Horowitz and Looney 2014a, pp. 125–132).

Controllers and interfaces can represent an avatar's acoustical behavior. And as an avatar moves through virtual spaces and interacts with virtual materials, the acoustical behavior of the static and moving objects of the game space around it may also be affected. Controllers and their interfaces can automate the selection of music best suited to a game's level or an avatar's actions and objectives at any stage of a game. Additionally, they can load and play specified dialogue for player and non-player characters designated for interaction between characters of either type. The expressive power of the sound controller and its interface is broad and deep as it delivers artistically selected sets of sounds designed for specific game states, allowing a player to hear only the sounds selected for their unique journey, while possibly never playing back much or most of the sound files programmed into a game.

The work of this chapter is to describe the uniquely *aleatoric* features of the sound controller and its interface used in game development and other forms of interactive media production. Sound design inherits much from music the idea of authorial control; a sound designer often "composes" with consideration of sound fundamentals such as volume (like dynamics in music), equalization (similar to pitch), reverberation and directionality (also known as pan). Since at least the introduction of musique concrète and tape-based composition (Pejrolo and Metcalfe 2017, pp. 4–6), sound design production has resulted in what is much like a "composition" or "performance." Interactive media invites a listener or game player to enter the composition with a presence, often a touch of a button or movement of the body, to affect the performance at an unpredictable instant during game play. The sound designer, in consideration of the game design, must anticipate the "chance" event of the player's interaction at aesthetically meaningful moments. Game design provides incentives in the game play to increase the probability of player interaction during intended game states. But as games give the player many conditions for interaction, each condition may give a rationale to modify fundamental qualities of an original sound recording. These modifications expand the combinations of sonic variations for any sound event. For game sound design, whether instrumentally derived with the conventions of music composition and performance, or whether recorded, processed and edited in the compositional tradition of musique concrète, the playback of sound is subject to the "random" whim of the player's interaction. This feature in game sound design I will identify as *aleatoric*, a term used to describe some of the mid-twentieth-century music of Pierre Boulez and John Cage (De Leeuw 2005, pp. 33–36).

Aleatoric features in sound design are those aspects of the sonic presentation of a creative work that are not precisely predictable by either the designer or the game player. As the software of game development has evolved, it owes much of its functionality to the software intended for linear media production. Game sound design software provides nearly all the features that shape the properties of a sound's playback as found in linear media sound design and music production software. As this chapter focuses on a software tool type used primarily for game development, we will not discuss the features of the sound controller and its data interface that are functionally identical to their linear media production software cousins. So, there will not be a discussion of how a game sound controller creates a filter for equalization or reverberation, nor will we discuss how a sound file might be routed through a bus by a game sound controller. Instead, we will focus on the aleatoric features of the game sound controller and its interface that are uniquely useful to the design of games or other interactive media where a sound is played by a game play algorithm or by a choice of the player at an indeterminate instance. The aleatoric features of interactive media do address how a sound can be given one of various combinations of equalization or reverberation if the game play conditions required for that combination are true.

The reader will also discover that this chapter does not explicitly identify a specific software. There are several reasons for this choice. First, among the handful of fully developed sound controllers commercially available, they all do similar things with varying degrees of control and detail, though the names they give for the tasks they perform and the internal interface and software objects they create are different. Some of these controllers do more tasks than others with differences in how they organize and label these tasks. They each provide distinct metaphors in their graphical user interfaces that enable the game sound designer to perform her work. Additionally, there are proprietary game sound controllers that game companies have developed for their own productions. But despite the variety of game sound controller software one might encounter, this chapter focuses on the features that are universally found and most expressive for the game sound designer, and that are not features shared with linear sound media production software. The objective of this chapter is to introduce the learning professional artist to a way of thinking about game sound design that is necessary to optimize the aleatoric characteristics.

2.1. The Architecture and Functions of a Sound Controller

Interactive media presents some special conceptual challenges for sound design. A work of interactive media is not fixed in duration, though a recorded sound is. With interactive media, the objects or combination of objects that collide or vibrate in the game space emit sound from forces that are often variable, random and unpredictable. These aleatoric characteristics of game sound that audio professionals are trained to control, give game sound literal or metaphorical meaning. Even if the recorded sound of our swag looting avatar's footsteps described above may in fact not have been a recording of an actual footstep, when we see a foot colliding with the floor synchronically with most any sound with similar dynamics and materials, the player connects the apparent visual collision as the cause of the sound (Chion 1994, p. 63).[1] That same sound file, processed through one or many filters, can be heard algorithmically by instructions held in the controller. It may be used again for other types of footsteps, or even other apparent collisions of visual objects. Thus, in the context of visually realized game objects such as feet and floors, sound files playback synchronically with aleatoric properties as well. A variety of materials may be used to create many recordings of the same type of collision to express the sound of different materials at the point of contact in the game space. Similarly, changes of footstep speed at the points of collision in the game space may produce increased volume levels and changes in pitch. With so many possibilities for unpredictable sound events such as footsteps, sound design for interactive media becomes a combinatorially complex task for study and design, and a challenging art form to master.

Like any art form, sound design must embrace an aesthetic system that integrates with other aesthetic systems within an overall interactive design. Other systems within the game include graphic, narrative, navigational, textual and animation aesthetic systems. This is not an exhaustive list, but a fundamental one. Shaping an aesthetic system for sound requires the designer to consider what the anticipated player has previously heard, and therefore already knows. This presumed knowledge includes two fundamental realms of sonic cognition. First, the game sound designer must consider those recognizable sounds the listener has observed in nature and that will be remembered in the player's mind. These memories form a culturally defined and idiosyncratically accepted notion of "naturalism" in the sound domain. The game player has likely heard a footstep, probably many. And, the player will likely have heard footsteps on a variety of materials displaced by several different states of locomotion—walk, shuffle and run to name a few. The game player compares the sonic likeness of the collections of footsteps in a game with those in his memory. Thus a "naturalistic" sound might resemble one heard previously in the unmediated experience of the player's personal history. Second, the game sound designer must consider the memories of previous mediated expressions of a particular sound used in the game. The implementation and construction of meaning made by recorded and edited sound creates conventions that, despite deviations from a physical simulation of the "natural," can become a "preferred" sound that matches the game player's memory of a previously recorded sound from some other game, or work of audiovisual media, such as a film or television show. With these two categories of previous knowledge—memories of unmediated and mediated sound—the game sound designer records, edits, processes and programs the playback of a sound such that the player will recognize a meaning caused by the sound's perception in the context of all the other components of a game or work of interactive media.

We mentioned earlier that interactive media is fundamentally aleatoric. For the sound designer, the fact that a sound-triggering event can happen in one or many places in the game space makes for a problem of semantic assignment. Every footstep has a meaning larger than the physical fact of the collision of a human appendage against a flat surface. Beyond a mere recording of a generic event, a sound designer creates a sound that suggests the consequences of an avatar's or non-player character's movements through a space. Coming back to our avatar searching for golden doorstops, the sound designer might ask if the player is aware that a misstep will cause its avatar's death. In linear media, a designer can concentrate on creating one set of footsteps for this decisive moment in a narrative. But in interactive media, the sound designer does not always know if the character will enter this room fully informed of the risk of death. Therefore, the recording of the footsteps might not have acoustic features that would suggest the avatar moves across the floor with existential uncertainty. To provide a distinct set of death-defying footsteps, a game designer may create a special set of footsteps that will play only under the condition that the avatar had already experienced some game event that informs the player of the risk of its avatar's death. Now the sound designer must create another set of footsteps (perhaps with creakier floors with tiptoe steps) for the same room and for the same avatar. The example we describe is a simple one. But the reader now should extrapolate how it is to design sound where there are exponentially expanding combinations of game states the sound designer could consider. It should not be a surprise to learn that game sound designers record and implement thousands of sounds and parameter settings for games to fit the idiosyncratic game states of each game. And, it should not be a surprise then to learn that for any one player's journey through a game from start to finish, many of those sounds will never be played during a player's game because the player never experienced the game states required for some sounds to play.

We mentioned earlier that the game sound designer practices her art with a constant problem of semantic assignment. She must ask, will a particular recorded sound work for all the game states I will assign it to play? Or will I have to record another sound to replace it if a different game state occurs? Once the game sound designer answers these questions, there is a tool that allows her to implement the sound design with aesthetic consistency. *Middleware* (the audio engine) uses its sound controller to take a wide range of game states as inputs to trigger playback of a particular sound that meets a game state's requirements.

As the player proceeds through the game, the game state frequently changes. When there is a change in game state, we may call that an *event*. Game events encompass the entire domain of all possible changes of any measurable value in a game. Many of them do not involve sound. But some do. It is the role of the sound designer to determine as many sound-triggering events as the game requires to achieve a desired aesthetic experience in the game. The resulting design could be sparse with very few sounds and only one channel of output to the audio monitors. Or, a game sound design could be very dense with tens of thousands of recordings to reference with dozens of sound recordings playing simultaneously at a given moment during game play over eight channels of surround sound output. For any of these sounds to play at the programmed instance, an event must be defined that will trigger the sound and its parameter settings (Horowitz and Looney 2014a, p. 165).

Let's imagine the first footstep taken in a new game level is on an impermeable stone floor. Hence, we define such a collision of our avatar's virtual flesh with virtual stone as an event. This event will have some unit-measurable parameters that describe its features. For example, the speed of travel of the foot may be very rapid. We could give it a number, say three units in the game space per second. In addition to a rate of travel, we might also consider a parameter to describe the weight of the avatar and its load. If the avatar has a bag full of golden doorstops, then the weight of the avatar plus its load might be high, say nine out of ten. With these parameters, the event of the footstep will *trigger* the playback of a footstep sound and its parameter settings that has been matched for all the properties we have discussed so far: speed (three), weight (nine), floor material ("stone"), foot material ("flesh").

The game player may expect the designer to consider how the values of these parameters will shape the footstep sound. Will the sound "match" a similar footstep remembered from previously heard recordings? Will the sound "match" a memory of a footstep remembered from those heard in the "natural" world? We have made use of these words, "natural" and naturalistic with intentional semantic distance so as not to give the impression that the primary purpose of game sound design is to simulate the properties of sound as experienced or known in the natural world. Instead, we refer to any frameworks of naturalness as socially constructed (O'Keefe 2011, p. 47) by the context of the game play and the memory of the player.

The mechanics of manipulating objects and causing collisions within a game world with reference to the observed behavior in the "natural" world, is often referred to as a game's *physics* (Horowitz and Looney, p. 152). A game sound designer can use the controller to create a game physics that simulates the natural world, or that deviates from it. But, like the Laws of Physics we recognize in the "natural" world that explain observable consistent behaviors and properties of sound signal impulse, emission and reflection, a game sound designer can implement her own set of laws for game sound. To accomplish this creative task, the sound designer programs the controller's real-time digital signal processes to alter a sound's fundamental parameters during a triggering event. Once the sound plays back with these filters, its altered characteristics are restored. The original file remains waiting for the next triggering event and the next parameter settings (Collins 2008, pp. 95–96).

The sound controller also works for events that will play musical cues. The event of our golden doorstop-laden avatar crossing from one level to the next may trigger a musical recording to play. Likewise, the speed of movement that our avatar uses to flee from the robots could also trigger a change in the tempo or add an instrument to a currently playing music cue. We should notice now that events and triggers are two concepts fundamental to the idea of sound controllers in interactive media, and that these events are affected by parameter values defined by the game design. There can be a multitude of combinations of values that define each game state. But only the values that occur within the range of the selected parameters of a game are those that might set off an event trigger. These values are caught by an event *listener* (Horowitz and Looney, p. 164), and actuated by what is sometimes called an event *handler*.[2]

Thus far, we mentioned this relationship between event and trigger as focused mostly on the actions of the player through his avatar. The game sound designer will try to anticipate most every possible event that the player could trigger. This anticipation requires that the designer place event listeners to sense when a trigger is "tripped." If the interactive media we are discussing were totally analog, like medieval theater performed without any electrical devices, a "listener" might be a drummer, among the team of performers, whose eye watches an actor's movement, probably synchronizing her hands on her drum so that each drum clap matches the faux-gestures of an actor in faux-battle. So-called silent cinema employed behind-the-screen sound effects artists and musicians that watched ("listened" for) the action of a movie as it projected through to the backside of the screen. On a predetermined visual cue in the action, the musician or sound effects artist would create a designated sound with an instrument or some other object to strike, to augment the "silent" experience of the film with sound, synchronized to the movement and collisions of objects on screen (Bottomore 1998, pp. 129–131). But as most interactive media are now found in digital devices, we define a "listener" as a block of computer code that runs on a given frequency (usually some number of frames per second). A listener's job is to periodically monitor some specific measurable values of the game state. The listener waits for an assigned parameter to reach a specific value. If it does, it triggers another block of computer code, sometimes called a *handler* that "handles" this specific change in the game state with a command, such as the playback of an assigned sound (Horowitz and Looney 2014a, p. 165).

Every game object has a set of properties described by a finite set of values, usually numbers, words or Boolean (true, false) values. If we create a trigger for an object, such as a collision of our avatar's foot on a pebble resting on our virtual stone floor, the pebble and foot each have at least a parameter preset to "false," that describes if any object other than the floor has just collided with it. The foot and pebble also have a position in the virtual space described by numbers that are distances on the *x*, *y* (and *z* for three-dimensional games) axes relative to the world origin (0,0,0) of a game's space. When the listener "listens" for changes in the selected game object collision properties, it listens for the event of a specified value to change to "true," such as the collision of a foot with a pebble. When such an event occurs, the distance parameters between the location of the collision and all other objects, including the game player's viewport, may be set to affect the fundamental properties of the footstep recording during playback of the pebble bouncing across the stone floor as our avatar's foot triggers it with a kick. At the event that the collision value changes to "true," the trigger is tripped, and a sound event occurs. The game player hears a scrape of a rock against a stone floor as an embedded or linked sound plays for the duration of the recording with any filters applied that could change its fundamental properties: volume (gain), frequency (equalization), directionality (pan) and reverberation during playback. The sound will play until it is finished, or until another trigger may cause the sound to stop.

At the instance of the hypothetical footstep event we have imagined thus far, a game may have one or many states whose parameters may change the values of the fundamental properties of a sound files during its playback. The four fundamental properties are altered by digital processes, such as a filter. Thus, a sound is often recorded with "neutral" values for these fundamental properties so that the playback of the file is optimally mutable to settings that fit the input parameters of the many game states a player may encounter. A recorded footstep with a peak average of -12 to -18 decibels, recorded with a frequency range that closely resembles the range at the original *point of emission* event, and with no audible reverberation from any reflections in the recording space would be optimally neutral, and often referred to as a *clean* recording (Yewdall 2011, p. 63). The controller can receive as input parameters data from the game state to alter the playback of the clean recording. As an example, a firm and rapid footstep on gravel for a passing non-player character may be programmed by the controller to reduce in volume as its distance from the *point of audition* decreases.

We mentioned two important terms here that should be clarified for further use as applied to computer games—*point of emission* and *point of audition*. Point of emission is straightforward. Sound waves are propagated as energy through air whose origin radiates from a point in space where a collision of objects occurs. This energy radiates and degenerates at a periodic rate into space. When a sound is "heard," by ears or microphones, we can identify in space a point of audition (Chion 1994, pp. 89–92) where the hearing occurs. While picture editing in cinema sometimes makes locating the "point" of audition uncertain and debatable (Kassabian 2008, pp. 299–304), computer games mostly present sounds as heard from the point in space where the camera (player viewport) exists in the x,y,z coordinate system of the game space.[3] If those footsteps we mentioned earlier belonged to an approaching non-player character, perhaps a few meters away from our player's point of audition, and on the far side of a thick hedge nearby, that hedge may be programmed to cause some obstruction of the higher frequencies of the approaching footstep between the points of emission and audition. Thus, in this example, the *game physics* for audio (rules of parameter behavior for audio) of the game, as implemented by the sound controller, may depress the higher frequencies of the footstep playback to simulate the sonic obstruction caused by the hedge. Similarly, a stone wall behind the hedge may trigger this same footsteps' sound to reverberate as if reflected off a closely positioned stone surface in the physical world. In the example, the wall and hedge are in positions relative to the player's point of audition, and thus the footstep sound could be directed for processing to simulate reverberation caused by the wall, and absorption caused by the hedge. In summary, the sound controller is set to dynamically change the properties of the footsteps' playback so that it references the expected effect of the position of other game objects and the player. The sound controller takes as inputs the parameters of the game state, which includes the position in space of any sound event and all the three-dimensional objects involved, including their mass, density, surface texture and speed. As output, the sound controller can produce in real time a new sound that has been altered by all the assigned processes triggered by the most recent game event (Horowitz and Looney 2014a, p. 126). Be aware, however, that the sound controller can be used more simply than described. Games that intend complex and granular sound design will aspire to maximize a controller's potential.

We now have an idea that the values of parameters that affect a sound's playback can occur dynamically, based on the sound controller's many possible configurations. It is also possible that the controller can save a group of settings of any of the playback properties for a sound file. Thus, instead of creating a listener (coded) that constantly triggers new events that will cause a handler to continuously adjust a sound's playback properties, a *group setting* may provide one configuration for all the properties of a sound's playback while the player remains in a particular game state or set of game states. For example, if a player enters a highly reverberative virtual space, like a cave with many reflective surfaces in every direction, the properties that govern reverberation can be set and fixed. Properties such as the volume of the reflective sound signals, the number of reflective surfaces and sound signals, and the delay of the reflective sound signals, to name a few among many—these properties may be set once regardless of the position of the sound emission, and applied consistently to all sounds propagated around a point of audition within a game's level.[4]

When the configurations of the playback of sound interpret the game state parameters consistently with a set of designer-determined rules, we can then say that the game physics govern the sonic world of a game much like it governs the animation and motion for a game's rigid body objects, their masses and the appearance of gravity.[5] The rules govern how recorded sounds play back in naturalistic ways. And, often, a game's physics can artistically deviate from "natural" physics to create exaggerated, subjective or playful rules that express unnatural features of the game world. From those rules, whether consistently applied, or thoughtfully broken, the player can then sense the meaning of sound as a cohesive experience that forms an *audio aesthetic*. An audio aesthetic, developed from a game's physics, allows the game player to have an experience enriched by psychological processes. Mood, emotion, attitude, memory, causality and identity, are psychological processes expressed sonically in music, opera and film soundtracks, and are evoked during game play as a result of dynamically processed sound design. The sonic expression of these ideas is shaped by thoughtful use of a sound controller.

The examples we have considered so far rely on a commonly understood model of game physics that roughly simulates sound in the natural world. We use a commonly understood model for clarity. A game sound designer may elect to deviate substantially from this model as a game's sound aesthetic requires. We should be careful to not assume that "good" sound is naturalistic because it simulates nature. Sound as occurs in nature is far more confusing and chaotic than game sound, and does not occur with the intention of artists and technicians creating meaning for others to comprehend in their game experience. The meanings conjured by the player hearing a sound in a game, cue a player's "operant behavior" at the instance of a sound event (O'Keefe 2011, p. 53). The value of using a naturalistic game physics and aesthetic for game sound design is that it contextualizes the player in the game world with meaningful and remembered sound experiences from the natural world and from other games that also assume a naturalistic sound design.

2.2. Metaphors of the Game Controller Interface

In the discussion so far, we have focused attention on the ability of the sound controller part of middleware to integrate individual sounds into a game engine to optimize a game's aleatoric acoustics. For game design, there are interfaces to consider that enable the controller. Let me first clarify my use of the term *interface*. For many readers, this term is defined by the field of Information Architecture as the graphical features of software that allow users to execute commands or apply values to properties using iconic shapes, windows, buttons, fields and checkboxes. This definition considers the user interface. Additionally, we shall consider a broader use of the term as expressed in Software Studies—"software, or hardware-embedded logic, that connects hardware to software . . . specifications and protocols that determine relations between software and software" (Cramer and Fuller 2008, p. 149).

Let's address the hardware first. There is the handheld game controller, or other haptic device, the player manipulates with his body. Then, the game controller actuates game objects on screen during game play, or modifies on-screen settings shown in control panels that allow the player to make more game-specific choices. These player-facing interfaces deployed to levels within a game, allow more combinations of the parameters of sound events. For example, a player chooses an avatar that has a distinct mass. The mass value can influence the fundamental properties of the avatar's footsteps, speed or strength for launching weapons. Each of these features impacts how a sound should play at a collision. The desire to more efficiently integrate expanding combinations of sounds and processes in real time shapes the implemented design of programmer-facing interfaces of sound controllers in middleware (Brandon 2007a, p. 68).[6]

In 2007, Brandon published an extensive review of middleware for the publication *Mix* (Brandon 2007a, 2007b, 2007c). In the decade that followed, middleware developers traded innovations and development trajectories to keep pace with contemporary game design, game engines and console technologies. A review of the surviving software can be generalized as an optimal aleatoric model as appears in the diagram in Figure 2.1. Each of its components will be fully elaborated in the ensuing discussion. Let's review the key components of the model more formally than we

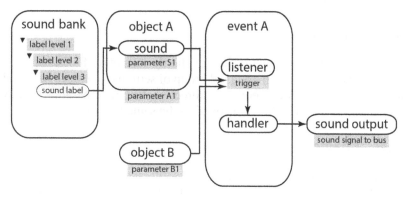

Figure 2.1   A Data Mapping Model to Visualize the Flow of Data to Trigger a Sound Event.

did in the previous section. We should acknowledge that the names of these components may or may not be the same found in any named sound controller software in a middleware package. However, the functions that these named components perform are usually found in most commercial and proprietary game sound controllers, even if they go by different names.

2.2.1. Event

For game design, an event is a meaningful change in a game's state as registered by changed data in memory, data that are usually parameter values. Game design generally requires that events are anticipated and designed in advance of their occurrence (Horowitz and Looney 2014b, p. 68). For sound, an event is the playback of a sound resulting from a game state where a selection of values for parameters falls within a specified range. We should recognize that an event occurs only when the required conditions to trip its *listener* are "true." The conditions of an event can change to "false," limiting the duration of the event and restricting if any coded function is listening for the event. *Events* initiate and terminate the playback of a sound file's signal to an output bus. They can also change the settings of a sound during playback.

2.2.2. Listener

An *event* requires a component that "listens" or observes the values of the parameters that can make the conditions to trigger an event. These *parameter* values are often dynamic. The set of values that cause an event to occur may be called a *trigger*. Sometimes there is one *listener* with multiple triggers. One listener may check for "true" conditions, each of which can trigger an event. Another might check for "true" conditions that will terminate an event.[7]

2.2.3. Handler

An *event* has a handler function that explicitly calls other functions to play back sounds, often with an assigned group of settings for each of a sound's properties (e.g. volume, frequency, directionality, reverberation). Sometimes listeners and handlers are coded as the same function.

2.2.4. Parameter

Each game object has parameters that describe its current state in a game. These parameters are often numerical, textual (string) or Boolean (true, false) values of a game object's properties, such as position in the *x, y* and z axes or its visibility, among others. User-defined properties are also possible as some games have idiosyncratic parameters that reflect the game's unique rules, such as a Boolean value that would indicate "true" or "false" if a player's avatar has ever encountered a Taser-wielding patrol robot. In this example, a "true" value may cause ominous music to play the second time a player encounters the robot. But if the player had never seen the threatening robot before, the game designer may elect not to foreshadow danger with an ominous music cue, and thus decide that music should not play, or perhaps decide that an innocent musical cue should play instead.

2.2.5. Trigger

A *listener* "listens" or observes values of *parameters* of game objects, such as the proximity of positions of vertices of two meshes (to detect a collision), or the passage of game play time taken by the player while occupying a level or room. A *trigger* is a parameter in the listener function that trips an event *handler*.

2.2.6. Sound Bank

The sound controller allows the game sound designer to create distinct hierarchic collections of sounds, and their combinations of settings, saved into a database. A "sound" may or may not be a stand-alone source file. A "sound" in the sound bank is often a link to a sound file on which the controller can apply unique property settings. It is possible that one sound source file can have many combinations of settings, each referencing the same linked sound source file, and each bearing metadata collections of settings. The sound bank stores the digital processing instructions and conditions for playback (Brandon 2007b, pp. 66–67).

These principal components of the middleware interface comprise a game sound designer's means to program sound behaviors. The interface connects the sound controller to the game engine and is essential for applying the aleatoric properties of a game sound design aesthetic. Let's use the diagram for our footstep example. Let object A represent a three-dimensional mesh model of a foot. Each of its vertices contain location values in an x, y and z coordinate system. The foot may have a placeholder for an undetermined footstep sound file. Think of the placeholder as a variable to be algorithmically provided a pathname to an actual digital recording. We don't know yet what material the foot will step on. Object B could represent the floor surface on which the foot will step. The listener (code) will monitor the values of the parameters A1 for object A, and B1 for object B. What might those parameters represent in the case of a footstep?

The parameter could be a Boolean value of "true" or "false" set by a collision detection function that looks to see if the vertices of the mesh of the foot and the floor are near each other. If they are within a narrow range of proximity, then the collision detector will set the parameter value to "true" until the listener detects neither object vertices are near each other. While the collision parameter is set to "true," the event listener notices. Its trigger detects the collision parameter's change to "true." Once the trigger is tripped, the listener calls the event handler, which sends a command to play the sound through an output channel of the game's host device. But how shall the sound play? Let the sound itself have at least one parameter we call in the diagram "S1." S1 could measure step type, such as "walk" or "run." Now we have minimum parameter values to playback the assigned sound programmed through the interface.

The model and explanation above illustrate a simple flow of data that the controller manages. An event programmed in the controller can listen for other parameters in the game state of most any object. For example, some events may listen for time triggers. If our golden door stop hunting avatar has been perusing for too long in one level, say for a minute, then the tasing robot patrol could be programmed to show up and harass us. Such a rule will require a listener whose trigger will be tripped by an event clock that would begin counting down when our avatar enters a level. When the event clock expires creating a parameter value of zero time, the trigger trips, and the event listener will call the event handler. In addition to causing the tasing robot patrol to approach, its attached sound or sounds will play. However, since the approaching robot may begin its pursuit near the limit of audible distance from the point of audition of our avatar, the parameter value of volume for the approaching robot will likely be low. Dynamic control across all output channels of volume, directionality, frequency and reverberation results in playback as output.

The input conditions in the game state constantly determine the output playback results using dynamic digital processing managed by the interface of the sound controller. A real-time system that allows for dynamic parameters must trip the triggers of listeners with regular frequency, as much as once each "frame," to maintain smooth animation for moving game objects. Changes to the parameters of the sound output can occur with the same frequency, usually sixty times each frame for console and PC games or ninety times for VR head-mounted displays. Mobile and Web games vary in frequency because of data traffic on networks. Applications of groups of settings that are not dynamic may allow the trigger to be tripped "on" by one event listener, and tripped "off" by another. The event listener that terminates the sound relies on whatever conditions that the game physics and aesthetic require. The initial description of the middleware systems was provided by Brandon in 2006 (pp. 57–60). Current middleware explored to develop the model shown above remains consistent with his investigation, but with expanded features.

2.3. *Sound Bank* Categorization As "Naturalistic," Personal and Cultural Expression

Footsteps are a useful example to demonstrate how a sound controller monitors a game state and plays back a single sound triggered by an assigned condition, processing it in real time to fit the parameters of the game state. As we previously evolved the example of footsteps, the reader may have intuitively considered a condition that might cause a great variety of footstep sounds. The many surfaces with which our imagined foot may collide will cause the game player to expect distinctions among footsteps. A game world's variety of materials that collide to trigger sounds give reason to record as many audio assets as possible to represent a variety of features with desired detail. Additionally, a footstep event can be broken down into smaller sub-events, each with unique recordings. A footstep sometimes consists of two collisions. The heal strikes the ground separately from the toe or ball of the foot. Moreover, different types of steps may require that the heal collision occur before the toe collision, or vice-versa. Thus, if greater detail in the control of sound is desired, the sound designer turns to the controller to categorize sounds in granular ways for many more variations of sound events.

In the case of footsteps, the controller can create organizational hierarchies within what we have called *sound banks*, where sound recordings and variations of sound recordings differentiated by groups of property settings can be organized under a designer-defined hierarchy. The sound bank we refer to would be a collection of links to sounds (not sound files) associated with an assigned event category, such as footsteps. A sound bank allows for each sound to be tagged with categorical labels that descend hierarchically into more granular specificity. So, for our footstep examples, there could be a primary label category that defines the "Floor material" on which the step occurs. Then, the "Shoe material" the foot wears could be defined below the "Floor material" tag. Then, there could be a set of variations of step types inferring the "Step weight," such as "Tiptoe," "Shuffle," "Walk" or "Run." Lastly, the recordings themselves could be differentiated with a variety of "Step speeds" for each "Step type." The spatial context of the sound could provide relevant data for a reverberation process for a "Step space" tag. One sound in a sound bank of many footstep sounds might be found by following the path from labels using the hierarchy described above: Stone/Moccasin/Tiptoe/Slow/Cave, or Wood/Boot/Run/Fast/Hallway. In these examples, a sound event becomes a five-parameter set of values to trigger the event: the playback of a step with the designed digital processes as output that interprets the parameter (text string) values.

The need for a bank of sounds that uses hierarchical (tree-like) ordering becomes obvious when one considers how the aleatoric property of inter-active media creates exponentially expanding combinations. Each combi-nation of parameter values presents a possible sonic hearing in the game play if the game state warrants it.

The reader might think that the order of the hierarchy in the sound bank that governs these parameters is arbitrary. Some methods of organizing creative assets optimize the expressive capacity of game sound (or any other aspect of game design). Categories of sounds in a sound bank are often ordered in *ascending order of variability* with Boolean values con-sidered first. Organizing categories and subcategories of parameter vari-ables by an ascending order of variability does not change the number of possible combinations of parameter values. The product of the count of values of each of i parameters will be the same regardless of the order in which they are applied to any one sound file. But some parameter values can eliminate others because of their dependency on the value of another parameter, usually a Boolean parameter. Boolean parameters allow only two variations ("true" and "false"), which can cancel the use of another parameter. For example, a footstep weight is affected by the weight of the character. If a character can carry a load, then a "load weight" param-eter could have several values such as "light," "medium" and "heavy." The result increases the character weight, each parameter could change the volume and speed of a step's sound. The value of other variables will depend on if the load variable is "true" or "false." A "false" value for the load would eliminate all three possible "load weight" values and elimi-nates from the set of all possible combinations of process settings those values that would be affected by a weight increase of the character due to a load.[8] A "false" value would prune the combinations by three times the count of all possible parameter value combinations. For sounds that have many possible parameter values, this pruning can save production time and space in memory.

Finally, to demonstrate how organizational and combinatorial think-ing can be an integral part of the creativity of the game sound designer's process, we will consider arguments advocating for either order between the two last levels in the sound bank hierarchy. Should we consider the speed of locomotion above or below the weight of a footstep? As is often the case, there are several perspectives to consider: the avatar character design, the game level design, computational efficiency and the physical interface that the player uses to control its movement through the game. Let's tackle a few of these.

While the avatar moves, we know that the player embodies its persona. The physical extension of the character's behavior is actuated by the game controller that the player holds in his hands. Ground movement along the horizontal plane is generally caused by continuous pressure on a vertical stick, mouse ball or one of four keyboard buttons. Varying the speed of movement requires an occasional instantaneous push of a button, rather than continuous pressure. Varying the type of step by choosing what type of locomotion, such as crouch, walk or run, for example, may be changed by an occasional instantaneous push of a button that has precedent over direction or step type, like the caps lock has precedent over alphanumeric keys on a keyboard (Skalski et al. 2010, p. 227). But can our avatar change from a run to a crouch without first slowing to a walk? This is not usually possible. (Try running in a crouched position.) If our character must slow to a walk before crouching or crawling, then the type or force of the footstep is dependent on the speed of the footstep manipulated by the handheld game controller. Therefore, speed will likely have more frequent variability than the force or type of step. Obviously, this argument makes assumptions that might not be true for every avatar character in every game, and our conclusion might not be true for every game controller button assignment map. Nonetheless, a reasoning through the game design by examining character design and user interface mapping will lead the game sound designer to organize a sound bank for optimum expressivity using the logic of the handheld controller mapping. However tedious this discussion, the reader should take away the idea that the sound bank hierarchy design is an important part of the sound design. Later in this chapter, we will discover how these hierarchies are implemented in several popular games so that the reader can realize how hierarchy design in the sound bank affects the sound design in specific game play.

The example of creating a footstep event sound bank thus far addresses variation of physical characteristics, such as the material on which a foot can step. But what about cultural variations, such as the way one walks in a social space? If our golden door stop filching avatar steps into a cemetery, will his step be the same? Would it worry about disrespecting the dead as it unearths any buried golden doorstops? We have chosen to refer to our avatar so far without identifying its gender. In this imaginary game, do female avatars walk differently than males? Do the genders in this world wear different shoes with different materials requiring gender distinct gaits in their step? Are the points of collision on the avatar's shoe concentrated on a small surface with gendered rhythms, as occurs with high-heeled shoes? Or are the points of contact widely dispersed across the foot as occurs with a sandal? These are cultural determinations of the game world that the sound designer in concert with the game designer will choose to resist, accept, ignore or transform. Whatever the design decision, the choice may result in another layer in the hierarchy of a sound bank.

The same logic of categorization could be applied when considering the personal states of the avatar itself. Our golden door stop burglar may be a complex humanoid with a spectrum of personal motives and emotions. If our avatar has a partner on its journey—perhaps a pet, a child, a friend, a parent—might its footsteps change when approaching that partner to reflect the interpersonal circumstance of the relationship? Our avatar might scold a child or a pet, or might apologize to a parent or friend for some misdeed. Will these actions and their associated emotional states require another variation of recordings or parameter settings to add to the hierarchy of linked files in the sound bank of footsteps? Depending on the game or interactive project, a set of personal categories may be required. And if it is required, would it be necessary to create unique recording sets for each avatar or some of the non-player characters in the game? There is no universal answer. The degree of granularity in the organizational hierarchy of a sound bank is a function of the complexity of the game design requirements. With the middleware sound controller and its sound banks, the game sound designer can define the degree of granularity in relation to the game's underlying aesthetic systems.

2.4. Some Typical Applications of the Game Sound Controller

Our discussion so far has described the game sound controller's function and architecture. We introduced the sound bank as a primary aleatoric feature of the sound controller's interface and as a hierarchic data manager for accessing sounds for playback by the game engine. In the data mapping model (Figure 2.1), we should notice that labels for sounds within the sound controller's interface are defined by the semantics of natural language to describe a sound object. "Footstep" has been our frequently used example under which many materials and modes of stepping are considered. Labels like "footstep" can describe physical, cultural or personal features of a sound. They can also describe the objects that collide to make a sound, and often the avatar or non-player character name and its state are carried in a hierarchy label. While a sound bank of footsteps can show how a game sound controller can use labels that adhere to objective and naturalistic uses of sound effects, subjective perceptions can also use the hierarchic interface of labels in the sound bank. Subjective labeling hierarchies may even describe sounds for a surrealistic game world, as we will see.

Before we can consider subjective hierarchies for sound, we must be aware that a game engine can also record a player's progress through the game. Thus, every event that can trigger a sound has a context of data that represents what the player knows, where he has been and how long he has been playing. The pairing of an event context with a point of audition is sometimes significant. Suppose we hear a door opening mysteriously. If we hear it through our avatar during game play, we may ask ourselves, "Is that door opened for me to enter?" But if our avatar is fleeing from an approaching assailant, then if that same door opens, it instead may cause us to ask ourselves, "Is the other side of the door a safe place to hide?" The sound design of the door creak could be different if the door opens during the flight from an assailant than from a chance discovery of an open door without an urgent need for escape. We can define an event context as the unique combination of all related parameters in a game, including those that represent a player's progress through the game. The pairing of an event context and its point of audition may affect the organizational hierarchy of a sound bank and allow for another way to label sounds to express game-specific meanings. But we have exhausted the footstep as an example from our imaginary game, and must now discuss how the sound controller may have been implemented in specific video games the reader may have played.

This research of popular video games has not unveiled exactly which middleware was used in the games discussed below—that information is often unavailable. The names for sound bank hierarchy levels, their sounds, or the names of events, triggers or parameters are unknown to us. While game credits sometimes reveal the software, they don't reveal the alphanumeric data processed during production. However, to better understand how to implement sound design using most contemporary game audio middleware, the reader may find it useful to consider the fundamental model (described in the diagram in Figure 2.1) of the sound controller and its interface as applied to these game sound design case studies

2.5. Naturalistic Haptic Sound in *Red Dead Redemption*

The most familiar and obvious categories of diegetic sound are those whose emission is caused by vibrations of explicitly visualized objects. Consider those touched by an avatar. For example, in a third-person open-world game space, such as that found in Rockstar's *Red Dead Redemption* (2010), we see our avatar's arm extend with a weapon in hand. This is the arm of John Marston, who cocks, loads, aims and pulls the trigger of an early twentieth-century weapon. When we lift the gun, we hear and see the weapon's metal parts moving in our avatar's hands. The cause and effect of the sound we hear are immediately obvious as they respond to the inputs of the physical game controller that moves the on-screen hand. Our implicit haptic senses know that sound emanates from objects we handle when they collide, even slightly. And when we move as John Marston through the space of the town, our footsteps displace the gravel beneath our feet. Our on-screen movement, often an ambling performance of our body, shoulders pulled back, one leg lifting over the other, cowboy style, matches the sound of this event—leather boots, one after the other, stepping in the gravel and dirt of the *Red Dead Redemption* West. Even the footsteps of Marston's horse are expressed by the ambling or galloping of the third-person camera that trails behind.

Haptic sound effects, those that are triggered by things that the player feels through the handheld controller, often reference a naturalistic game physics circumscribed around a point of audition where the avatar exists. The audio physics synchronize with the animation physics of our avatar's appendage touching an on-screen prop. The degree of "naturalness" in game sound is often established in the first levels of play to create a pattern of sound behaviors that the player will recognize in later levels. In the early levels of game play, a sound's event context is minimal. The game player mostly holds in mind sound behaviors remembered from the natural world, and sound behaviors remembered from other works of audio or visual media. We must recognize that artistic constructions of naturalism in game sound are not actually "natural" sound, but negotiated representations of sound physics with the game sound aesthetics entrenched in the player communities and heavily informed and reinforced by contemporary cinema sound. While sound design may be built around the experience led by what is seen in the game play viewport, sounds are still stylistically processed, edited and mixed by digital audio workstation software. Sound in the natural world is governed by laws of nature. Natural sound, much like nature, is chaotic, unfiltered, unfocused, unbalanced and, in general, a bit messy to the ears when placed in the context of the rendered visuals of a video game. What may seem like natural sound is an idealized sonic expression tethered to the causality controlled by a game mechanic. Naturalistic sound seems easily recognized. The colliding or vibrating objects belie the cause of their movement, the material they are made of, and the materiality of the spatial design of the surrounding objects that reflect a sound emission. The examples given so far of John Marston's gun handling and trail ambling are naturalistic by this definition because we observe them without apparent subjective filtering. And while we hear them subjectively through the ears of a nearby observer embodied in the camera that follows him, we have no reason to believe that they would be heard any differently by any other character standing nearby.

In the sound bank of the controller, however, we still can organize sounds around a causal actor. A sound designer could use "Gun Marston" (the noun object that causes the sound followed by its adjective owner) as a label at the top of the hierarchy, followed by "Handling," followed by a group of action labels at the same level to choose from, usually in a drop-down menu, such as: "Draw," "Cock," "Twist," "Spin chamber." As our player watches, Marston will hear himself and others creating similar sounds. There is a benefit to creating distinct effects for the player's avatar—Marston's sounds will stand out from the others. Thus, we let the sound bank's hierarchies reflect this distinction. The sound bank hierarchy design can reflect the objectivity of the naturalist aesthetic while also embedding hapticity in the selection of sound effects around a causal agent, such as Marston.

2.6. Subjective Surreality in *Grand Theft Auto V* and On-Screen Avatar Death

The audio characteristics of the physics of a game determine how a sound should be processed. We discussed earlier that the fundamental properties of volume, frequency, reverberation and directionality are each defined in the sound controller by triggers, events and processes in response to the physics of the game. But to what extent are the physics deviating from the natural world if sound design is not a simulation? A conventional first concern is to design a game physics that references the natural world. The second more artistic concern is to clarify if, when, and how the game will deviate from the physics of the natural world. If such deviation is part of the game aesthetic, then the game designer must establish those non-naturalistic game states for the player that are aesthetically justified and observable. We should also anticipate that a game design could attempt to deviate totally from any reference to the natural world altogether. Sometimes intentionally "unnatural" sound design is justifiable because it is simply more enjoyable for the intended player.

The intoxication of the protagonists in Rockstar's video game *Grand Theft Auto V* (2013), is one such example. In the *Grand Theft Auto V* single-player version, the player chooses one of three protagonists as an avatar who at times can intoxicate himself with a bounty of sense-altering liquid and solid substances available on several of the game's missions. The sound design during intoxication deviates from the "naturalistic" by imposing conventional cinematic sound processes that signify an intoxicated state. These processes reference conventions established in the second decade of the sound era with films like Hitchcock's *Notorious* (1946), where we subjectively experience being poisoned through the sound effects mix representing the subjective point of audition of Alicia Huberman (Ingrid Bergman). The intoxication sound processing effects for the inebriation of *Grand Theft Auto V's* avatar Michael Townley and others, enjoy a similar treatment. Environmental sounds are heard through a point of audition of the intoxicated player with reverberation using long sonic delays. All the effects are compressed in the higher frequencies, and fewer sounds enter the mix. The long delays seem to stretch time in sync with the stretching image, imitating Salvador Dali's painting *Persistence of Memory* (1931), with its melting clocks. Sound gets a similar treatment using a stylized, perhaps cliché, reverberation effect. The drunkenness in *Grand Theft Auto V* dynamically shifts the sound design in varying degrees of inebriation as the player sips more booze or inhales more marijuana. But as time passes, the effects dissipate, signifying recovery. These are dynamic shifts between the natural and the surreal. The variable to determine how much deviation from the natural, and how long the effect should last, depends on how many shots of booze or hits off a joint the player takes.

These variations of sound effects from the naturalistic to the narratively motivated surrealistic as described here, could easily occupy variations in a sound bank hierarchy for labels of subjective diegetic effects. We can imagine the hierarchy to organize a group of such sounds even beyond what we find in *Grand Theft Auto V*. "Intoxication effects" could be one of the higher-level labels followed by "Drinking" and "Smoking." Under "Drinking" we might include "Wine," "Beer," "Whiskey." And under each of those we could have variations of processes that alter ambient sound or other diegetic effects to express the number of drinks imbibed, and therefore the degree of intensity to apply the intoxication effect settings. For the characters that can experience intoxication, we might insert a level label in the hierarchy for each character name, such as "Townley," under the level "Drinking." We would do this to consider how sensitive each character is to drunkenness and how potent is the beverage. Heavier and bigger people sometimes have higher tolerances to alcohol. But no matter the mass of the person, each character may have interesting idiosyncratic degrees of alcohol tolerance that could be considered in the structure of the sound bank hierarchy. Below the names of the characters might then follow acousmatic sounds (those with ambiguous origin) to represent a character's personal variation of alcoholic hallucination.

From this discussion, the reader might erroneously conclude that deviation from the naturalistic is triggered by the player's entry into a subjective state of perception, one encapsulated by an avatar character. Instead we should recognize that deviation from a game's construction of the naturalistic is a convention that the designer might use for familiarity or clarity of game play or a game's narrative. There are games that begin with less natural or unnatural audio whose physics never shift toward *the* "natural."

Similarly, the transformations from life to death, or to mission failure, are depicted as a convention in first-person and third-person shooter games. A player's avatar often diminishes in vitality due to a trigger of some number of body strikes by an enemy's blows or projectiles.

As an avatar nears on-screen death, animation often slows down and becomes blurry, shapes stretch, colors reduce to black and white and the lens of view speckles with splashes of blood-colored red. The death screen effect is not naturalistic, but subjective, approaching the surreal. As the avatar enters a near-death state, sound and image transform into an other-worldliness. Sound in on-screen avatar death often loses its regularity of pitch or volume, and gains a cavernous reverberation with fewer sounds at lower volumes, each sounding distant. The sonic transition to death or fail-ure has a diminishing relation to a naturalistic representation of the game space. While the player-characters in *Grand Theft Auto V* die or fail, they leave consciousness and control of the game with a subtle other-worldly pause from the gangster society. Other games provide more overt and sur-realistic player-avatar death transitions. In each, the player enters a simpler unnatural aesthetic to signify loss of consciousness and agency. The sound controller can reflect these shifts, as triggers from the game state can cause sound property settings to switch to those combinations designated for near-death game states. Reverberation and frequency compression is often increased. Directionality of sound orients to the center. Volume decreases as the end of life arrives.

2.7.   Acoustic Narrative Causality in *Thief*

Michel Chion describes how sonification of image action allows the viewer of audiovisual media to render in the imagination a space of experience for other senses (Chion 1994, pp. 109–111). David Bordwell evolves the discussion to include sound's narrative functions (Bordwell, Kristin and Smith 2016, pp. 293–300). Video games readily inherit much of cinema's aesthetic function for sound when game play outcome is altered by the hearing of a sound in the game space. This narrative technique, sound as lure for action, was noticeably used in the first release of Konami's video game *Silent Hill* (1999). Sound as trigger was evolved in Eidos' *Thief* (2014). The properties of a sound event and its location in the game space can change the player's game state, shaping the player's sense of space, affecting the narrative outcome of his mission, alerting listeners embedded in non-player characters. The player learns, by on-screen tips and by experimentation in the game space, that when Garrett, the thief avatar of the player, steps on carpets or other sound-dampening materials, the stealth score and ability will increase. The higher stealth score will better enable Garrett to steal an object and evade guards. A sound bank of the sound controller might address this design with a hierarchy starting with "Stealth surfaces." The next level might be labeled "High stealth" with sibling levels to choose from labeled "Medium stealth" and "Low stealth." For "High stealth" we could include the recorded sounds to play for a footstep event on carpet, snow or other soft absorbent material. Each of these may be triggered by a footstep colliding with a floor material, as we would expect. But each step collision could also trigger a change in the stealth-related scores of Garrett. Thus, a step on a tin roof could cause the guards in *Thief* to notice Garrett. If he steps on carpet, perhaps no guard will hear Garrett's presence. In this example, the sound bank is organized with a level that might also link to a character property, "Stealth." A footstep would change not only the sound played but also a property in the avatar state.

2.8.   Sonifying Objectively, Visualizing Haptically in *Spore*

In the game *Spore* (2008), developed by Maxis, the player can design their avatar as a simple multiple-celled organism. The game requires the player to move the avatar with the game controller through a liquid sea of other organisms. In the beginning levels of the game, the player feels the game controller vibrations when its organismic avatar touches other game objects. But the player sees in the monitor the avatar's movements and collisions in the third person. The sound design provides an objective and naturalistic aesthetic as well. But while the player moves through the game space with the locomotive constraints of the organism, he hears and sees as a human player, not as an organism. Thus, the haptic triggers of on-screen touch enabled by the game controller cause sounds to play that a microscopic multi-celled avatar creature would not likely be able to hear in nature. One might think that hearing, as the game represents it, is a unique feature of all living things. But that thought contradicts the evolutionary logic of the evolving avatar. An evolving organismic avatar would also evolve its hearing. But the game's sonic attributes do not change with organismic evolution. Throughout the game play of *Spore*, we experience sound objectively, regardless of the hearing abilities of the evolving organism avatar. Thus, the avatar's own point of audition is ambiguous. As the player progresses through the game, we see and hear as a third-person objective viewer, uncertain or unconcerned about what the avatar sees or hears. Instead, the player must convert third-person point of audition information into first-person haptic input through the handheld controller for the avatar to survive the threats of other organisms seeking the avatar as prey.

This objective-subjective split creates an "in-game" schizophonia (Collins 2013, pp. 23–26) that could affect the sound controller's implementation. Schizophonia is a much used cinema sound technique originally developed by Canadian composer R.M. Shaefer. Typically, a sound is recorded and then edited and mixed in such a way that the playback of the recording is completely disassociated from the recognizable original sound event. Spore makes use of this idea. We can imagine that the label hierarchies of the sound bank would describe sounds that reveal very little about the avatar in Spore, and instead describe the objects and materials of the game space. The hierarchy of the sound bank will likely have many distinct categories that have very few levels. Each category might describe collisions of materials such as "Claw strikes wood," or an ambient drone such as "Wind blows trees" or "Water flows rocks," or a category of utterances under a creature label such as "Howl," "Sigh" or "Grunt." The sounds themselves could be labeled with descriptions that indicate their intensity of energy, their duration, their rhythmic tempo or their number of varied repeats of the same sound type. Since sound effects need not be identified with specific characters but instead with materials and processes in *Spore*, the hierarchy can be structured with great breadth, few levels deep.

2.9. Layered Instrumentation in *Journey*

Few games use music as a primary interface for game play. In Thatgame-company's game *Journey* (2012), music provides an extensive component of the game play and player interface. The game play imposes some important rules. Music is always playing. Also, spoken or written words in languages known in the player's world, such as English or Chinese, do not exist. The "language" of *Journey's* world is constant musical expression. Occasionally, a language of inscribed and sometimes glowing rune-like characters is found on a few game props, or they float above the character at moments of contact with important objects or non-player characters. The game play objective is to move the player's armless robed avatar to a metaphysical mountain where the robed avatar will find an awaiting spiritual-like master character. To get there, the robed avatar must travel into deserts, through hills and passes, meandering through ruined civilizations and evading marauding automatons. To heighten the sensuality of the wordless experience, music responds to the animated aerial physics. Like many musical scores for media, they are layered with distinct instrumental parts, each assigned for layers of a score that express melody, harmony, rhythm or accent using specific instrumental sounds we might recognize as strings, percussion, horns and all other sections of the modern symphony orchestra. Each region of play within the game space triggers a score consisting of looping parts that progress melodiously as the robed avatar advances through a region of the game space. The game's musical score uses a recognizable key from the twelve-tone scale of Western musical tradition. The triggers in the game regions are usually tripped when the robed avatar finds and touches special objects, passes through tunnels or doors, or mounts a precipice or architectural spire. Once tripped, the triggers express an accent chord or trill in the same key as the score, and they advance the melody closer to a musical climax where chords harmonize and resolve. Often a trigger will introduce a new layer of instrumentation to the music mix, creating a denser musical expression and a more complex harmony or rhythm.

We should imagine the sound controller's role in *Journey* as the player progresses. Triggers are mostly tied to points in architecture or landscape. Thus, the game's levels become the uppermost labels of the sound bank hierarchy. Within each of these levels, the sound labels will describe a set of instrumental layers that must turn "off" or turn "on" the constant flow of the musical score. One notable accomplishment of *Journey* is synchronization. Without much complaint, a game player's ear and eye can accept a relatively wide margin of synchronization error between sound effects that play simultaneously. The game player's ear allows nearly no such error for synchronized music. Unlike the discussion so far, where sounds triggered by the sound controller will play on demand usually against an atonal and arrhythmic ambient level sound, percussive or melodic music cues that play simultaneously must play in musical time so that the intended notes line up and play precisely together. A mistaken start of a musical layer by even a beat can make music sound unbearable, destroying the intended experience or at least creating unintended meanings. Therefore, for *Journey*, or other similarly designed games, the score's tempo must play a role in the sound controller so that any trigger that calls a musical cue must begin only where the musical time signature will allow. The introduction of the concept of musical tempo does not change the sound controller model we have evolved. It does however create an imaginary conveyor belt that will allow a triggered sound to join the movement only at a permissible instant. Therefore, the event handler must include a mechanism that identifies the occurrence of each instance at which a sound it controls can start to play.

Not all musical sounds are controlled so precisely in *Journey*. Most of the "touches" the robed avatar makes are with glowing objects or with non-player characters. These touches set off instantaneous musical expressions that are not regulated by the musical time signature of the playing score. They occur exactly at the moment of collision when the robed avatar bumps the object or non-player character. But, these touches play a tone, or series of tones, that occur in the same or compatible key signature. The sound controller allows the designer to categorize a hierarchy of levels to describe the categories of touches; "key signature" and "tempo" are among the least mutable given that they remain the same throughout most game regions. Next, object and non-player character names might appear on the hierarchy level below. Then each of these may be categorized by their type. For example, some non-player characters are kind and helpful travelers and trigger similar musical cues that connote that character's demeanor. Other characters are dangerous and trigger ominous musical phrases. Within this label grouping, we might find the musical phrases themselves, labeled by their energy expressed in the sound file's original volume, tempo or beat count of the recording.

2.10. Physical Gesture Triggers for Dialogue in *L.A. Noire*

Games whose play depends largely on dialogue present a new problem for the model we have illustrated so far. The past examples have emphasized object collision triggers that cause sound events. For games that use a lot of dialog, avatar gestures in the presence of non-player characters often trigger the play of dialogue recordings. As we will discover, this problem allows the sound designer to shape the game world's expression of social atmosphere through oral inflection, word choice, speech tempo and volume as expressed by the spoken word. In Rockstar's and Team Bondi's *L.A. Noire (2011)*, the protagonist Cole Phelps is a late 1940s-era Los Angeles police officer who begins his journey on a beat that covers a vast stretch of the game's vintage Los Angeles map. If a player only pursues the mission objective, he may not explore the open-world features of the game, nor notice the subtlety of the sound designer's work with non-player characters and dialog. If the player wanders around the streets of Los Angeles, it won't be long before non-player characters pass him, approach him and react. If the player's Cole avatar merely lets a non-player character pass at a "socially appropriate" distance for a stranger, the non-player character will sometimes utter a passing comment about a personal thought. That thought might have some cryptic relevance to one of the narrative threads or missions of the current game level, or not. The non-player character may also express an impression or gossip about Cole or about police in general. The play of a dialogue line seems to be triggered by Cole's proximity. But the choice of the line is triggered by his recent previous behavior. If Cole steps closer to the non-player character, the dialogue reaction can become more personal, sometimes negatively judgmental and reactive to Cole's confrontational gesture. Cole can persist and step even closer, even confront or block the passage of the character. The non-player character may cycle through a few lines that reject Cole, or it may not respond at all. If Cole behaves politely, the lines are not as rejecting.

The sound controller's event management system will be important for Cole's dialogue interaction with the passing Los Angeles pedestrians. Each non-player character name or type (there are so many, they may not all have names) might be found at the top of the sound bank hierarchy. Under their names, we might find three variations on Cole's trigger gestures toward the non-player characters: "Pass," "Approach," "Confront." We notice that the level of action intensity expressed by Cole increases in the list order provided. We also should notice that the semantics of these labels limits the kind of interaction the game interprets for Cole. Given this list, Cole's action toward the non-player characters in this example suggests that as he approaches each non-player, his action consistently leans toward "Suspect" or "Accuse" rather than something else, like "Greet" or "Flirt." With triggers labeled as social interaction verbs, we then can name the dialogue recordings that react to Cole "passing" non-player character X, "approaching" non-player character Y or "confronting" non-player character Z. Cole's speed, distance and impact on collision provide parameter value triggers in relation to any given non-player character that causes the non-player character to speak a line. A residual memory of Cole's previous behavior suggests subtle artificial intelligence (AI) at work. The AI determines whether to choose dialogue from subcategories such as "Suspect," "Accuse," "Greet" or "Flirt."

2.11.   AI Triggers in *Bioshock Infinite's* Elizabeth

In Irrational Games' *Bioshock Infinite* (2013), the character Elizabeth is part of the super-objective for the game player's avatar, Booker DeWitt. To complete the game through DeWitt, the player must find Elizabeth then escape with her from Columbia, a mythic pseudo-Christian utopia that rationalizes its racially segregated class-divided society through prosperity-focused political and spiritual dogma. DeWitt eventually discovers Elizabeth locked in a tower guarded by a powerful giant flying mechanical sparrow-shaped monster named Songbird. While Elizabeth is a non-player character, the complexity of her programmed behavior while relating to DeWitt or the environments and obstacles they must overcome presents a remarkable challenge for the sound designer. The AI that brings Elizabeth to "life" is revealed in the sounds that she triggers. The fact that *she* autonomously triggers sound, rather than DeWitt, the player's avatar, makes this example distinct from the others. Artificial intelligence programming requires a few fundamental features that together give the observer the sense that an independent object has its own goals, thinks about the input provided and input remembered, decides what to do, then takes an action based on what it knows. First, Elizabeth appears to have a goal—self-preservation and escape. Her movement and words suggest she knows she is trapped in a tower and wants to survive. Her reaction to the arrival of DeWitt suggests at first that she sees him as a threat. She attacks him with books. But once he presents a key that could lead her out of the tower, she realizes she can trust him, so she invites him to follow her. On their escape journey she moves following impulses to discover and delight in the objects she finds around her. She also keeps an eye on DeWitt as they both search around for the right way to escape.

The sound design team was tasked with giving Elizabeth a liveliness of audible touch and speech. To do this, they had to mirror the emotional triggers developed by the AI programming. When we play *BioShock Infinite*, we discover that Elizabeth is a companion character whose super-objective is the same as the DeWitt's. This fact frees Elizabeth to pursue smaller sub-objectives along the path to freedom from Columbia. Clues to obtain her sub-objectives are found in positions that are always between where DeWitt stands and where DeWitt wants to go (Abercrombie 2016). When DeWitt seems to go in the "wrong" direction for completing the game's narrative path, or if she senses the arrival of a threatening character or situation, she expresses her doubts and warnings aloud. Thus, Elizabeth is programmed to react to DeWitt, and to react to other things that she senses—things that she likes and wants, and things that she doubts or that threaten her.

So how might the sound designer have used the sound controller to integrate the AI? We must notice that the input data to trigger sounds caused by Elizabeth had to come from her behavior rather than the player through DeWitt. On first consideration, it might seem that all the categories and their hierarchies of sound or dialogue should be very similar to our discussions so far of John Marston in *Red Dead Redemption* and the non-player characters of *L.A. Noire*. After all, Elizabeth moves and talks, as do *Red Dead Redemption's* Marston and *L.A. Noire's* pedestrian population of late 1940s Los Angeles. But what makes the sound design implementation distinct is that Elizabeth is not controlled by a handheld game controller like Marston is; she has a complex and robust memory of possible movements and utterances that use conditional logic and game state data values to trigger sound. We will give an example of conditional logic to illustrate sound design for reactive and event-memory AI for the sound controller and its interface.

Conditional logic uses two types of block statements—the if-then block and the switch-case block. The if-then block considers the truth value of a specific statement as a condition for executing a set of instructions contained in the block. If-then blocks can be grouped in a sequential chain of conditional statements that will be tested, of which the first one that proves to be true will execute its own subset of instructions, ignoring those that follow. If-then blocks can also be nested inside of the set of instructions executed when an outer if-then statement proves true. Given what we mentioned about Elizabeth earlier, some fundamental if-then blocks are apparent for reactive AI. For example, if DeWitt is looking in her direction and standing nearby, then she need not raise her voice to get his attention before or during her programmed utterance. Let's suppose he is looking in her direction. So, whatever expository dialogue she must say, she will say it conversationally low, reacting to DeWitt's proximity. During the escape from the tower scene of the game, she is programmed to react to the mysterious behaviors of statues and crumbling architecture activated by Songbird, her guard. We must imagine the set of instructions to find a nested if-then block. If stimuli from Songbird is true at a given instance and Elizabeth is still in mid-speech, the game allows a reactive pause of the speech for an instant to react to Songbird, otherwise it does not pause the speech. With this programming structure in mind, the sound design team must consider how they will direct the voice actor for recordings of Elizabeth during the tower escape scene. The recordings of Elizabeth's voice must include silent pauses so that the controller can take an event like Songbird's intrusion and stop the speech in a predetermined place so that the utterance does not sound clipped. So even though the sound bank hierarchy might resemble the broad and deep labeling structure one might expect for any character's possible sounds, the many combinations of events that trigger the playing and stopping of these sounds are coming from a conditional algorithm instead of the player's handheld game controller.

Another feature of AI is that the part of its algorithm that can trigger audio also can access a stored value in memory of previous events. That memory can change the set of possible utterances from Elizabeth. During the journey out of Columbia, some lines she will only say once. Therefore, we must keep in mind as a sound designer all the possible event contexts and their point of audition pairs where she could express a one-time-only line. We must be aware of where she could be physically placed (e.g. near, far) in relation to DeWitt (close enough to hear her), and how other things she might say after saying such a line should or should not be affected by already having said such a line to DeWitt. Here again, we consider the sound bank hierarchy and the possible branches in the structure for this kind of design. Sibling branches could be labeled to consider if DeWitt has heard the one-time-only line: "DeWitt heard X" and "DeWitt heard not X." Below this branch would be recordings of any dialogue that should express Elizabeth's different inflections that indicate she knows that DeWitt has heard or has not heard information about X.

We mentioned that the switch-case blocks also were related to Elizabeth's AI and the sound it triggers. Unlike conditional block statements that test the truth of statements for a value or combination of values, switch-case statements take as input one member of a finite set of constants. For each value in the set, a set of instructions is executed. With the relationship between Elizabeth and DeWitt, we can illustrate a possible implementation of reactive AI from the previous example. When DeWitt finds Elizabeth, she must decide if she will escape with DeWitt and trust him. While she is hearing the series of scripted lines that DeWitt expresses as he gets oriented to the room in the tower where he finds her, he is free to move farther from or closer to her. She too has scripted lines. But if he is close, she need not yell. If he is far, raising her voice would show that she senses his distance, and can modulate her vocal energy and volume depending on his distance from her. But since voice recordings must be recorded at a fixed energy, the voice talent had to express and record the same line with varying energy to accommodate a distance parameter between her and DeWitt. So, the programmer must develop a variable that can be "clamped" to a finite set of constant values when a switch-case statement is used. Let's call this variable *distance*. We can give the switch-case block variable distance a value of 1, to represent all values when Elizabeth is close, and 2 for when she is far. When the switch-case block receives a 1 in its distance variable, it will call the event handler to play the lower energy recording of Elizabeth's line for the case when she is near DeWitt. When the switch-case statement receives a 2, it will call the event handler to play the far and loud recording of Elizabeth's line. Both recorded dialogue lines have the same semantic content, but one is quietly expressed by the voice actor, the other, loudly. With these known possible values, the game sound designer can prepare combinations of settings in the sound bank label hierarchy to fit each of the required possibilities of the AI programming. The discussion of Elizabeth's AI describes the simple implementation of reactive and event-memory AI.

2.12. Conclusion

Game sound designers ultimately deliver a large set of sounds for any given game. If the sound design is rich with detailed variations for many possible paths through the event structure, many sounds will never be heard during any given player's path through the game. This fact strongly implies that the aleatoric characteristics of game design are what differentiate it from other types of sound design for media. One tool that gives game sound designers expressive power is the game sound engine, with its sound controller and interface. Standing out from its various features, its interface and the sound bank (or equivalent) is critically important in responding to the many triggers that a game sound designer may designate for collisions, clock changes, non-player character reactions, AI programming, game progress through space or other quantifiable measures. Thus, as game sound designers learn and create new interactive media for their games, they will benefit from developing an awareness of programming principles and the structures they infer. The labeling hierarchies game sound designers organize in the sound controller will reflect their understanding of a game's underlying logic and overall design. The "fun" part of game sound development, selecting sound event triggers and recording and editing sounds to respond to them, must be followed by that more cerebral and sometimes tedious step of interpreting the conditions that trigger sounds and creating the categories of sounds that can play once the triggers are tripped. We should now remember how hard it is to make game play fun. As we insist that our games provide increasingly complex and immersive experiences to satisfy our expectation of pleasure, the more the game designer must also invent an aesthetic with the technology that keeps the player engaged in an insatiable goal to finish or win.

Notes

1. Michel Chion calls this epistemic relationship *synchresis*—to believe that a synchronized image collision is a sound's cause.
2. The concept of an event handler is most often used in Java, Javascript, C# and C++, all languages used in the API of commercial game engines. But the idea is the same. The instruction caused by an event trigger is the handler.
3. There are many games where the player hears sounds (voice-overs, sound effects and music) that "emanate" from the mind of the player's character as sonic fantasy or memory. But the player's mind is not the "virtual" physical space of the game that I refer to here. It is also possible that the player character's mind can be programmed to filter and augment the pseudo-naturalistic sound of the game space, justified by subjective aesthetics. Even in this case, the game physics for sound will have a role in referencing (not simulating) nature through its application of processes to shape the fundamental sound properties.
4. Applying a fixed setting to a sound in a game is one among several ways to minimize a sound's use of calculations performed on the CPU or GPU. Fixing (or "baking") a group setting allows the game to respond to new events more rapidly. But, the resemblance to accurate reverberation may stray somewhat from simulating the natural world.
5. The idea of a consistently applied audio framework implementation has gained attention and use with the availability of more powerful processors and physics-based sound synthesis systems (Raghuvanshi et al. 2007, p. 68–69).
6. It should be noted that contemporary middleware is a relatively recent software innovation that solved a more laborious and limited sound design workflow. Horowitz and Looney note that as game design became more complex (and more aleatoric),

middleware became an urgently needed tool for the game sound designer (Horowitz and Looney 2014a, p. 124).

7. The concept of the listener, handler, parameter and trigger are common in coded game audio controlling systems that use C++, Java or other object-oriented languages. However, some middleware implicitly adopts these concepts, often without revealing these named relationships.

8. We can generalize the count of all possible combinations with product notation: $\angle n_i$ where n is the size of the set of acceptable values for the ith parameter. If we reduce any ith parameter by i-1 values, the total size of all combinations reduces by a factor of i.

References

Abercrombie, J., 2016. Bringing BioShock Infinite's Elizabeth to Life: An AI Development Postmortem. Also Known as Building the AI for BioShock Infinite's Elizabeth, Game Developer's Conference 2014. *YouTube Video*. Viewed Apr 25, 2018 www. youtube.com/watch?v=wusK-mciCVc&t=2369s

Bordwell, D., Kristin, T. and Smith, J., 2016, *Film Art: An Introduction*, 11th edition. New York: McGraw-Hill.

Bottomore, S., 1998. The Story of Percy Peashaker: Debates About Sound Effects in the Early Cinema. In: Abel, R. and Altman, R. (Eds.), *The Sounds of Early Cinema*. Bloomington: University of Indiana Press, pp. 129–142.

Brandon, A., 2006, Game Audio Integration in *Mix*. Mar, 30, no. 3, pp. 55–60.

Brandon, A., 2007a, Audio Middleware in *Mix*. Mar, 31, no. 3, pp. 68–70.

Brandon, A., 2007b, Audio Middleware, Part 2 in *Mix*. Apr, 31, no. 4, pp. 66–68.

Brandon, A., 2007c, Audio Middleware, Part 3 in *Mix*. May, 31, no. 5, pp. 82–85.

Chion, M., 1994. *Audio-Vision*. New York: Columbia University Press.

Collins, K., 2008. *Game Sound: An Introduction to the History, Theory, and Practice of Video Game Music and Sound Design*. Cambridge, MA: The MIT Press.

Collins, K., 2013. *Playing with Sound: A Theory of Interacting with Sound and Music in Video Games*. Cambridge, MA: The MIT Press.

Cramer, F. and Fuller, M., 2008. Interface. In: Fuller, M. (Ed.), *Software Studies: A Lexicon*. Cambridge, MA: The MIT Press, pp. 149–153.

De Leeuw, T., 2005. *Music of the Twentieth Century: A Study of Its Elements and Structure*. Amsterdam, The Netherlands: Amsterdam University Press.

Horowitz, S. and Looney, S., 2014a. *The Essential Guide to Game Audio: The Theory and Practice of Sound for Games*. Burlington, MA: Focal Press.

Horowitz, S. and Looney, S., 2014b. How To: Master Class: A Guide to Game Audio Middleware. *Electronic Musician*, Jul, 30, no. 7, pp. 64–72.

Kassabian, A., 2008. Rethinking Point of Audition in *The Cell*. In: Beck, J. and Tony Grajeda, T. (Eds.), *Lowering the Boom: Critical Studies in Film Sound*. Urbana, IL: University of Illinois Press.

O'Keefe, L., 2011. Sound Is Not a Simulation: Methodologies for Examining the Experience of Soundscapes. In: Grimshaw, M. (Ed.), *Game Sound Technology and Player Interaction: Concepts and Developments*. New York: Information Science Reference, pp. 44–59.

Pejrolo, A. and Metcalfe, S.B., 2017. *Creating Sounds from Scratch: A Practical Guide Music Synthesis for Producers and Composers*. New York: Oxford University Press.

Raghuvanshi, N., Lauterbach, C., Chandak, A., Manocha, D. and Lin, M.C., 2007. Real-Time Sound Synthesis and Propagation for Games. *Communications of the ACM*, 50, no. 7, pp. 66–73.

Skalski, P., Tamborini, R., Shelton, A., Buncher, M. and Lindmark, P., 2010, Mapping the Road to Fun: Natural Video Game Controllers, Presence, and Game Enjoyment. *New Media and Society*, 13, no. 2, pp. 224–242.

Unity3d, Web Page Entitled "Audio". Viewed Mar 24, 2018 https://docs.unity3d.com/Manual/Audio.html

Unreal Engine, Web Page Entitled "Audio and Sound". Viewed Mar 24, 2018 https://docs.unrealengine.com/en-us/Engine/Audio

Yewdall, D., 2011. *The Practical Art of Motion Picture Sound*. Burlington, MA: Focal Press.

3

Designing Game Audio Based on Avatar-Centered Subjectivity

Jonathan Weinel and Stuart Cunningham

3.1. Introduction

Since the early days of computer games, the graphics, sound and narrative capabilities of the medium have steadily advanced. Sprites and early forms of wireframe 3D with flat shading gave way to accelerated OpenGL 3D in the 1990s, which was soon augmented with various effects such as colored lighting, bump mapping, fog, transparencies and lens effects. In game audio, early "chip sounds" were superseded by digital sampling, eventually leading to multichannel CD-quality sound effects and musical soundtracks that respond dynamically to events in the game (Collins 2008). In the current state of the art, video games within genres such as the first-person shooter (FPS) are able to depict highly realistic three-dimensional environments, which the user can navigate interactively with an avatar.[1] With the latest virtual reality (VR) equipment, this drive towards realism continues, as designers seek to immerse users in virtual environments that engulf the senses.

In immersive simulations, first-person perspective (FPP) is often used, which provides the user with an "eye-view" from the subjective perspective of an avatar. FPP video games do this by utilizing a virtual camera, which is typically located on the head of an avatar, thereby allowing graphics to be generated based on the virtual perspective of the avatar. In effect, the camera acts as a substitute for the eyes of the avatar, allowing us to view what the avatar would be seeing from its location in the virtual world. Similarly, three-dimensional audio is calculated based on the location of the avatar relative to the virtual sound sources within the level. This allows the amplitude of sound sources to change based on their relative distance from the avatar and for environmental effects such as reverberation to be applied, for instance. Using such techniques provides the user with the illusion of seeing and hearing the virtual environment from the perspective of the avatar, which contributes to the user's sense of embodying it. This sense of embodiment may also lead to a feeling of "presence," the sensation that the user has when he or she feels that they are really there in the virtual world (Slater and Wilbur 1997). Following Slater and Wilbur's definition, this sensation can be distinguished from "immersion," which describes the capabilities of technology to envelop the senses of the user. For video games, this feeling of presence, along with the achievement of flow (Csikszentmihalyi 1997, p. 39), may be among those features that make the experience particularly compelling, and perhaps partly explains why genres such as the First Person Shooter have gained such popularity since the 1990s.

Immersive technologies, such as the latest VR devices, augment FPP by bringing the user yet closer to the visual and aural perspective of an avatar, by mounting a stereoscopic display on the head of the user, which can be orientated in correspondence with the avatar. This impression is used by VR games and 360 VR videos in various ways. For instance, *Affected the Manor* (Fallen Planet Studios 2016) places the user within a terrifying haunted house that he or she must navigate through. The 360 VR video *Clouds Over Sidra* (Arora and Pousman 2015), by contrast, situates the viewer within a refugee camp, with the aim of promoting awareness of the struggles faced by people in the camp. Along similar lines, the *Autism TMI Virtual Reality Experience* (The National Autistic Society 2016) portrays the experience of a person with autism using FPP, to generate awareness of autism. Meanwhile, *The Machine to Be Another* (BeAnotherLab 2016) explores the idea of using VR to see through another person's eyes, by switching the viewpoints of two individuals using live cameras. Each of these examples utilizes FPP to create the illusion of seeing an environment from the alternate perspective of an avatar, which may be fictitious or based upon a real person. Yet the last three examples also show how VR is not only being used to provide a sense of being *there* within the environment, but also how it is being used to provide a sense of actually being *someone* (the avatar).[2] This concept—the idea of communicating the experience of a virtual avatar—is a key theme that we shall explore in this chapter.

For media that seek to convey the experiences of a virtual avatar, we must consider how this can appropriately be achieved through their design. In video games, VR applications and 360 videos that use FPP, there are two distinct approaches that may be followed. The first of these is to employ cameras and microphones, or their virtual equivalents, to capture the *objective* patterns of light and sound in the approximate location where the head of the avatar would be.[3] For the user, the approach replicates the incoming patterns of light and sound that would be received by the avatar in the given situation. The second approach must also do this, but takes further steps to represent the *subjective* perception of the avatar, by using various techniques to represent how the visual or auditory information is processed. The key difference between these two approaches can be clearly highlighted if we imagine a hypothetical scenario in which an avatar has consumed a hallucinogenic drug. In this situation, the avatar would be in an altered state of consciousness (ASC), which substantially changes their subjective perception of the world around them (Ludwig 1969). Using the first *objective* approach to FPP, the patterns of incoming light and sound that are external to the avatar would be represented, thereby giving no indication of the hallucinations or altered perception experienced by the avatar. By contrast, the second *subjective* approach would represent the visual and auditory experience of the avatar including the hallucinations, which might be shown by increasing the intensity of colors or using sound to represent auditory hallucinations. The key difference is that the subjective approach takes into account the perceptual processing of the avatar.

The second of these approaches, in which graphics and sound seek to represent the *subjective* visual and auditory perception of an avatar, is our main focus in this chapter. This approach, which we shall refer to as "avatar-centered subjectivity," may provide a means through which the user can gain a deeper sense of connection with an avatar, since it communicates more aspects of the avatar's subjective perception.[4] For example, using this approach we may represent how an avatar directs attention towards incoming sensory stimuli; has certain types of emotional reactions; or perceives sensory hallucinations that arise internally within the brain. Such information may allow the user not only to understand what it is like to be *where* the avatar is, but also to understand what it is like *to be* the avatar.

In order to explore the concept of "avatar-centered subjectivity," in this chapter we first discuss a variety of prototypes that demonstrate approaches to sound design based on subjectivity. These prototypes focus in particular on the concept of ASCs, such as the intoxicated experiences and hallucinations that people may have on psychedelic drugs. Such experiences provide an interesting way to explore notions of subjectivity, because they include experiences of hallucination that are more easily distinguished as "subjective." Through our discussion of these projects, we illuminate some approaches for sound design in relation to subjectivity, within the interactive context of video game engines. Following this, we then describe a framework for "avatar-centered subjectivity," which draws upon various theories of consciousness and cognition, drawing notably on the approaches of information processing.[5] Through this discussion, we aim to define the concept of "avatar-centered subjectivity" in terms that will point towards its generalized use for a variety of purposes, such as interactive sound design for video games and other networked communication tools that utilize avatars.

3.2. Prototype Systems for Representing ASCs in Video Games

3.2.1. *Quake Delirium*

In order to explore ways in which representations of ASCs could be designed in a video game through practice-led research, our Affective Audio research group (www.affectiveaudio.net/) created several prototype systems.[6] The first of these was a development of Weinel's *Quake Delirium* (2011), a modification of the video game *Quake* (id Software 1996). Using a Max/MSP patch and various scripts, *Quake Delirium* enabled the use of a MIDI mixing device to adjust various graphical and game parameters such as field of vision (FOV), drunk mode (causes the camera to sway drunkenly), fog density/color, game speed, stereo vision (stereoscopic effect for red and blue 3D glasses), gamma and red hue. By manipulating these parameters with a haptic device, fluctuations in the appearance and speed of the game could be introduced, reflecting the way in which an ASC may cause visual disruptions and alter perception with regards to the flow of time (Figure 3.1). The modification allowed the sequence of these changes to be automated, while the same fluctuating parameters were also used to control various sound-generating processes in a Max/MSP

Figure 3.1 Screenshot from *Quake Delirium*: A Max/MSP patch is used to automate distortions to graphics and sound, in order to represent a hallucinatory experience that varies in intensity over time.

patch. The latter caused filters and other audio effects to follow the visual changes, thereby producing a corresponding electroacoustic soundtrack. This connection between the soundtrack and the visual distortions was conceived as a way to reflect the synesthetic aspects reported in hallucinatory experiences.[7]

Quake Delirium EEG (Weinel et al. 2015a) expanded the system by utilizing a consumer-grade EEG headset (a Neurosky Mindwave), as a controller for the fluctuating parameters. Here, the main intention was to explore how a brain-computer interface (BCI) might provide a form of "passive" control,[8] linking the biosignals of the player to the simulation of hallucination, and thereby providing a deeper sense of connection between the player and the avatar. As part of the project, we explored the user experience of the system. Here we found some challenges in providing the user with a meaningful sense of connection and interactivity via the biofeedback, perhaps due to the quality and tangibility of the EEG signals as a controller source. Similar issues were also highlighted in a related study exploring the user experience of *Psych Dome*, an EEG-controlled audio-visualization based on the visual patterns perceived during hallucinations (Weinel et al. 2015b). Despite these shortcomings, we suggest that future systems could utilize improved biofeedback devices and build upon the basic approach demonstrated by *Quake Delirium EEG*, in order to yield effective results.

3.2.2. *Auditory Hallucinations Project*

While some existing representations of ASCs in video games[9] may rely on intuitive notions of what a hallucination might entail, we can argue that if improved accuracy is sought, we should refer to research regarding the experiences people actually describe during ASCs. Towards this aim, we carried out a large-scale analysis of nearly two thousand experience reports gathered from an online database, which described intoxication from a variety of substances such as LSD, MDMA, amphetamines, alcohol and others (Weinel, Cunningham and Griffiths 2014). Using these descriptions we were able to isolate key features that people described during intoxicated experiences of auditory hallucination, and assemble a large number of qualitative descriptions to use as a resource for designing audio.

Using the qualitative descriptions, we designed sound materials based on different types of auditory hallucinations, such as auditory-verbal hallucinations, hallucinations of noises and of music. These sonic materials were categorized in relation to Hobson's (2003) three-dimensional "AIM" model of consciousness, with the axes of "activation," "input" and "modulation." "Activation" describes the level of brain activity; "input" describes the origin of sensory inputs; and "modulation" describes how events are recorded to memory. The "input" axis is of particular interest for our purposes here, and ranges from "external" to "internal." "External" sensory inputs originate in the surrounding environment, whereas "internal" sensory inputs are generated within the brain. According to this model, dreams and hallucinations can be considered as a predominantly "internal" form of sensory input.[10]

Utilizing the concept of the input axis, we designed sounds based on auditory hallucinations that corresponded with "internal" sensory input, and digitally manipulated sounds based on "external" environmental sensory inputs, allowing us to traverse the "input" axis of Hobson's (2003) "AIM" model of consciousness through sound. Along the midpoint of this axis, where the perception of everyday sounds becomes enhanced or distorted due to intoxication, we used a variety of digital signal processing techniques such as reverb, delay, flanger, EQ, time stretching and spectral processes to give an impression of warped perception. This was undertaken by using the qualitative descriptions as a guide to decide which type of sound manipulation process to use. Developing a trajectory that was explored earlier in Weinel's (2012) PhD work, in which he composed electroacoustic music based on altered states of consciousness (see also Weinel 2016), several of these demonstration pieces were also assembled into miniature electroacoustic compositions.

Among these, Cunningham's *LSD No.1* (2014) was performed at several international electroacoustic concerts, such as *Étude de l'objet: Bilder und Klangbilder* held at the Kurt-Tucholsky-Literaturmuseum in Rheinsberg, Germany (2017). The piece was produced in response to a specific account of LSD hallucination gathered from our study:

> I was hearing voices in the wind when I had the window down, and I heard bizarre, distorted, nearly demonic voices coming out of the radio if I turned it on. I had my driver's license and car registration out on the dashboard, and I was ready, if I got pulled over by a cop, to tell him or her that I was way beyond anything that could be tested and to just arrest me and have it finished. But I made it home. I was on the back side of the acid peak by this time, but still really high.[11]

This account features descriptions of both "external" ("real") and "internal" (hallucinated, or "unreal") experiences of sound, which could be designed using corresponding layers of digital audio. An "external" layer of sound was constructed using a recording made in the interior of a car during a journey. This layer includes engine hum and road noise, interspersed with the sound of gears being shifted and car turn indicator clicks. Car radio sounds are introduced approximately halfway through the composition, which were made by recording the sound of an analog radio being tuned, introducing static and occasional clips of radio stations. Band-pass filtering was also applied to these, to reflect the typical frequency response of a car stereo. An "internal" layer of sound, which describes the subjective aural experience of hallucinations, was also constructed using recordings of male and female voice actors. A degree of artistic license was applied to elaborate on the basic description. In this case the word "acid" was recorded several times, to provide the ". . . voices in the wind . . ." component of the text. These sounds were processed with a high-pass filter, augmented with short, frequent delays, placed at a low level in the mix and alternately panned. The "demonic voices" on the radio were recorded by having the actors shout statements that might be relevant to the situation, such as "wind that window up," "acid," "who do you think you are," and "listen to me." Multiple sections of these recordings were overlaid to create a sense of confusion and distress. The voices were all band-pass filtered and processed with distortion. Several were pitch shifted into lower registers, to create a slower and more sinister sound. Finally, towards the end of the piece, the radio tuning sound was reintroduced and the sound of the radio being switched off is heard, as the listener is returned to the normality of the car interior. In this way the structure of the piece reflects a transition through external and internal layers of sound.

While *LSD No.1* and the other demonstrative compositions we created as part of this project were fixed-media outputs, the separation of materials into the layers of "external" and "internal" sounds allowed us to begin thinking about ways in which a real-time system could be developed, which parameterized sensory "input" as a basis for interactive sound design. For example, an extension of this project could map layers of "external" and "internal" sounds to separate channels of audio within a game engine, and mix between these in real time when a player avatar becomes intoxicated.

3.2.3. ASC Simulation

Building upon the previous projects, the first iteration of our *ASC Simulation* demonstration patch was conceived to explore other ways in which an interactive system could represent auditory hallucinations through real-time mechanisms (Weinel and Cunningham 2017). A prototype was created in the Unity game engine, consisting of a simple three-dimensional game environment with several colored cubes, each of which emit different sounds such as synthesized tones. This was used as a basis to explore three prototype mechanisms for representing auditory hallucinations using real-time approaches, which were based on ASC features identified from the previous study.

The first of these mechanisms looked at the idea of "selective auditory attention." This concept is often illustrated with the example of the "cocktail party effect," in which amidst a noisy room of people talking, it is possible to selectively focus on a single conversation. Broadbent (1958) carried out important early work in relation to selective auditory attention, arriving at his "filter theory of attention," which considers how attention directs incoming auditory inputs, promoting some for higher-level processing, while others are filtered out. Subsequent theories such as Treisman (1960) developed this idea further, suggesting that unattended sources are "attenuated." More recently Bregman's (1994) theory of "auditory scene analysis" (ASA) further developed our understanding of how complex aural sources can be separated into different auditory streams, distinguishable from one another as separate objects, though potentially constituted by complex acoustical phenomenon. For some ASC reports we looked at in the previous study, absorbed attention towards certain objects was pronounced, and hence we were interested in modeling this effect for an avatar using a real-time process. Our design selects which of the sound-emitting cubes in the demo game is being attended by the avatar, based on which of the cubes the avatar is looking at.[12] When the avatar attends a cube, the amplitude for all other cubes is reduced in accordance with a time-varying envelope, as shown in Figure 3.2. This mechanism therefore demonstrates an approach for interactive sound design based on the concept of selective attention.

The second mechanism our *ASC Simulation* demonstrates is that of "enhanced sounds." This aims to model a feature described in reports of intoxication, where sounds are perceived as more bright, intricate, detailed, enjoyable or interesting than usual. As illustrated in Figure 3.3,

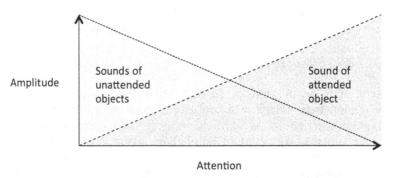

Figure 3.2 The "Selective Auditory Attention" Mechanism of the *ASC Simulation* Project Changes the Amplitude of Sound Sources Based on the "Attention" of the Avatar: Note that as the avatar looks at a sound source, all others fade out, so that only the object currently within attention is heard.

we split the source material for the sound-emitting cubes into three copies, each of which were then processed with EQ to create "dull," "medium" and "bright" versions of the sonic material. This then allowed us to manipulate an "enhancement" parameter, which applied cross-fading playback between the three versions of the material, thereby eliciting sounds from the cubes that became gradually brighter in correspondence with the degree of "enhancement." This demonstrated another specific effect that could be used as part of a package for representing intoxicated states using real-time processes.

Our third mechanism related to reports from our auditory hallucination study, in which people described disruption to their experience of the spatial location of sounds. As shown in Figure 3.4, we developed a real-time demonstration that caused the sounds being emitted from the

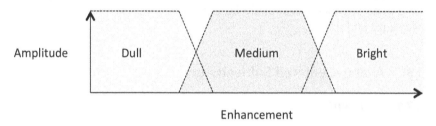

Figure 3.3 The "Enhanced Sounds" Mechanism of the *ASC Simulation* Adjusts the EQ Parameters of All Sound Sources, by Cross-Fading Between Different Versions: Note that as the "enhancement" meter increases, the source material mixes between "dull," "medium" and "bright" versions of the source material.

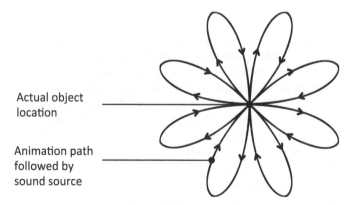

Figure 3.4 The "Spatial Disruption" Mechanism of the *ASC Simulation* Causes Sound Sources to Move in an Oscillating "Flower" Pattern Around the Central Point Where the Associated Object is Located in Three-Dimensional Space.

cubes to follow oscillating patterns of movement around the actual physical object, which remained stationary. Using this technique, the sounds of the cubes became detached from where they should be in terms of their spatial location, therefore demonstrating one possible real-time approach for representing spatial disruptions due to intoxication.

ASC Simulation is the most recent in a series of prototypes utilizing practice-led research as a means to explore ways to represent ASCs using video game development platforms. Through this project we have demonstrated a variety of specific techniques that could be used for representing effects of intoxication described in the experiential reports that we analyzed. However, ASCs are a very specific type of subjective experience, and therefore this work could also be considered as a subset of a larger project in the field of "avatar-centered subjectivity." With this in mind, in the next section we will seek to define "avatar-centered subjectivity" as a more general concept that could be utilized for representing ASCs, affective states and others.

3.3. Avatar-Centered Subjectivity

3.3.1. Concept

The concept of "avatar-centered subjectivity" proposes that we can model various features related to the conscious state of a particular avatar, such as may be used in video games or networked communication tools (e.g. those that utilize virtual avatars, such as social media, live chat and Second Life). By parameterizing various features of the avatar's cognitive state, and modeling specific perceptual systems, we can make this information available to inform how graphics and sound are rendered. In doing so, we can aim to communicate various aspects of the avatar's subjective experience to the user, such as sensory experiences or emotional states.

To develop such systems in ways that meaningfully correspond with human perception, we can inform their design based on research from the domain of cognitive psychology. For example, in this chapter we have already referred to Hobson's (2003) "AIM" model of consciousness, and theories of attention such as Broadbent's (1958) "filter theory" and Bregman's (1994) "auditory scene analysis." These and other established theories could be used to develop models of avatar-centered subjectivity with increasing levels of complexity. Eventually, by modeling multiple systems that exchange information collaboratively, we could conceive of designing highly complex VR systems that provide a "theatre of consciousness" (to adopt the metaphor used by Baars 1997). While such designs are a more long-term goal for this research, here we shall illustrate a basic system

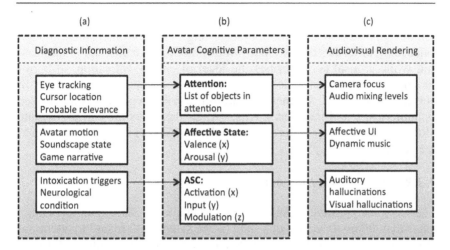

Figure 3.5 Diagram Illustrating the Proposed Concept of "Avatar-Centred Subjectivity" as a System That Utilizes Theories of Attention, Affective State and ASC: "Avatar Cognitive Parameters" (b) are acquired via "Diagnostic Information" (a) processes, and are used to inform "Audiovisual Rendering" (c) in a video game or VR application.

that could be developed and utilized for the purposes of interactive sound design with relative ease.

Our proposed system for "avatar-centered subjectivity" is illustrated in Figure 3.5. In this figure there are three main groups: "Avatar Cognitive Parameters" (b) store information regarding the attention, affective state and ASC of the avatar, which is acquired via "Diagnostic Information" (a) processes. The "avatar cognitive parameters" then become available to inform various aspects of real-time "Audiovisual Rendering" (c). In the following sections, each group will be discussed in further detail.

3.3.2. Avatar Cognitive Parameters

The core of avatar-centered subjectivity is group (b) labeled "Avatar Cognitive Parameters," which provides a set of attributes that describe the conscious state of the avatar. Here these attributes are limited to the categories of "Attention," "Affective State" and "ASC," though other models or categories could also be used. Following our discussion of Broadbent's (1958) "filter theory" and Bregman's (1994) "auditory scene analysis," "Attention" describes the game objects that are currently within the attention of the avatar; this could be modeled as a list or collection of game objects. "Affective State" describes the valence and arousal values of the avatar, in accordance with Russell's (1980) circumplex model of affect. Russell describes mood and emotion according to a two-dimensional model: valence (x) describes the extent to which emotions are pleasant or unpleasant, while arousal (y) describes the level of energy activation, in correspondence with systems such as the autonomic nervous system (ANS). In our proposed model, valence and arousal can likewise be parameterized as numerical x and y values. Lastly, "ASC" describes the conscious state of the game character according to Hobson's (2003) AIM model, with x, y, z values.

3.3.3. Diagnostic Information

The "Avatar Cognitive Parameters" (b) group provides the main set of attributes that define the conscious state of the avatar with regard to attention, affective state and ASC. This information needs to be acquired through some systematic means, which here we have suggested through group (a) "Diagnostic Information." This group provides methods through which the required parametric data can be acquired. Firstly, as demonstrated through our *ASC Simulation* prototype, a simple way to get attention information is by analyzing what the avatar is looking at (cursor location). If eye tracking is available, this could also be used to identify which objects the user is actually looking at, as explored by Hua, Krishnaswamy and Rolland (2006), Sundstedt (2010, p. 5) and Turner et al. (2014). Additionally, in the context of a game narrative, predictions could be made about which objects are likely to be of most relevance to the avatar; for example, an avatar searching for a door key might be drawn to looking at door key objects more than other items in a scene. Through a combination of methods such as these, it is possible that a list of attended objects can be provided, which is updated on an ongoing basis.

For identifying the affective state of the avatar, we propose that the movement of the avatar could be analyzed. Where an avatar is moving rapidly, this is likely to produce a high arousal state, due to the interaction between physiological condition and affective state. Conversely, an avatar that is not moving would likely have low arousal. Another approach could be to analyze the surrounding soundscape. For example, we can conceive in a war game that a soundscape filled with explosions may suggest a more fearful affective state for an avatar than one containing calm environmental sounds such as birds singing. We could also examine the narrative of the game, since events in the story of the game could be defining features that suggest what emotional state the avatar would be in. This would provide a useful way to determine the valence state of the avatar and could be accompanied by other in-game variables such as level of health and overall progress on the current task or mission. Using such information can provide "Diagnostic Information" (a) that informs the affective state component of the "Avatar Cognitive Parameters" (b).

Lastly, we need to consider how the ASC state of the avatar can be determined. Where states of intoxication are concerned, this could also be analyzed based on events in the game, such as the consumption of substances that cause an altered state of consciousness. Here we might use time-based envelopes with attack and decay parameters to model how the effects of a drug intensify over time and eventually wear off (for further discussion, see also Weinel 2016). Once we know what type of ASC a character would be in, we can define values for the AIM values of the "Avatar Cognitive Parameters" (b) accordingly, following Hobson's (2003) discussion of how these change during different states. Similar approaches could also be used for dream states; for instance, an avatar that remains undirected by the user for a period of time might start to daydream, or possibly fall asleep, precipitating corresponding changes to the AIM values.

3.3.4. Audiovisual Rendering

Once the "Avatar Cognitive Parameters" (b) are available within the game engine, they can be used to inform "Audiovisual Rendering" (c), which defines how various aspects of graphics and sound are rendered in order to communicate the subjective experience of the avatar in real time.

With regard to graphics, for instance, attention information could be used to focus the camera that represents the avatar's visual experience on attended objects. In the sound domain, "attention" could be used to adjust the mixing of audio levels for attended objects in real time, using an approach similar to the one we prototyped in the *ASC Simulation*.

With regard to "affective state," valence and arousal information could be used to inform the design of UI elements such as a visual impression of the avatar's facial expression, and/or other symbolic information regarding emotional state. In sound, these values could also be used to control parameters of dynamic music, so that the music changes emotion in accordance with the character to follow his or her affective state or to trigger the playback of a character monologue about their situation. In the case of the adaptive music, this would most likely be implemented by using the "Avatar Cognitive Parameters" (b) as controllers for a dynamic music interface, such as those provided by FMOD or Wwise.

The "ASC" values of the "Avatar Cognitive Parameters" (b) could also be used to control aspects of how graphics or sound are rendered. For instance, utilizing the "input" (y) parameter of the AIM model, we could design a system that modulates between "external" and "internal" graphical elements and three-dimensional geometry, effectively moving the avatar between real and unreal virtual environments. For the purposes of sound design, this input parameter could be used to gradually distort the "external" sonic materials, and introduce "internal" sonic materials related to auditory hallucinations. This may be achieved using approaches we began to explore through layers of sound in the *ASC Simulation* project, which were discussed earlier in this chapter. Using such techniques, as the avatar enters an ASC, real-time sound design could reflect that state by modifying the presentation of the virtual soundscape, and by introducing unreal sounds that reflect dreamlike or hallucinatory states.

3.4. Conclusion and Future Work

Through the course of this chapter we have defined the concept of "avatar-centered subjectivity." This concept is intended for use in video games, VR applications and networked communication tools, in order to provide a means through which to enhance the communication of information regarding the subjective experiences of avatars. This concept emerged from, and is informed by, our practice-led work developing several prototype systems for representing ASCs. Though our work in this area has focused predominantly on ASCs so far, the concept of "avatar-centered subjectivity" can be used to explore a whole range of cognitive systems, in order to model the ways in which subjective perception can vary for individuals under different circumstances. Therefore, it may be useful for looking at not only ASCs and states of intoxication, but *any* type of subjective perceptual state.

Working towards this goal, in this chapter we have outlined "avatar-centered subjectivity" as a general concept, which we propose points towards a system that could be useful for a much broader range of avatar designs. At its core, the concept of "avatar-centered subjectivity" is a system for communicating information regarding subjective experiences. Although we illustrated this primarily with regard to video games, this capability for communication could be useful for many networked communication tools in order to enhance understanding between individuals. In a world where our interactions are increasingly mediated through digital environments, this capability for enhanced communication of subjectivity could yield more "human" interactions that recover aspects that become lost in purely text-based forms of communication.[13]

As we have discovered, the concept of "avatar-centered subjectivity" can be used to inform various aspects of design including graphics and sound. Using interactive sound design techniques, with relative ease, we can immediately begin to design media that represents subjective experiences of attention or ASC. Communicating affective states may be trickier, but music already has an established use for indicating and stimulating emotion, and thus provides one obvious means through which to develop an affective interface. Using the system we proposed here, it would be relatively easy for a composer and games designer to implement an interactive system where dynamic music responds specifically to the emotional state of an avatar. Therefore, the essential concept of "avatar-centered subjectivity" can be implemented in the short term with relative ease. Moving forward, there may also be scope for "avatar-centered subjectivity" to inform other sensory stimuli that might be used in video game and VR scenarios, such as tactile and olfactory representations, as well as elements of game play or challenges.

In the long term we hope to see the basic concept defined here expanded upon. The next steps for researchers will be to design robust systems that implement different approaches to "avatar-centered subjectivity," and subject these to rigorous user testing.[14] As discussed, these can be devised by drawing upon the wealth of knowledge that is available in fields such as cognitive psychology. Therefore, while we outlined several systems that can be used, there are others we could look at; for example, we could model memory by using Atkinson and Shiffrin's (1971) distinctions of short-term memory and long-term memory, and consider how items in short-term memory may move to long-term memory through repetition (for example). Or, we might utilize Baddeley and Hitch's (1974) concept of "working memory," which describes a temporary memory system used for information currently being processed. It is easy to imagine how these systems could be useful for modeling avatar memories in games that require complex character psychologies, such as *L.A. Noire* (Team Bondi 2011) or *Max Payne 3* (Rockstar Studios 2012). By providing multiple systems, we could begin to supply a tool kit for designers, with different modules that can be activated as needed depending on requirements. By developing these we can begin to provide FPP simulations in video games and VR that closely correspond with our subjective human experiences. As the sophistication and pervasiveness of VR and FPP simulations increases, we believe that research into "avatar-centered subjectivity" is going to be a growth area of critical importance.

3.5. Online Resources

Additional media relating to this chapter can be downloaded as a. zip file from http://bit.ly/Avatar_Subjectivity.

Notes

1. Throughout this chapter the term "avatar" is used to refer to both fictional and non-fictional characters or persons depicted either in synthetic three-dimensional worlds, or via cinematography such as first-person perspective (FPP) video footage used in 360 VR videos.
2. Of relevance to this area of research is Slater's (2009) work, which seeks to expand the discussion by suggesting the definitions of: "place illusion" (PI): presence, or the feeling of "being there"; and "plausibility illusion" (Psi), the belief that what is occurring is actually happening.
3. While we refer to this as an *objective* approach to make the distinction clear, it should be acknowledged that locating these patterns in relation to the avatar affords them some basic features of subjectivity. Yet, as will become clear through the course of our discussion, we feel that there is room for these representations to take into account many more aspects of the avatar's subjective experience.
4. For example, Weinel, Cunningham and Pickles (2018), argue for the concept of "deep subjectivity," in which the representation of emotions and non-aural or non-visual aspects of sensory experience increases the information available to the user for discerning the subjective state of an individual. In the chapter, they propose "deep subjectivity" as a design approach that may provide the possibility of improved empathy between the user and the avatar.
5. For a general overview of "information processing" in relation to other approaches in cognitive psychology, see Galotti (2014). For the purposes of the projects discussed in this chapter, the use of this paradigm is pragmatic since it can readily be used as a basis for design within video game engines.
6. For each of the projects discussed in this chapter, various materials have been provided in the accompanying package of supporting files. These include beta source code, example sounds, videos and other related documentation which can be downloaded from http://bit.ly/Avatar_Subjectivity.
7. For example, the experience of visual hallucinations that respond to auditory stimulation is discussed in Bliss and Clark (1962, p. 97).
8. As discussed in (Zander et al. 2010, p. 185), "passive" BCIs utilize brain activity outputs without voluntary control in order to enrich human-computer interaction through the use of implicit information.
9. For a comprehensive review of video games and other audiovisual media that represents ASCs, see Weinel (2018).
10. It should be noted that Hobson's (2003) concept of "external" and "internal" inputs certainly allows for the subjective perceptual processing of both types of information within the brain. What is important in this distinction is where the *input* of the information comes from. In cases of dreams or hallucination, sensory information received from the surrounding environment is reduced, and internally generated sensations seem to play a more significant role. Of course, though these sensations seem to be generated within the brain while a dream or hallucination is occurring, they inevitably involve systems such as memory that we may reasonably presume have been shaped by interactions within the surrounding "external" environment, prior to the event.

11. This quotation appears in the data collected as part of our study, which is available within the supporting documents (AH_Descriptions_Draft.pdf). The original source can also be found online (www.erowid.org/experiences/exp.php?ID=8232).

12. This is achieved in Unity using "ray casting," where a line is projected along the avatar's "line of sight" (represented on-screen with a crosshair), and the objects that intersect with that line can be identified.

13. For example, the limitations of digital communication technologies in relation to trust are explored in an episode of Douglas Rushkoff's "Team Human" podcast series, in an interview with William Softky and Criscilla Benford (Rushkoff, Softky and Benford, 2017).

14. For a discussion of approaches for the testing and evaluation, see Cunningham, Weinel and Picking (2016).

References

Arora, G. and Pousman, B., 2015. *Clouds Over Sidra* [360 video]. *Within: Extraordinary Stories in Virtual Reality* [website]. Viewed Nov 29, 2017 http://with.in/watch/clouds-over-sidra/

Atkinson, R.C. and Shiffrin, R.M., 1971. The Control of Short-Term Memory. *Scientific American*, 225, no. 2, pp. 82–90.

Baars, B.J., 1997. *In the Theatre of Consciousness: The Workspace of the Mind.* New York: Oxford University Press.

Baddeley, A.D. and Hitch, G.J., 1974. Working Memory. In: Bower, G.H. (Ed.), *The Psychology of Learning and Motivation: Advances in Research and Theory*, Vol. 8. New York: Academic Press, pp. 47–89.

BeAnotherLab, 2016. *The Machine to Be Another.* Viewed Nov 29, 2017 www.thema chinetobeanother.org/

Bliss, E.L. and Clark, L.D., 1962. Visual Hallucinations. In: West, L.J. (Ed.), *Hallucinations*. New York: Grune & Stratton.

Bregman, A.S., 1994. *Auditory Scene Analysis: The Perceptual Organization of Sound.* Cambridge, MA: The MIT Press.

Broadbent, D., 1958. *Perception and Communication.* New York: Pergamon Press.

Collins, K., 2008. *Game Sound: An Introduction to the History, Theory, and Practice of Video Game Music and Sound Design.* Cambridge, MA: The MIT Press.

Csíkszentmihalyi, M., 1997. *Flow and the Psychology of Discovery and Invention.* New York: Harper Perennial.

Cunningham, S., 2014. *LSD No.1* [fixed stereo electroacoustic composition]. First performed at: *Concert of Electroacoustic Miniatures* Mexican Centre for Music and Sonic Arts (CMMAS), June 21. Morelia, Mexico.

Cunningham, S., Weinel, J. and Picking, R., 2016. In-Game Intoxication: Demonstrating the Evaluation of the Audio Experience of Games with a Focus on Altered States of Consciousness. In: Garcia-Ruiz, M.A. (Ed.), *Games User Research: A Case Study Approach*. Boca Raton: CRC Press.

Fallen Planet Studios, 2016. *Affected: The Manor.* Samsung Gear VR.

Galotti, K.M., 2014. Cognitive Psychology: History, Methods, and Paradigms. In: Galotti, K.M. (Ed.), *Cognitive Psychology in and Out of the Laboratory*. Los Angeles, CA: Sage, pp. 1–23.

Hobson, J.A., 2003. *The Dream Drugstore*. Cambridge, MA: The MIT Press.

Hua, H., Krishnaswamy, P. and Rolland, J.P., 2006. Video-Based Eyetracking Methods and Algorithms in Head-mounted Displays. *Optics Express*, 14, no. 10, pp. 4328–4350.

id Software, 1996. *Quake*. MS-DOS, GT Interactive.

Ludwig, A.M., 1969. Altered States of Consciousness. In: Tart, C.T. (Ed.), *Altered States of Consciousness: A Book of Readings*. New York: John Wiley & Sons, pp. 9–22.

The National Autistic Society, 2016. Experience Too Much Information in Virtual Reality. *The National Autistic Society* [website]. Viewed Nov 7, 2016 www.autism.org.uk/get-involved/tmi/virtual.aspx

Rockstar Studios, 2012. *Max Payne 3*. Xbox 360, Rockstar Games.

Rushkoff, D., Softky, W. and Benford, C., 2017. Episode 52: William Softky and Criscillia Benford "Recalibrating for Trust." *Team Human with Douglas Rushkoff* [podcast series]. Viewed Sept 5, 2017 http://teamhuman.fm/episodes/ep-52-william-softky-and-criscillia-benford-recalibrating-for-trust/

Russell, J., 1980. A Circumplex Model of Affect. *Journal of Personality and Social Psychology*, 39, no. 6, pp. 1161–1178.

Slater, M., 2009. Place Illusion and Plausibility Can Lead to Realistic Behaviour in Immersive Virtual Environments. *Philosophical Transactions of the Royal Society B*, 364, pp. 3549–3557. doi:10.1098/rstb.2009.0138

Slater, M. and Wilbur, S., 1997. A Framework for Immersive Virtual Environments (FIVE): Speculations on the Role of Presence in Virtual Environments. *Presence: Teleoperators and Virtual Environments*, 6, no. 6, pp. 603–616.

Sundstedt, V., 2010. Gazing at Games: Using Eye Tracking to Control Virtual Characters. In: *ACM SIGGRAPH 2010 Courses*. doi:10.1145/1837101.1837106

Team Bondi, 2011. *L.A. Noire*. PlayStation 3, Rockstar Games.

Treisman, A.M., 1960. Contextual Cues in Selective Listening. *Quarterly Journal of Experimental Psychology*, 12, no. 4, pp. 242–248. doi:10.1080/17470216008416732

Turner, J., Velloso, E., Gellersen, H. and Sundstedt, V., 2014. EyePlay: Applications for Gaze in Games. In: *Proceedings of the First ACM SIGCHI Annual Symposium on Computer-Human Interaction in Play*, pp. 465–468. doi:10.1145/2658537.2659016

Weinel, J., 2011. Quake Delirium: Remixing Psychedelic Video Games. *Sonic Ideas/Ideas Sónicas*, 3, no. 2, pp. 22–29.

Weinel, J., 2012. *Altered States of Consciousness as an Adaptive Principle for Composing Electroacoustic Music*. Unpublished PhD thesis. Viewed Aug 31, 2017 www.jonweinel.com/media/ASC_Commentary_38.pdf

Weinel, J., 2016. Entoptic Phenomena in Audio: Categories of Psychedelic Electroacoustic Composition. *Contemporary Music Review*, 25, no. 2, pp. 202–223.

Weinel, J., 2018. *Inner Sound: Altered States of Consciousness in Electronic Music and Audio-Visual Media*. New York: Oxford University Press.

Weinel, J. and Cunningham, S., 2017. Simulating Auditory Hallucinations in a Video Game: Three Prototype Mechanisms. In: *ACM Proceedings of Audio Mostly 2017*. London. doi:10.1145/3123514.3123532

Weinel, J., Cunningham, S. and Griffiths, D., 2014. Sound Through the Rabbit Hole: Sound Design Based on Reports of Auditory Hallucination. In: *ACM Proceedings of Audio Mostly 2014*. Denmark: Aalborg University. doi:10.1145/2636879.2636883

Weinel, J., Cunningham, S. and Pickles, J., 2018. Deep Subjectivity and Empathy in Virtual Reality: A Case Study on the Autism TMI Virtual Reality Experience. In: Filimowicz, M. and Tzankova, V. (Eds.), *New Directions in Third Wave HCI Vol. 1: Technologies*. Switzerland: Springer HCI Series.

Weinel, J., Cunningham, S., Roberts, N., Griffiths, D. and Roberts, S., 2015a. Quake Delirium EEG: A Pilot Study Regarding Biofeedback-Driven Visual Effects in a Computer Game. In: *IEEE Proceedings of the Sixth International Conference on Internet Technologies and Applications 2015*. Glyndwr University, North Wales. doi:10.1109/ITechA.2015.7317420

Weinel, J., Cunningham, S., Roberts, N., Roberts, S. and Griffiths, D., 2015b. EEG as a Controller for Psychedelic Visual Music in an Immersive Dome Environment. *Sonic Ideas/Ideas Sonicas*, 7, no. 14, pp. 85–91.

Zander, T.O., Kothe, C., Jatzey, S. and Gaertner, M., 2010. Enhancing Human-Computer Interaction with Input from Active and Passive Brain-Computer Interfaces. In: Tan, D.S. and Nijholt, A. (Eds.), *Brain-Computer Interfaces: Applying our Minds to Human-Computer Interaction*. London: Springer, pp. 181–199.

Presence and Biofeedback in First-Person Perspective Computer Games

The Potential of Sound

Mark Grimshaw-Aagaard

4.1. Introduction

I have just bought a house with my wife that we moved into a few weeks ago. Throughout my life, I have lived in many houses and apartments and have often wondered at the speed with which I adapt to my new surroundings, making them a part of the world in which I live. Typically, this is a matter of the perceptual modalities accommodating themselves to, and cognition taking ownership of, the new domestic ecology; an ecology of touch, as my feet acquaint themselves with new floor surfaces and temperature gradients, of space, as I learn to navigate the angles and doorways in the dark, of smell, as my nose accustoms itself to the cooking of our new neighbors, and of sight as the visual ephemera of my 50-odd years of life, the artworks, worn-out rugs, furniture, books and so on are positioned in their new abode, nestling amidst those walls that are freshly painted and those that are still testament to the previous owners' tastes.

Above all, though, it is through perception of the set of sounds attached to the new sensory world that I am able to act within that world. This is not just because the ability to use the hearing modality utilizes a sense I concentrate on and that is special to me because I am a sound man who is particularly attentive to the sounds of my environment, but because this approach, I believe, is innate to all hearing humans. A particularly noisy refrigerator stands to my left as I sit and write at the kitchen table, an electrical device that is not working properly and that, with its constant, insistent humming and all-too-frequent liquid burbling, my wife has labeled the aquarium (indeed, there are brightly painted, fish-shaped magnets adorning its door). A heavy creaking on the ceiling above my head informs me that one of our new neighbors is pacing up and down on his wooden floor. Occasionally, the front door to the building opens and then shuts with a muted bang either preceding footsteps, usually running and lighter if children ascending the stairway to one of the other apartments or following the heavier tread of those descending the stairway on their way out. Often, voices can be heard and it is possible to locate such sounds either within the building or at its exterior, widening my sonic salience horizon beyond our immediate home and those of our neighbors. Other such exterior sounds include those of the wind in the bushes and trees of the building's courtyard, of clicking and clattering footsteps on the old cobbled street beneath our front room windows, or of infrequent vehicles driving up or down one of the two passageways bounding the building. This being late at night and during the run-up to Christmas, there is the almost constant sound of merry, singing Danes making their way from bar to bar to bar. Other sounds are within or caused by my self. A faint tinnitus rings in my right ear, a bone cracks as I shift weight, my breathing is occasionally audible, and there is the almost constant scratching and stabbing of my pen as it scrawls across the paper, crossing my t's and dotting my i's.

Before I begin, a word about particular terminology and what, at first sight, might appear to be idiosyncratic phrasing. Although I explain these in greater detail in what follows, I think it useful to the reader to be first given a brief introduction. To begin with, I deal with the self as a concept that is differentiated from the nonself (i.e. that which is not the self and is identified as such as part of the process leading to presence) this is the reason for my use of, for example, "my self" as opposed to "myself." (The latter might be best viewed as a conscious expression arising from the subconscious formation and unity of the former.) Second, I specifically and carefully distinguish between concepts such as "world" and "environment" (all too easily conflated in the literature on computer games, Virtual Reality [VR], and presence); "world," in my thinking, is that set of objects and events available to be sensed by any one individual across any sensory modality and "salient world" is the part of that world that we are aware of through sensation whereas "environment" is a perceptual construct derived in part from that salient world and that provides the place for presence. Third, I distinguish between "sound wave" and "sound" such that the latter is an emergent perception that is not necessarily dependent upon the former.

In the scenario sketched out above, all the auditory sensations are sounds that are perceived as of me, from me or around me. The more proximal the sound wave sources are to me, the more efficient I am at identifying their motivating causes and in locating the sources (being of me, and therefore familiar, or from me and around me; I make use of the frequent experience of similar sounds in the perception of new sounds). The further away these sources are, the more colored the sound waves are from the part of my world that is more absorptive and reflective of such waves and thus the more vague and uncertain the identification and location of source becomes. Still, all sounds are perceived in relation to me and so are unique to me; those sounds that are of me give me the sense of my self, those that are around me provide a sense of the external world, the nonself, while those that are from me are an interface between my self and that nonself.

Very rarely is our hearing modality alone used. More typically, it is used in a multimodal context to inform us about the world and our place in it. As far as we are aware, hearing is correlated mainly with vision in establishing information about the world and our place in it. However, there is increasing evidence of more extensive connections between perceptual modalities whereby one modality affects (and even effects) the perception attributed to another. This, however, is a chapter focusing on the modality of hearing and so, while I do account below for other modalities' contributions to the ultimate perception of sound while also accounting for the influence of other factors such as reasoning, experience, emotion and so forth in the main, I approach the question of presence from the direction of the perception of sound. My reasoning is simple: more than any other perception, sound is the perception that not only informs us about objects and events[1] in the immediate and more distant and often unseen world but also is the result of a modality that is omnidirectional and which allows us to locate those objects and events with varying degrees of precision at any point in the space around us. In so doing, hearing is the modality nonpareil that confirms and locates our selves in the world and so affords our feeling of presence.[2]

The main themes of the chapter are the role of sound in the establishment of the feeling of presence in computer games (particularly first-person perspective games) and the potential of biofeedback in the context of sound to enhance that presence beyond what is (arguably) possible now. In order to begin to expand upon these themes, I continue with a discussion of what sound is. I then move onto a discussion of presence and what presence is in computer games (and the role that sound plays in it) that proceeds from the sketch of presence in real-world environments with which I began this chapter. Finally, I then present a more speculative section dealing with ideal biofeedback devices and how they might be used to exploit the manner in which we perceive sound and so increase our feeling of presence in computer games.

4.2. Sonic Virtuality

In various writings over the past few years, I have enumerated and described various past and current definitions of sound and have detailed objections to each definition's scope and application while also identifying fundamental flaws in these definitions that are particularly egregious.

I therefore do not repeat the same tasks here and instead direct the reader to some of these past writings (e.g. Grimshaw-Aagaard and Garner 2015; Grimshaw 2015). I use this section to briefly present a definition of sound that arose in response to the objections and flaws noted in other definitions and to show how this new definition can be of use regarding the topic of this chapter. Where necessary, though, I do briefly present some of those objections and flaws where they support my argumentation for the presented definition of sound.

The definition of sound I present here is one devised by myself and Tom A. Garner, the exposition of which is detailed in our book *Sonic Virtuality* (2015). This definition is:

> Sound is an emergent perception arising primarily in the auditory cortex and that is formed through spatio-temporal processes in an embodied system.

Briefly, in this definition, sound is conceived of as a perception rather than as a sound wave. The definition accounts for the effects of cross-modality in the perception that is sound and accounts for the effects of space and time on the brain's relationship with the body's embodiment in a sensory world and the role that this has in the emergence of that perception.

Regarding the first contention, I point to two important pieces of evidence to support it (there are more as detailed in the sources above). The first is the McGurk Effect in which two videos are shown each of a mouth articulating a syllable, for example, "ba" and "fa." (Numerous examples of the McGurk Effect can be found online including a BBC Horizon television program.[3]) When the audio on the video of the mouth articulating "fa" is replaced by that from the video of the mouth articulating "ba," one cannot help but hear "fa" as per the visual cues of the "fa" video. An acoustician might ask the question: how can we hear something different when the sound (i.e. the sound wave) remains the same? Using the principle of Occam's Razor, though, the simplest explanation for the quandary is to suggest that the acoustician's definition is flawed; a sound is not, in fact, a sound wave (the standard view of sound from physics) and a sound wave is merely one contributing factor to something else that we then perceive (and this is what Garner and I call sound). Thus, in the McGurk Effect, the sound does not remain the same but changes as its balance of modal components changes and this is why we hear something different. The second piece of evidence arises from neuroscience. There are many examples of brain imaging studies of subjects viewing an event in a film (for example, a hammer falling on an anvil) where the activity shown in the auditory cortex in the absence of the audio track is strongly similar to that shown in the presence of the event's expected audio track (e.g. Bunzeck et al. 2005;

Hoshiyama, Gunji and Kakigi 2001; Voisin et al. 2006). In conjunction with the results of the McGurk Effect, I suggest that this demonstrates the role of cross-modality in the perception of sound: what we perceive as sound depends on other perceptual modalities, in this case vision. Additionally, the effects highlighted by brain imaging would support Garner's and my contention that imagined sound (auditory or aural imagery see Baddeley and Logie 1992) is indistinguishable from sound perceived in the presence of sound waves, and thus the perception that is sound does not necessarily require the presence of sound waves.

The second contention in the definition that there is an embodied, spatio-temporality to the perception that is sound can be defended and illustrated from a number of positions and directions of which, here, I present just a few. That sound takes place over time is self-evident and is an element of many different definitions of sound including the standard view: "*Oscillation* in pressure, stress, particle displacement, particle *velocity*, etc., *propagated* in a medium . . . " (ANSI/ASA S1.1–2013, my italics). In addition to this, the definition also accounts for other temporal effects on the emergence of sound as perception: effects from the past of experience, knowledge and memory, all those aspects of a lived life that inform the present perception.

It is the spatial aspect of the definition, though, that has more relevance to my argument on presence, which follows this section. Acousticians and psychoacousticians subscribe to and study a phenomenon known as auditory localization. In this, humans (and other experimental subjects such as owls and ferrets) are assessed as to their accuracy in determining the location of a sound wave source. According to this field of research, and the theory that comes with it, our stereophonic hearing achieves localization mainly by dint of time difference and intensity difference between the sound wave arriving first at one ear and then the other (assuming a lateral sound wave source[4]). In *Sonic Virtuality*, Garner and I propose another understanding of localization that accounts for various anomalies and which I have since developed further in the study of presence. How does one account for the ability of the brain to adapt to changes in the human auditory system and yet still, over time, be able to localize sound wave sources with a reasonable degree of accuracy (see, for example, Slattery and Middlebrooks 1994; Kacelnik et al. 2006)? How does one account for the fact that we are able, seemingly effortlessly, to localize sound at locations other than the sound wave source (the most obvious and widespread example being the use of loudspeakers and/or headphones in conjunction with a moving image displayed on a screen from which no sound waves issue—see the concept of synchresis [Chion 1994])? The answer proposed in *Sonic Virtuality* is a different understanding of what it is to localize sound: we localize the perception that is sound on the most likely source in the external world whether there is a sound wave involved or not. This accounts for experience and learning (post-natal and onwards) and embodiment in an external world and it is this understanding of localization that I have used in my thinking on sound in the role of presence. In this conception, the localization of sound is a means to probe the external world of sensory objects and events and so a means to hypothesize about the sensory spatio-temporality of that world.

4.3. Presence

"Immersion" and "presence" are two terms that are often conflated in the literature on computer games. Although I too, in the past, have been guilty of this conflation, today I prefer to make a useful distinction between the two. In this, I follow theoretical positions put forward in the field of VR (see Slater 2003, for example). Thus, immersion is an aspect of specific technology that is more or less capable of sensory fidelity to a reference world of sensory objects and events (i.e. the real world as opposed to a virtual world). This is view is mirrored later by Mestre stating that: "immersion is achieved by removing as many real-world sensations as possible, and substituting these with the sensations corresponding to the VE" (2005, p. 1). Technology, in other words, can be immersive to a certain degree and this level of immersion can be objectively assessed as a measure of its relative sensory fidelity. Presence, as Slater states, is the subjective "human reaction to immersion" (ibid. p. 12). I deal in more detail with aspects of immersion and presence elsewhere (e.g. Grimshaw-Aagaard 2017, 2019) and so, for now, am content to state that this differentiation is useful in the context of this chapter particularly when I discuss the potentially immersive technology of biofeedback below.

As noted in the introduction above, I also conceptually distinguish between "world" and "environment." Again, the reasons for this are presented in greater detail elsewhere (e.g. Walther-Hansen and Grimshaw 2016; Grimshaw-Aagaard 2019) and so here I present the briefest of expositions of this distinction. Environment is a perceptual construct that is drawn from a set of perceptual hypotheses that themselves are increasingly refined and increasingly usable models of a salient world (a term Walther-Hansen and I introduced to describe that subset of the external world that not only is immediately available to be sensed but which is being sensed in the here and now). The successful perceptual hypothesis is that which is sufficient in order to be able to act in the external world, and this is what I call the environment. What is important here is that Walther-Hansen and I described the creation of this environment as the means by which I distinguish between my self and nonself and that this distinction, once defined, is what then allows for the possibility of being present in that environment and being able to act within the external world (which is represented by the concept of the nonself). The purpose of presence, therefore, is to be able to act within and on that part of the nonself that is salient to the perceiver, and the process of acting can be encapsulated within the concept of engagement. One must first feel present and able to act before engaging with a world, real or virtual.

A simple flow model from sensation to environment might involve the following steps that improve in speed and efficiency with age and experience:

1. Perceive sensations from the immediate external surroundings (the salient world).
2. Use cognition to fashion these perceptions into hypothetical models of the salient world.
3. Test each model as it is fashioned against the sensations, refine it, test again, eventually selecting one (under time constraints required by the necessity of action) as the workable environment.

This process of creating an environment as a model of the salient world is the process that distinguishes between self and nonself in effect, the environment is a metonym of the nonself that is the salient world and is dynamic according to changing sensory circumstances. Once this environment exists, the conditions for presence have been attained: one must first have a model of the salient world in order to be able to act within it.

It should be noted that it is perfectly possible to be absent, that is, not present, despite an external world rich in sensory objects and events (Waterworth and Waterworth 2014, p. 589). Presence requires attention or saliency, and thus the salient world must be dynamic in order to maintain attention; stasis is not conducive to presence. In the real world, this dynamism derives from my self engaging with the world and other actors engaging with their worlds whose projected sensory reach overlaps mine, and it can be conceived of in like manner for virtual worlds such as computer games. I return to the need for a perpetual salient dynamism below but will state, for now, that a lack of engagement with the world will decrease the dynamism of the salient world and that, thus, there is less need to continually model an environment in which to be present.

Thus far, I have avoided defining presence other than using Slater's description of it as a subjective response to immersive technology. Slater is discussing presence in VR so, unless one greatly increases the scope of the definition of immersive technology to the point of uselessness, it is not alone a definition fit to describe presence outside VR (assuming that the presence aimed for in VR is that experienced when engaging with the real world). Slater further constrains his definition to VR by stating that presence is "the extent to which the unification of *simulated* sensory data and perceptual processing produces a coherent 'place' that you are 'in' and in which there may be the potential for you to act" (op. cit., p. 2, my italics). This definition encapsulates definitions of presence provided by others (e.g. IJsselsteijn, Freeman and de Ridder 2001; Waterworth and Waterworth 2014) viz. the feeling of being in a place and the possibility to act within it.

Implicit in Slater's latter definition is a proportionality between level of immersion (of the technology) and the "extent" of presence, and this equation is also expressed by IJsselsteijn, Freeman and de Ridder who state that presence is enhanced by "more accurate reproductions and/or simulations of reality" (op. cit., p. 180). Assuming one can be present in a virtual world (see below for further thoughts on this), while I agree that the feeling of presence is that of being in a place and having the potential to act in that place, I disagree with the proportionality between level of sensory fidelity (of immersive technology) in relation to the sensations provided by the real world. One can either be present or not; there is no halfway house whereby one can be part-present in a virtual world while being part-present in the real world. I liken the switch in presence from real world to virtual world and back (or expansion of one world to incorporate part of another) to being in an art gallery and moving across the threshold from one room to the next by passing through doorless entranceways. Here, I make use of the concept of environment defined above. One is present in one's environment, and this environment, as a model of the salient world, provides the means with which to act within that salient world. This environment, as noted, is dynamic as the salient world is dynamic; it is continually modified or even refashioned as new sensory data is available that triggers new perceptions when combined with cognition (e.g. experience, knowledge, memory). It is a feedback loop and, in its dynamism, it maps out a shifting presence horizon that, in the case of an art gallery, may encompass just the one room but that, as the saliency horizon is breached through, for example, a child screaming in an adjoining room, can grow to encompass a now expanded salient world: the presence horizon delimiting the environment no longer maps out the sensory boundaries of the one room but includes sensations from another. This concept of a presence horizon explains how it is possible to act within the virtual world of a computer game but still have one's attention caught by the ringing of a telephone in the real world—there is no blocking out of an external world, merely a shifting of attention and focus.

Clearly, though, one must have a workable definition of presence if one is attempting to create virtual worlds that have the potential to evoke presence. An assumption, then, is that what the designers of virtual worlds such as computer games (those games that are claimed to be immersive) are attempting to produce is the same sense of presence as experienced when present in the real world: the feeling of being in a place (the perceptual environment) and the possibility to act within the world it models. One must then ask if, despite marketing hype and its claims for immersion/presence in computer games, it is indeed possible to be present in environments of such virtual worlds in the same way as we are present in environments of the real world. On the basis of current computer technology used to fashion virtual worlds, one would have to conclude no; while there may be some fidelity of the sensory data in virtual worlds to the sensory data of real worlds, it still falls short of complete fidelity not only in each sensory modality's quality but in the limiting, in the main, of sensory modalities solely to vision and hearing. (One must not only think of missing modalities such as touch [if this exists at all in computer games, it is a rudimentary tactility], smell and taste but also of other sensory modalities such as mass/gravity and temperature.) One is then forced to ask: can a deaf person be present in the environment they have fashioned of the real world if one sensory modality is lacking? I do not know the answer to this but would tentatively suggest that the answer is yes, but that the feeling of presence is different, the missing sensory modality being compensated for by increased use of the other modalities. This then brings us back to presence in the environments of virtual worlds: if the technology is missing sensory modalities, can one achieve a sense of presence albeit one that is different to that felt in the real world with all sensory modalities available? On the belief that my speculation about a deaf person's feeling of presence is correct, I suggest that the answer to that particular question is yes. An acceptable and usable environmental model of the salient world (real or virtual) can be produced even in the context of a limited number of sensory modalities.

To close this section off, I quote from *The Psychopath Test*, a journalistic investigation of psychopathy: "I remembered a time I perforated my eardrum on a plane and for days afterward everything around me seemed faraway and hazy and impossible to connect to. Was that foggy sensation a psychopath's continual emotional state?" (Ronson 2011, p. 100). When I read this a few days ago, I was immediately struck by its connection to my thinking on presence (although I would not yet go so far as to suggest that there is a connection between presence and psychopathy). I was reminded of the frequent times I had been struck by a sudden, intense and unexplained ringing tinnitus in one ear and the sudden decrease in my presence horizon that resulted; things do indeed seem "faraway and hazy and impossible to connect to" when suffering this temporary condition until—and this I have noticed before—I moved back from concentrating on the inner world of my tinnitus by compensating for the decrease in hearing sensitivity with other senses, particularly vision. In this way, I have been able to maintain my presence in a dynamic environment and have been able to feel that I am in that place with the potential to act within what it models.

4.4. Biofeedback, Sound and Presence

The theoretical framework mapped out above and that I use here can be summarized thus:

- Sound is a perception that is multimodal.
- Sound as a perception is localized on objects and events in the external world that we are aware of (the salient world) as a means to probe that world and hypothesize about its structure.
- Immersion is an objective measure relating to the degree of sensory fidelity produced by virtual world technology according to a reference standard (the real world).
- Presence is a subjective (perceptual) response to the salient world either as modeled by the immersive technology of a virtual world or that is part of the real world
- It is possible to be present in worlds that are deprived of some sensory modalities (such as those of computer games) but this presence is different than when all sensory modalities are active; missing senses are compensated for by increased reliance on available senses.
- Environment is a perception and is a usable hypothetical model of the salient world.
- In constructing this environment, we distinguish between self and nonself (the salient world), and the localization of sound is an important means with which to accomplish this.
- We are present in the environment and, as a model of the salient world, we are thus able to act within and upon that salient world.

In what follows, I use this framework to map out a means to include biofeedback devices in the immersive technology of virtual worlds such as computer games, the intention being to make more efficient the environment creation process and thus to provide a space in which to be present and from which to be able to act within that virtual world. I begin by defining what is meant by biofeedback and by describing some of the current state of the art of biofeedback in relation to sound within computer games.

With the definition of sound presented in the section above on sonic virtuality, it could be argued that we already make use of biofeedback to form the perceptions of sounds in virtual and real worlds. These perceptions are formed in response to objects and events in the salient world and so represent a form of biofeedback in that our model of the world in which we can act, the environment, formed from these and perceptions from other modalities, is dynamically modeled by cognition in response to the sensations our embodied mind is subject to. As an example, using a first-person perspective computer game (although other computer games follow this model too), the virtual world is created anew at each frame relative to our actions in the game (or head movements using VR headsets) and these actions are driven by new and changing objects and events presented by the game engine that enable us to continually remodel the environment in which we are present.

Here, though, I wish to deal with the use of explicit biofeedback devices (as part of the immersive technology of virtual worlds such as computer games) the data from which are used to process audio artifacts or to create them from scratch—these artifacts leading to the formation of new perceptions of sound that then produce new data for the devices to supply feedback to the game engine and so on. A biofeedback device, in this context, is a sensory device that registers some of body's psychophysiological state and frames this within a particular protocol, producing data on that state that can then be used externally for particular purposes. Such devices might make use of electroencephalography (EEG), electromyography, heart rate, galvanic skin response, eye movement, facial and bodily gestures and so forth. Many of these devices have already widely been used in computer games (such as Microsoft's now defunct Kinect or Sony's Playstation Move, among others) and other motion sensors have long been used in a range of other applications such as sound installations and household burglar alarms. Here, though, I wish to concentrate on the use of EEG to process audio artifacts in response to player psychophysiology, the purpose being to achieve efficient environment formation, thereby providing the conditions necessary for presence.

EEG devices have long been used for analog sonification of brainwaves, such as works by the composer Alvin Lucier (e.g. Lucier 1965) where EEG signals were used to trigger the sounding of various musical instruments in live performances. Sonification and EEG devices are increasingly used in the digital domain (i.e. the data from the EEG control digital sound cards) for the purposes of monitoring and/or diagnosing brainwave patterns and brain disorders (e.g. Hermann et al. 2002; Vialatte et al. 2012), and EEG devices continue to be used for sonification in purely creative purposes (e.g. Deuel et al. 2017). EEG devices can be interfaced with game engines for various purposes such as creating adaptive game play to increase engagement with the game (e.g. Ewing, Fairclough and Gilleade 2016), and they have been used to process game audio files according to measurements of arousal (e.g. Garner 2013; Garner and Grimshaw 2013).

The work by Garner in particular demonstrates the practicality of assessing brainwave data via EEG and then mapping elements of the data to various audio parameters (equalization, position, volume, etc.) according to the assessed arousal level of game player. With a game engine that can process audio in real time, audio can be produced that is designed to arouse when the player is judged to be too calm or that is designed to calm when the player is judged to be too aroused. Garner continues to work in this direction (e.g. Garner 2016; Garner and Jordanous 2016 [this last exploring the possibilities of an in-game "synthetic listener" responding to player-generated audio events]).

I now move into more speculative territory by first mapping out a framework for the use of biofeedback for presence purposes. This concentrates on audio artifacts and sound, my contention, as stated above, being that it is the auditory modality that is most important for the establishment of presence; but, as a generalized framework, I see no reason why it could not be applicable to other modalities too especially when our understanding of perception is increasingly cross-modal. It also concentrates on the use of the EEG as the means for biofeedback as I view the environment, in which the player is present, as a perceptual construct. Second, I briefly peruse what the future might hold should the technology for biofeedback and the interpretation of psychophysiological data from biofeedback devices continue to develop.

If one were to map out a chain of process from sensation to presence and engagement following what I have discussed above, it would look like this:

Figure 4.1   From Sensory World to Engagement.

However, I have noted above that, if one is to be present in an environment, one requires a salient world that is perpetually dynamic in its saliency. The sensory dynamism can come from other actors emitting sensations that impinge on our saliency horizons but it can also be through our engagement with the world in which we become an actor having an effect on the salient world we model. There is, therefore, feedback to be added to the model thus:

In the real-world scenario sketched out at the start of the chapter (and concentrating solely on sound), the sensory world of Figure 4.2 can be viewed as the pool of available sound waves all of which might well be

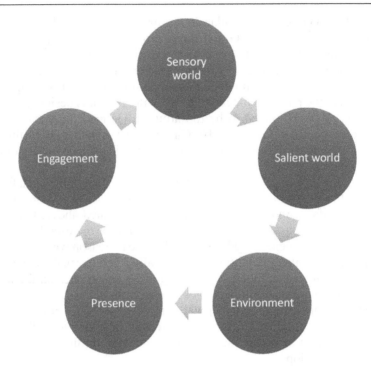

Figure 4.2   From Sensory World to Engagement Modeled as Feedback System.

within my sensory horizon but only some of which will be within my
saliency horizon. Those audio sensations that are within my saliency horizon are those that I either choose to become attentive to (through focus,
often initially via another modality such as vision) or that I am made aware
of (through a differentiation from the ambience, via sudden loudness, for
example). From this salient world, the environment is created that, as a
usable perceptual hypothesis (a model) of the salient world, provides the
place in which I can be present. In other words, as I sit at my kitchen table,
the environment in which I am present comprises perceived sounds (and
other perceptions) that are modeled on sound waves from the salient world
but that also include representations of what the sound wave sources are.
In a sense, sound as a perception is a symbol composed of several factors
(sound wave, if present, other sensory information and cognition), and
this multifaceted perceptual artifact mirrors our everyday language for
sound (see Gaver 1993) in that sound is represented in language not by
frequency, intensity, phase and so on but by the sound wave source (vehicles, drunken Danes, etc.) and all that that denotes and connotes. These
perceptual symbols, sounds, are what provide the self the means to act

upon and within the nonself of which the salient world is part. It is not just the formation of these perceptual symbols that aids in fashioning the environment but also, importantly, the fashioning of the perceived space of that environment through the localization of the sounds onto likely sound wave sources modeled within the environment. Thus, presence.

At this point in Figure 4.2, should I choose to act in the salient world, I am engaging with this world. Engagement takes many forms, such as the decision to put pen to paper, but, vitally, it closes the feedback loop by pushing fresh events out into the external world. In auditory terms, because the sound waves of my engagement are of me or from me—my breathing, the scratching and stabbing of my pen (note the everyday language here)—this can have the effect of shrinking my saliency horizon to the task in hand; the writing of this chapter and the environment adapts accordingly to remodel this dynamic saliency. My salient world and its horizon shrink with focused engagement.

The use of EEG devices for biofeedback purposes in computer games is still in its infancy. Partly this is due to a still-developing understanding of how to interpret in a meaningful manner the data retrieved from the EEG, and to thereby use it in a meaningful manner. In Tom A. Garner's work briefly mentioned above, it was only possible to gain an impression of arousal during game play. EEG as a technique is not well suited to deep brain monitoring whereby one might gain an impression of emotions, for example, or responses in the auditory cortex to specific sound waves. Part of the problem is that any object or event in the salient world of the game must compete for attention with other objects and events, and so it can be difficult to ascertain precisely what it is that causes a particular fluctuation in EEG data. Another issue is that, if one wishes to monitor the brain using EEG in as unobtrusive a manner as possible during game play, one must use commercial EEG headsets (such as the Emotiv EPOC) that are less than satisfactory when compared to the clinical devices but are easier to use. EEG can be combined with other devices—eye or facial gesture-tracking, heart rate monitors, galvanic skin response and so on— but this risks an over-supply of data that is difficult to manage and assess in combination; in any case, such multiple devices are cumbersome and time consuming to set up for technicians not to mention for the home-based gamer.

However, let us assume that at some point in the future, there will exist the ideal biofeedback device (EEG [or some other form of brainwave monitor] for my purposes) perfectly integrated with computer game engines. Although I have above suggested that both real worlds and virtual worlds are implicitly feedback-based, the explicit use of biofeedback devices such as EEG as part of the framework presented above must impinge upon and take account of several factors. In the first instance, an assessment of the data collected from the EEG should be used to make dynamic the salient world of the computer game. In terms of the auditory modality, then, new audio artifacts must continually be sounded and saliency horizons must shrink and expand by means of new, foregrounded audio artifacts, the sources for which the player is encouraged to engage with. As an example of this latter method, in early work, I identified a mode of listening I called navigational in which an audio artifact functions as a form of beacon, aiding the player in navigating the game space (Grimshaw 2008). In a first-person perspective computer game, audio need not be specifically designed to function as a beacon for it to function as one: for example, almost any audio with high enough salience and perceived as coming from off-screen (localized in the sonic virtuality sense) is likely to provoke the player to explore in that direction.

Second, and perhaps most important, the data from the EEG device must be able to be interpreted correctly according to the psychophysiological state of the player as regards an assessment of presence aided by evidence of engagement with the game. Once interpreted, is the player present or not? The data can be used to produce new audio artifacts and to position them within the space of the game world. Here, though, is the problem that remains to be solved: precisely how does one assess presence from brainwave data? It might well be that the tracking of arousal noted above can be combined with an assessment of engagement by using the player's in-game actions to interpret the EEG data, the hypothesis being that presence is judged to have been achieved when the dynamism of the engagement can be seen to affect the dynamism of the EEG data. Without further work, though, it remains to be seen if this hypothesis is correct and so can be used with a gaming system to provide the conditions necessary for presence in virtual worlds.

Notes

1. It is through the sensation of events that we become informed and create knowledge, not only about them but about the objects behind the events. In this, I follow a constructivist line. I discuss this further below when dealing with everyday language.
2. This standpoint does not discount the contribution to presence of other modalities, and, indeed, below I account for a deaf person's feeling of presence.
3. www.youtube.com/watch?v=G-lN8vWm3m0 (Viewed Dec 22, 2017).
4. The pinnae of the anatomical auditory system also aid in the localization of sound wave sources in the vertical domain (and, to some extent, contribute to resolving in-front or behind ambiguities) through their filtering properties. This frequency spectra-based filtering is helped by minute (nutating) or larger movements of the head in order to create frequency differentials between the outer ears as the sound wave impinges on

the auditory system. Clearly, the use of headphones militates against this movement, as does the use of stereo loudspeakers, while surround sound systems can provide the necessary cues and VR headsets, tracking head movements, should be able to process the audio data appropriately.

References

American National Standard, 2013. *Acoustical Terminology*, ANSI/ASA S1.1–2013.

Baddeley, A. and Logie, R., 1992. Auditory Imagery and Working Memory. In: Reisberg, D. (Ed.), *Auditory Imagery*. Mahwah, NJ: Lawrence Erlbaum Associates, pp. 179–197.

Bunzeck, N. et al., 2005. Scanning Silence: Mental Imagery of Complex Sounds. *NeuroImage*, 26, no. 4, pp. 1119–1127.

Chion, M., 1994. *Audio-Vision: Sound on Screen*. Translated by Gorbman, C. New York: Columbia University Press.

Deuel, T.A. et al., 2017. The Encephalophone: A Novel Musical Biofeedback Device Using Conscious Control of Electroencephalogram (EEG). *Frontiers in Human Neuroscience*, 11. Viewed Jan 15, 2018 http://journal.frontiersin.org/article/10.3389/fnhum.2017.00213/full

Ewing, K.C., Fairclough, S.H. and Gilleade, K., 2016. Valuation of an Adaptive Game That Uses EEG Measures Validated During the Design Process as Inputs to a Biocybernetic Loop. *Frontiers in Human Neuroscience*, 10. Viewed Jan 15, 2018 www.frontiersin.org/articles/10.3389/fnhum.2016.00223/full

Garner, T.A., 2013. *Game Sound from Behind the Sofa: An Exploration into the Fear Potential of Sound & Psychophysiological Approaches to Audio-Centric, Adaptive Gameplay*. Unpublished PhD Thesis, University of Aalborg, Denmark.

Garner, T.A., 2016. From Sine Waves to Physiologically-Adaptive Soundscapes: The Evolving Relationship Between Sound and Emotion in Video Games. In: Karpouzis, K. and Yannakakis, G.N. (Eds.), *Emotion in Games*, Vol. 4. Berlin: Springer, pp. 197–214.

Garner, T.A. and Grimshaw, M., 2013. The Physiology of Fear and Sound: Working with Biometrics Toward Automated Emotion Recognition in Adaptive Gaming Systems. *IADIS International Journal on WWW/Internet*, 11, no. 2, pp. 77–91.

Garner, T. and Jordanous, A., 2016. Emergent Perception and Video Games That Listen: Applying Sonic Virtuality for Creative and Intelligent NPC Behaviours. In: *The 2nd Computational Creativity and Games Workshop*, June 27. Paris.

Gaver, W.W., 1993. What in the World Do We Hear? An Ecological Approach to Auditory Perception. *Ecological Psychology*, 5, no. 1, pp. 1–29.

Grimshaw, M., 2008. *The Acoustic Ecology of the First-Person Shooter: The Player Experience of Sound in the First-Person Shooter Computer Game*. Saarbrücken: VDM Verlag.

Grimshaw, M., 2015. A Brief Argument for, and Summary of, the Concept of Sonic Virtuality. *Danish Musicology Online—Special Issue on Sound and Music Production*, pp. 81–98.

Grimshaw-Aagaard, M., 2017. Presence Through Sound. In: Wöllner, C. (Ed.), *Body, Sound and Space in Music and Beyond: Multimodal Explorations*. London: Routledge, pp. 279–298.

Grimshaw-Aagaard, M., 2019. Presence, Environment, and Sound and the Role of Imagination. In: Grimshaw-Aagaard, M., Walther-Hansen, M. and Knakkergaard, M. (Eds.), *The Oxford Handbook of Sound and Imagination*. New York: Oxford University Press.

Grimshaw-Aagaard, M. and Garner, T.A., 2015. *Sonic Virtuality: Sound as Emergent Perception*. New York: Oxford University Press.

Hermann, T. et al., 2002. Sonifications for EEG Data Analysis. In: *Proceedings of the 2002 International Conference on Auditory Display*, Jul 25. Kyoto.

Hoshiyama, M., Gunji, A. and Kakigi, R., 2001. Hearing the Sound of Silence: A Magnetoencephalographic Study. *NeuroReport*, 12, no. 6, pp. 1097–1102.

Ijsselsteijn, W.A., Freeman, J. and De Ridder, H., 2001. Presence: Where Are We? *Cyberpsychology & Behavior*, 4, no. 2, pp. 179–182.

Kacelnik, O. et al., 2006. Training-Induced Plasticity of Auditory Localization in Adult Mammals. *PLoS Biology*, 4, no. 4, pp. 0627–0638.

Lucier, A., 1965. *Music for Solo Performer.*

Mestre, D.R., 2005. *Immersion and Presence*. Unpublished article. Viewed Jan 29, 2018 https://scholar.google.fr/citations?hl=en&user=l9RLpn4AAAAJ&view_op=list_works

Ronson, J., 2011. *The Psychopath Test: A Journey Through the Madness Industry*. New York: Riverhead Books.

Slater, M., 2003. A Note on Presence Terminology. *Presence Connect*, 3, no. 3. Viewed Jan 15, 2018 http://publicationslist.org/data/melslater/ref-201/a%20note%20on%20presence%20terminology.pdf

Slattery III, W.H. and Middlebrooks, J.C., 1994. Monaural Sound Localization: Acute Versus Chronic Unilateral Impairment. *Hearing Research*, 75, no. 1–2, pp. 38–46.

Vialatte, F.B. et al., 2012. Audio Representations of Multi-Channel EEG: A New Tool for Diagnosis of Brain Disorders. *American Journal of Neurodegenerative Disease*, 1, no. 3, pp. 292–304.

Voisin, J. et al., 2006. Listening in Silence Activates Auditory Areas: A Functional Magnetic Resonance Imaging Study. *The Journal of Neuroscience*, 26, no. 1, pp. 273–278.

Walther-Hansen, M. and Grimshaw, M., 2016. Being in a Virtual World: Presence, Environment, Salience, Sound. In: *Proceedings of Audio Mostly*, Oct 24. Norrköping, Sweden and New York: ACM.

Waterworth, J.A. and Waterworth, E.L., 2014. Distributed Embodiment: Real Presence in Virtual Bodies. In: Grimshaw, M. (Ed.), *The Oxford Handbook of Virtuality*. New York: Oxford University Press, pp. 589–601.

Composed to Experience

The Cognitive Psychology of Interactive Music for Video Games

Hans-Peter Gasselseder

5.1. Introduction

Playing a video game is a two-way street. You play the game, and the game plays you. At the interface between the physical and the virtual realms, one's actions reciprocally affect the course of the game, as the narrative and dramaturgy in turn also define the behavioral traits of the player. Interactive music, too, aims to accommodate a similar dialogue by adapting and remodeling its structure from linear storytelling towards a nonlinear framework. The latter brings its own requirements to the composition process as for integrating challenge-based and dramaturgic motives within musical expression, while satisfying the aesthetic expectations of the audience. When attempting to juggle with these structural aspects, one will begin to understand that a prerecorded music loop cannot meet the above premises single-handedly. The non-responding quality of the loop would turn game play into an experience of stasis, which more often than not results in demotivating players, or pushing them to turn off the game. This brings us to the question, why exactly does this happen? Given those prerequisites, the player certainly would not transcend into the immersive, fantastic world that was enticingly promised in the marketing material of the game. If a music score can hamper the gaming experience to such an extent, it also begs the question about its massive potential. How would the experience be if the music seamlessly accompanied one's actions or, even better, motivated the gamer to perform them? Have we found the holy grail of supporting immersive experience in the game play?

This pretense frequently forms the main motivation of implementing an interactive music score into video games. And while it offers a perfectly sensible reason for branching out into audio programming, as with every powerful tool or skill set, it may do more harm than good in the wrong hands. That is why this chapter will introduce you not only to the most common techniques for transforming your music into an interactive system, but it will also inform you about the effects of its resulting soundtrack and its potential for a ludic experience of the player, which is actually what and who you should be writing for. You do not produce music for games to sound cool, epic or cinematic (and it is better for you not to posture as the next genre-defining composer because, as yet, neither of us is; though, if you want to be better, you are on the right path). With this in mind, music for games is not about showing off how great you are in turning your music cue into a chameleon, which changes every other second in an unpredictable manner. It is just as little about making your mockups stand out and sound like a real orchestra (though, that's great!).

Notwithstanding this, you are allowed to impress your peers, but only if you do so by making them enjoy the game you are working on. And that is the first and only objective: the experience of the player. This may seem more straightforward than it really is, but it is helpful to remind yourself that you are not writing music for the sake of music, but rather for an experience that is more than the sum of its parts. This paradigmatic way of thinking will invite ideas you would not have thought of initially (i.e. the functional application of pauses and silence as part of a music cue). Nevertheless, if we entertain the thought that experience is not just a constant but a moving target, an important question comes to mind: do you really know what kind of experience you want the player to have in your game? You may answer that the music should contribute to the enjoyment, performance, emotional and immersive experience of the player. But do you know how to write and implement music for these rather specific experiential concepts, as in having tested your peers with different versions of a music system to see how they would respond to it? It falls to the reader to answer this question for herself, but there is no need to feel guilty for not having conducted systematic studies on the subject. These experiments take lots of time and resources. So much so that sometimes even major game studios struggle with evaluating a music system that they have spent weeks on implementing.

Thus, contrary to the many technique-oriented introductions to the subject, this chapter is going to spend a significant part on the insights gained from some of the authors' experiments, which have since posed as a methodological template for a few follow-up studies on the subject (e.g. Williams et al. 2016; Prechtl 2016; Schrader et al. 2017). In this way, the reader will learn what compositional concepts require special attention during the design process of a music system and how to use them to either attain a similar effect or counter effect in the player by purposefully following or challenging these guidelines (e.g. as in the use of anempathic/counterpointal expression that aims to arouse emotional dissonance by scoring a scene opposed to the expressive content of the remaining modalities, i.e.

visuals). But before we delve into the depths of the functional concept, the reader will be quickly familiarized with core components and functions of music in games as well as a selection of the most important techniques for its implementation. The later sections will be dedicated to music's ability to alter attentional focus as well as involvement and the experience of being immersed into the game world and/or action. Gaining an understanding as to how interactive music can affect game play experience will allow the reader to adapt their own future system to accommodate the requirements of most games. The final section will review these findings and showcase some notes about how to (responsibly) apply the previously discussed techniques for ensuring that your players will have a blast.

5.2. Core Concepts and Implementations of Interactive Music

As we discuss the core concepts of interactive music, we will also meet some basic terminology that you may have been familiarized with from composing for film and other media. Oftentimes these concepts may have larger consequences within the interactive realm so that they deserve special attention when laying out your ideas to game producers, designers and programmers. First, some considerations as for the *genre* of a game may give the reader an idea what stylistic elements and scoring conventions future players have potentially been accustomed to (i.e. some major genres are action, adventure, strategy, simulation, logic; for a detailed review of genre and music style see Crathorne 2010). Second, for any music cue, it needs to be determined where music stands in the *diegesis*, that is whether music will originate through objects or music-making characters inside the game space (i.e. diegetic music; see *BioShock Infinite* (2013)). Despite the effectiveness of diegetic music, the most common use of music in games has its roots in non-diegetic contexts, that is, music that accompanies characters, objects, events or locations from the off and acts outside the perceptual reality these elements are situated in (cf. Stevens and Raybould 2011). As will be shown in later sections, non-diegetic music possesses the capacity to alter the apparent relationship between the aforementioned design elements on several hierarchical levels, which resembles an important function of how humans interact with others in real life (i.e. "theory of mind" is the ability to attribute mental states to oneself and to others; see Decety and Grèzes 2006; Gasselseder 2015b). Transdiegesis, a term originally coined by Jørgensen (2007), refers to a special kind of diegesis that occurs when audio acts as a communicative device between the game world and the player (this may also apply to music whenever a specific event is announced that requires a response by the player). As transdiegesis is believed to operate bidirectionally, the source of music may not stem only from the game world but also from the players' world (e.g. creating a personal playlist to which the game engine will adjust its visualization, as seen in futuristic racing games such as *Audiosurf 2* (Fitterer 2015)).

5.2.1. What, Who, When and Where: The Functional Components of Interactive Music

Knowing about the functional components of music enables a composer to subdivide a cue for its most relevant structural features, such as timing, tonal, affective and semantic/symbolic qualities. Further down the process, these parameters impact the balancing of music with other elements in the audio mix (i.e. sound effects, dialogue) and game play. At that point, depending on the identified parameters, it will be determined how these functional components will be implemented as an interactive music system. For the purposes of simplifying the identification of these components and their respective parameters, in the experience of the author, it has reaped tangible rewards to follow the questions what, who, when and where as a guideline. Thus, the paragraph below will explore the "four Ws" as for their functional counterpart in game music.

What does music relate to? When underscoring a relevant moment in the action, a music state may appear

- As a *marker* of significant events in terms of game play or narrative-dramaturgic development. Examples of game play events may be seen at the start, in intermediary states or at the completion of a challenge (note that musical markers of successfully finishing a challenge are crucial to game enjoyment and flow experience; more about that in the later sections).
- As a method to prompt *expectation* in the player, as in the informing about impending events (e.g. dangers). It is important to ensure that the link between the appearance and expression of music and game variable remains intuitively understandable to the player (though, the exception proves the rule in that some games may wish to set the player on the wrong track, as in the survival horror genre).
- To induce *tension*, whether of visceral nature (i.e. shock) or as a positive reinforcement (e.g. "fiero," which is a motivational state brought about shortly before overcoming an obstacle, as in the last few remaining strokes to knock out an opponent in a fighting game (cf. Stevens, Raybould and Mcdermott 2015)).
- As a *navigational* device that guides the player through the game world. For example, using slow, legato (i.e. connected) lines in the

melodic instruments supports exploration of the world, whereas fast, accentuated staccatissimo (i.e. distinct with strong accent) will motivate players to move in a more target-oriented manner (e.g. towards the next challenge, such as rescuing the hostages).

Who does music relate to? If targeted towards an individual or a group of characters, a music state may appear

- As a *theme* or *leitmotif*, which encompasses a recurring musical motive that is attached to the presence or mentioning of a specific character, situation/event, place, mood or dramaturgic motif.
- As a *reminder* motif or as a *programmatic modulation* of the theme that mirrors the development of the specified referent or its relationship to others (note that examples for this will be given as part of discussing relational differentials in later sections).

When and *Where* does music relate to? For differentiating multiple points in space and time, music may appear

- As an *anchor point* to inform the player about cultural, stylistic as well as game play-related factors, such as challenge difficulty and emotional qualities of a particular space.
- As a device of mediating *perspective* and *mental states*, such as altered states of consciousness. For example, the latter is frequently realized by applying filter effects onto the audio mix of a music state (see the Joker transformation scenes in *Batman: Arkham Origins* (Warner Bros. Int. 2013); for a further review of composing for altered states of consciousness in audiovisual media refer to Weinel (2018)).

5.2.2. Core Implementations of Interactive Music in Video Games

Typically, when implementing music in a video game you may choose to use either the music system that is part of a game development engine (e.g. Unity (2018), Unreal (Epic Games 2018)) or a third-party software tool that interfaces with this engine. As of the time of writing this chapter, several solutions aim to provide a more intuitive workflow by offering musically intelligent transition points, (e.g. ELIAS (2018)) as well as algorithmic and procedural (e.g. Melodrive (2018)) paradigms, but, due to their recent arrival, have yet to find their way into the mainstream (e.g. Wwise (Audiokinetic 2018), FMOD (Fireflight 2018)). Thus, the following paragraphs will introduce the reader to the most common core techniques of implementing dynamic music (note, in order to remain compatible with terminology used by other practitioners, this overview will orient itself primarily to Sweet (2014, 2016), Stevens and Raybould (2011), and McLeran (2017)).

Because the following techniques make use of prerecorded material, some important considerations have to be made with regard to the limited resources available to the audio engine. As uncompressed audio files are typically large, it is recommended to select a suitable audio compression format that is loopable as well as supported by your audio engine and the destination computing platform (e.g. Vorbis is a popular compression algorithm). Choosing an appropriate audio codec will ensure a minimum latency when streaming from the storage space before the sound is triggered (note that most audio engines allow to specify the prefetch length of data that will be initially accessed from RAM). Having ensured that your audio tracks meet those requirements, you are ready to explore dynamic (or adaptive) music systems.

5.2.2.1. Music Looping

The first and probably most familiar technique to the reader is *looping*. It works as the foundation of most other techniques involving prerecorded material. Due to the undetermined amount of time players spend in a particular state, it is important that the transitions between the beginning and end of the looping point are seamless and do not draw attention to themselves. To avoid listener fatigue, it is recommended to have a loop repeat for a finite number of times and subsequently fade out to silence. If you plan to rely exclusively on this technique, keep in mind that you may want your music to act in the background and thus avoid using recognizable distinct motives in looping points. Instead, you can opt to introduce a healthy amount of indeterminacy in your musical material so as to keep effects of musical expectancy as low as possible and cover up loop transitions.

5.2.2.2. The Horizontal Dimension: Re-Sequencing

Another common technique, horizontal re-sequencing, pertains to the connecting of different music cues depending on the actions of the player and various states in the game world (cf. Sweet 2014). Depending on how the transition is implemented, various compositional techniques will apply to ensure a sensible evolution from one state to the next. Similarly, the type of changing game states also impacts which transition you may want to use.

Transitions based on *cross-fading* are most apt for changes of states that root in the game world, such as location or weather. In terms of compositional style, it is advisable to keep musical parameters, such as harmony, meter, tempo and instrumentation/mix rather ambiguous at the point of contact between the two states. Sweet (2014) recommends to limit switching between music states to not more often than every 30 seconds as it may result in interruptions to the game play (more about this as well as so-called "local" and "global" goals will be discussed in later sections).

Phrase branching transitions are suitable for switching between states of the game world that share the same location. The technique waits for the currently playing music state to complete before moving on to the next state. This brings about the advantage of higher flexibility with regard to the compatibility of musical parameters between two music states. However, due to the dependency of phrase length, longer music cues are more likely to cause lag when moving to the next state.

Musical demarcation/synchronization branching come closer to reflecting avatar- and character-related states within game play when used appropriately. This technique enables two music states to transition into each other at musically meaningful divisions in the time-domain (i.e. beat, measure). Hence, the music system needs to be aware of the currently active music parameters of both states. While allowing for faster transitions, this type of branching runs the danger of interrupting phrases. Hence, changes in melody, rhythm and instrumentation should ideally complement each other.

Bridge transitions are particularly useful for accentuating a significant event within the dramaturgy of game play or cinematics. Rather than directly switching to the subsequent music state, this technique acts as a building block of its own that permits a more evolutionary shift towards the next music state. Also, here, attention must be paid with regard to the frequency of transitions (30 seconds, see above), though, this problem can be mitigated by assimilating start and end points of the bridge to the connecting points of the origin and destination music state.

Stinger-based sequencing represents the first entrance towards a more action-oriented approach to scoring. Rather than operating as a transition, stingers may be laid on top of a music bed so as to punctuate a specific action or event. The technique is frequently used during scripted sequences and is thus more appropriate for underscoring highly significant dramaturgic events in game play. Thus, the applied musical material often consists of short snippets of crescendi and accents whose expressive quality fits into a concurrently playing musical bed (e.g. tension and shock induced in *Dead Space 2* (Visceral Games 2011)). Alternatively, the technique can also be used to hint the player towards clues in the game environment.

5.2.2.3. The Vertical Dimension: Re-Orchestration and Parallel Form

Also referred to as a multilayered approach or vertical remixing, vertical re-orchestration makes use of stacking several looped tracks onto each other to be played simultaneously. These layers can be separated by criteria such as the density of the instrumentation or other musical functions, such as punctuation or percussive/rhythmic elements. To give an example for instrumentation, the first layer may consist of a slowly moving synth pad or drone, the second layer adds staccati strings while the third and fourth layer add brass and woodwinds as well as percussion, respectively. The number of layers can range from two up to five or more, depending on the resources available and control-inputs exposed by the game engine. These game variables permit for defining state changes, singular events or thresholds to trigger the entrance and exit of a layer. Frequently, these parameters relate to progress in game play and thus are used to accentuate the dramaturgy of the action as much as they act as a communication device towards the player (e.g. music hints about the number of remaining adversaries in the Penguin's museum map of *Batman: Arkham City* (Rocksteady 2011)).

One major challenge when producing content for re-orchestration is ensuring that all layers share the same harmonic, rhythmic, tempo and length parameters in order to appear as a coherent entity. Some more elaborate varieties of this technique utilize musical demarcation or synchronization rather than simple cross-fades for introducing and removing individual layers. A weak point of implementations found in some commercial games relates to slow response times to move down the line of hierarchy within the arrangement (i.e. removing a layer to reveal the one underneath). While this may partially serve in the interest to keep the player motivated to remain active within the current quest, it may be worthwhile to double check its detrimental effects as for maintaining congruence within game play. Overall, this offers the greatest potential to game play experience by means of modulating emotional arousal with varying expressive trajectories. Re-orchestration is particularly useful for varying between different behavioral traits of game play, such as stealth exploration and combat in the action genre. Specifically, it has been shown to effectively impact the emotional experience (particularly on the valence dimension, that is, pleasure-displeasure), absorption and spatial presence in game play (cf. Gasselseder 2014, 2015a). As for a hint about its technical implementation, one way to avoid memory bottleneck is by exporting all layers into a single multichannel file (to save simultaneous music streams) and to turn on their respective channels rather than playing back each layer from a separate file (cf. McLeran 2017).

Now that the reader has been familiarized with some of the most common principles of its implementation, it is time to move the discussion towards the effect of interactive music onto the attention system as well as its structuring function of experience by its very nature.

5.3. Interactive Versus Noninteractive Music: Shifting the Focus From Sounding to Acting Music

Interactivity comes in where music ends, which is not to state that music is inherently not interactive. More to the contrary, the contention of the abovementioned thought experiment is that music indeed is interactive by nature, but that it is so by providing an anchor point for the synchronizing of the aesthetic experience of the listener. Whether you might find yourself comforted by music matching your current mood, or like to listen to music while you are at work, there is a good reason to assume a functional equivalent in the aesthetic pleasure of music listening alone. Thus, it is less surprising to find music as a suitable reference for associating other stimuli that appear in conjunction with it. Walking through a shopping mall, sitting in a café, a movie theater or in front of your TV will make you inadvertently tend to synchronize the incoming visual stimuli in relation to their sonic counterparts. Hence, music interacts with listeners also in noninteractive formats whenever it triggers the recall of associations or supports schematic processing that might make you move your feet with its rhythm or walk at a pace that matches its tempo. Such unconscious behavioral responses are rarely attributed to a soundtrack that in fact may have prompted viewers to react with goosebumps or an agitated feeling when watching their favorite character fall into a deadly void or a zombie suddenly attacking from behind. The visual primacy in our perceptual processing will make us look at what we hear, rather than hear what's in plain view. Because one comes into the experience with a certain set of expectations, finding them met by music to varying degrees, one also finds oneself interacting with the music, which in turn, also affects one's responses to the content that one is currently concerned with, or, in the words of music psychology, to that to which you are selectively attending.

We may assume that listening to sound and music always involves a certain *active* component, even when it may seem that music would not be at the center of attention (a mode of listening that is usually referred to as "*passive*"). To demonstrate the interplay between music and attention, let us entertain another thought experiment: if an unfamiliar piece of music is interrupted with one's favorite tune, one is likely to pay attention to the change of musical background, even if primarily focusing on doing something else. In this case, while one may have listened to the music scene in a passive mode, one would immediately be able to detect the change and switch to "*active*" listening. But, as in almost all that makes us human, this covert processing does not always work to the benefit of our perception, such as in the case of failing to notice changes of objects, scenes ("change blindness") or their overall appearance ("inattentional blindness"). The seminal experiment on inattentional blindness by Simons and Chabris (1999) demonstrates the latter notion of our capacity for selective attention that holds even when a woman in a gorilla-costume walks through a dynamically changing scene—unbeknownst to the viewer who is tasked to focus on the total number of passes made by a basketball team wearing white t-shirts. Whereas rooted in the visual, the studies' main finding has been shown to apply also for the auditory domain in that the likelihood of detecting an unexpected object depends on its similarity to other objects and the difficulty of the monitoring task (see Dalton and Fraenkel 2012). More specifically for music, Koreimann and colleagues (2014) report related cases of inattentional blindness with their subjects failing to notice an incongruous guitar solo in a famous musical piece while being tasked with counting the number of drum beats.

Returning to the aforementioned example of switching between passive and active modes of listening, it becomes apparent that some kind of covert processing must have carried on while you were focusing on whatever else was laid in front of your eyes. Such covert processing does not exclusively affect the detection of familiar or similar objects or events, but it also contributes to the structuring of the experience of time per se. Hence, functional music, as we hear it in movies and games, holds the potential to change our perception of time flow or intervals that are inherently linked with the concurring stimuli of the overall presentation. However, previous studies have shown that the perception of synchrony may be independent of multisensory integration (Harrar, Harris and Spence 2016). This suggests a fast and slow system of multisensory processes where the former may be linked to the physical synchrony of stimuli and thus promote an automatic processing that is unaffected by temporal adaptations (i.e. the delay of the onset of an audio cue relative to the visuals). The slow system, in contrast, is perceptually driven and more likely to adapt to temporal changes of synchrony. Related results are echoed in several studies conducted by Gasselseder (2015a, 2015b), which indeed suggest a predominant role of the slower perceptual system when testing for different implementations of music stimuli in video games. One may think of the slower perceptual processing as a platform for setting the stage of the situation that is to be depicted. Hence, the sense of perceptual synchrony acts as a driving force for putting your mind into the world of the experience. Within this context the faster processing platform, which automatically integrates multisensory stimuli, may contribute to the sense of proximity of imminent, specific events that a user may ascribe to the experience. This is evidenced by sudden bursts of sounds triggering a scare reaction—an observation that is automatically processed due to its high value to survival in nature. Further work by Gasselseder (2014) suggests that these modes of processing also contribute to ascribing different meanings to the depicted scenario. Accordingly, different implementations of music in a video game result in varying ascriptions of relationships between characters. These ascribed relationships can even change in hierarchy depending on the articulations used as much as the potential of inducing emotional arousal to the user. Whereas low- to mid-arousal music may primarily affect attributions towards extra-protagonist characters (i.e. all characters except for the protagonist), high arousal music prompts more frequent attributions of character relationships towards the protagonist. This is not very surprising considering that the emotional arousal conveyed in music often indicates the relevance a given event or character may have to the protagonist (see Vitouch 2001; Hoeckner et al. 2011). However, the dynamics of changes of arousal also appear to affect attributions of relationship hierarchy. A somewhat more static underscore may convey meaning attributed to the realm of the extra-protagonist, whereas a dynamically changing implementation, that often moves between the extremes of arousal, is more likely to be attributed to the protagonist (Gasselseder 2015b). But what can we learn from that when composing music for interactive media? Assuming that our processing of music in the medium is governed by two distinct mechanisms of synchrony allows us to make assumptions about its potential effects on selective attention ("whom do we want the audience to attend to?"), time flow ("how much attention should be paid to these events?") and its inherently related offspring, namely, immersive experience and its overarching term of non-mediation.

5.4.   A Proposed Mechanism of Non-Mediation and Immersive Experience: Acting by Music Versus Listening to Music

Where there is nothing, there is imagination. As human beings, it is almost impossible to imagine nothingness. Shutting off sight, hearing, taste, smell and touch, as in depriving oneself from the sensory inputs of one's surroundings, will inevitably lead to psychedelic experiences ranging from the hallucinatory to the meditative as well as to the disturbing. These effects are explained by false prediction errors where new sensations are not properly integrated and result in misattributions as originating externally (cf. Mason and Brady 2009). Besides sensory deprivation, various affective states (e.g. high stress levels) have been shown to impact heuristic judgements as they pertain to errors in source monitoring with regard to the context in which an individual is situated. Previous work by the author also indicates that congruence in parallel information streams plays an important role in reality monitoring and perceptual hypotheses testing, which brings about the question as to its consequences for non-mediation (cf. Gasselseder 2015a, 2016).

The term "non-mediation" conjures an illusion that stems from nothingness. It outlines the uplifting of the border between the real and the virtual, where individuals respond as if the medium were not there (cf. Lombard and Ditton 1997). By removing something, it adds a new dimension of immediacy to the experience. While the primary understanding of non-mediation relates to illusions of embodiment, another facet is found in absorption and self-transcendence that may make the imagination of the user prone to be taken over by fictional settings as suggested within the realm of the media. Terms such as "immersive," "involvement," "absorption," "suspension of disbelief," "transportation" respective to "presence," "self-location," "flow" and "interactivity/possible actions" have been used interchangeably to denote these overlapping concepts within their specific application, such as simulations in virtual environments or video games. Despite their vague distinctions, common denominators can be identified in motivational states and their altered cognitive representations of situational factors, such as in imaginary aspects that also reside in absorption, suspension of disbelief and transportation. With self-location, flow and perceived possible actions, an embodied, sensory-spatial aspect adds to the experience.

The multiconstruct "immersive presence" (Gasselseder 2014) aims to consolidate these imaginary and sensory-spatial aspects of non-mediation within a unified framework that incorporates the notion of relational differentials during agency detection. In this view, immersive experiences arise from perceptual processes that juxtapose expected and incoming sensory data as a function of situational demands (cf. Bruner and Postman 1949; Popper and Fay 1997). Here, relational differentials operationalize an agent's current and future state as well as realm of interaction in the environment with regard to expected outcomes for the user. Ascribing purpose and relevance to surrounding events in relation to our own beliefs and desires appears to be a ubiquitous process of perception (cf. Zwaan 1999). Research on theory of mind supports this view in that activity of the mirror neuron system is only observed when actions are attributed to agents but not to non-agents (see Decety and Grèzes 2006). In order to assess a situation, these relational differentials are subsumed to a syntax or reference frame that determines the situational context, which is further projected to subsequent cue juxtapositions and awareness of the range of possible actions. Relational differentials may then be seen in connection to intrinsic motivation, which in turn is believed to support the experience of flow and the exploration of the environment. It is hypothesized that music achieves immersive experiences by altering relational differentials as a result of directing selective attention and retrieving schemata in function to varying levels of expression-congruency. In doing so, connotations based on prior experiences and cultural codes drive expectations and evaluative functions of music (cf. Cohen 2001; Gaver and Mandler 1987). Applying this information to relational differentials enhances the validity and predictive value of individual cues. The attribution of a reference frame and its associated situation model then emerges from sensations caused by corresponding audiovisual accent structures (cf. Boltz, Ebendorf and Field 2009; Cohen 2001; Petrini et al. 2011).

The first step in achieving immersive experiences through music in multimedia may be seen in the primal urge of humans to synchronize incoming stimuli (cf. Maasø 2000). When initializing selective attention and searching for salient cues in the environment, other senses are taken over by the superior temporal resolution of sonic dimensions (cf. Spence and Driver 1997). At this point, a first set of filters directs subsequent hypotheses testing towards congruent percepts (Bruner and Postman 1949). Synchronization ensures the assessing of audiovisual accent structures at contact with visual and other stimuli. If music and the remaining modalities are found to follow similar structural features causing analog sensations, multisensory expectations on emotional congruence towards the situation at focus are formed. If matching combinations of stimuli are found to be congruent to a hypothesis of perception, attention allocation to the media content is intensified. While at this point connotations of music are processed on an extramedial level, that is a conscious integration into relational differentials within the situational context of media reception (e.g. sitting in front of a PC and knowing that music is played back by speakers placed in the room), the emerging reference frame (e.g. defining challenge-based motives) is attributed to the situational context implied by the media content.

This process of "situational context localization" sets the stage for experiencing imaginary immersion by giving access to portrayed intentions and motivations (cf. Ermi and Mäyrä 2007; Wirth et al. 2007). At this point only expressive features reach the processing of relational differentials. Previous work by the author suggests that expressive features related to emotional valence may play a dominant role during extramedial processing (cf. Gasselseder 2014). This may be due to a basal matching process of synchronization that yet does not fully account for momentary changes. Accordingly, valence is less likely to change spontaneously; suggesting that the potential of music in modulating emotional valence, as, for example, by the means of minor keys and dissonance, may provide an efficient way of establishing mood and situation. The associated connotations are integrated consciously, meaning that the perceiving subject is still able to discern the presence of music as well as its surface features from the remaining modalities of the media content. This is of relevance for the attribution process of agency, which is negotiated between a sub-personal automatic level for action identification and a more conscious level for sensing agency related representations about intentions, plans and desires (cf. Decety and Grézes 2006). While this hierarchy is asserted for real-life social interactions, the situational context model proposes an inversion of this sequence for media reception. Hence, the conscious sense of agency preexists and is followed by covert automatic processing that couples premotor action with a virtual avatar. Compared to automatic bottom-up, the conscious top-down path processes information slower, making it more susceptible to information that is carried by the valence potential of music.

For the faster bottom-up path, however, a more efficient source of information may be seen in the changing levels of arousal potential in music. Having reached extramedial localization, the basal matching process of synchronization may be extended by momentary changes. The gained relevance of the latter allows for lower latency in action identification so that varying levels of music-expressed arousal take a dominant role in driving multisensory expectations on emotional congruence. If proportional to the arousal potential of remaining modalities, an increase of arousal in music leads to an intramedial localization of schemata recall. Arising connotations are now unconsciously integrated into relational differentials attributed to the situational context implied in the media content. However, the transition to intramedial localization is gradual insofar as it depends on the latency that music takes to follow the remainder of the accent structure. This latency determines the degree of attributing a particular event or action to an agent, such as the user itself. Note that due to constraints in terms of syntactic structure and scoring conventions, music expression rarely mimics on-screen action directly (Cohen 2001). Within premotor activation, however, music may affect expectations directed towards action readiness or "forward models" at a higher level that encodes global specifications of the action with the controlling and adapting to their goals and underlying motivations (cf. Decety and Grèzes 2006; Jeannerod 1997). This synchronization of premotor activity marks the point when intramedial localization has been reached and the self has become aware of its physical extension towards the possible realm of action and its location. Since an action may become an intrinsic motivator in its own right, it is more likely to be attributed to the self. Within the context of the flow model (Csikszentmihalyi 1991), the additional information provided by music affects the assessment of task demands, while also modulating self-perception of skills. For presence, previous studies have found a correlation between varying levels of induced arousal and self-location (e.g. Robillard et al. 2003). Moreover, forward models may also contribute to discerning one's own thoughts and emotions from others, providing the foundation of cognitive empathy (cf. Decety and Grèzes 2006). This discrimination may allow relational differentials to become emotionally contagious. Finally, schemata recall and emerging relational differentials are contextualized beyond those motivational ties that were ascribed to the usage situation (e.g. playing for fun in the living room). The situational context model thus operationalizes immersive presence as a mediated perspectivation of situational characteristics that are represented by the media content and its expressed meaning structures (Bruner 1986).

5.5. The Game Plan: Constructing a Functional Concept of Musical Interactivity for Your Game

Whereas the previous sections touched on issues that often emerge indirectly during the composition process, the present section will be concerned with the multitude of ludomusical strategies that aim to affect players by linking music to forms and content of game play. At the beginning of this process the overall design and character of a music system will need to be adjusted to a specific game project (i.e. whether interactive/nonlinear music is wished for in the first place). As outlined during the introductory section, linear music may be regarded as interactive to a degree and thus may provide a sufficient accompaniment to select genres ranging from casual to racing to logic. This is to assume that the horizontal dimensionality of a score (i.e. one music cue transitioning to another) adequately reflects the motivation, difficulty and overall aim or intended outcome of the challenge. Thus, a music cue that repeatedly appears throughout a game level can establish a supporting background (i.e. a unified flow from one moment to the next and deepened cognitive involvement) to challenges that demand similarly repetitious action or higher logic skills. Then again, for challenges with stronger ties to exploring a narrative, the same approach may result in listening fatigue, which will prompt players to turn off the music after having been exposed to a track multiple times. Consequently, the previously discussed components of non-mediation correlate differently with game play experience and performance depending on game genre (see Gasselseder 2015a). As music, too, has been shown to differentially affect these components, it is advisable to consider their supporting and detrimental effects to the player within the context of a specific game with great care.

5.5.1. Know Your Genre and Your Players: Interactivity Is Not a Constant

Rather than being touted for their absorption or spatial presence, racing games often profit from soundtracks supporting the experience of flow whereas role-playing and action-adventure rely more strongly on presence (see Weibel and Wissmath 2011). In game genres such as racing and other vehicle simulations that typically aim to recreate perceptually realistic scenarios, flow experience is enhanced by handing over music choice as well as control over other playback parameters to the player (i.e. "personalization" such as customizable volume and event triggers realized with a dedicated music player integrated into the user interface). This is to suggest a stronger diegetic, more familiar and thus "realistic" anchoring of musical stimuli within the game world. Other genres that similarly refrain from using a narrative as a primary motivational design element and that involve a higher level of abstraction, such as logic, board-games or reaction-time challenges, may invite better game play performance if accompanied by mood-matching music, as empirical evidence suggests outside the realm of gaming applications (see Franco et al. 2014). Ideally, in this case the music score would anticipate a priori or, going one step further, deduce the potential mood of the player (i.e. a relatively long-lasting emotional state) from all available data points inside the game. However, whether it is advisable to make music react to particular events (i.e. affect-matching; a relatively temporary emotional reaction) depends on the level of interactivity that a game offers. If within a challenge the indicators of progress are limited to the beginning and end of game play as much as to lack any feedback for reward and punishment, the requirements for a music system may be lower than for a game that ascribes instant rewards and punishments to an action while also constantly adjusting a high score. Indeed, both the lack and abundance of interactive complexity may need to be reflected within your music system. Consequently, a thorough genre-related consideration of mood- and affect-matching music represents one of the first steps of the design phase, as does a review of dynamic systems linked to rewarding and punishing game events.

5.5.2. Focus on a Core Experience: Immersive Is Not the Only Way

It is important to note that the insights above may not be considered as absolute rules, as it depends on the motivational design as much as the specific implementation whether music will have a detrimental or favorable effect on game play (e.g. emotional responses, challenge/competence), immersive experience (see previous section) and performance (success/failure;

in-game progress). Typically, not all of these experiential aspects can be accounted for to the same degree within one soundtrack. Instead the challenge is to identify the core experience of a scene or level as much as the overall arch of the game and try to balance these with the remaining aspects.

In this vein, a short excursion towards composing for states of flow may throw further light on the issue. In the literature, it has been suggested that flow acts as a transfer mechanism between immersive and game play experience with regard to gratification and the formation of game-related attitudes as well as motivation (see Örtqvist and Liljedahl 2010). Wherever themes like these comprise the backdrop of a games' proposal, you are well suited to examine musical options that optimize flow and, consequently, game performance. Among these options are often softer toned pieces that vary in their dynamics as realized either by a looping linear music cue that exploits the previously discussed phenomena of synchronization, or by a dedicated interactive mechanism that manipulates the vertical dimensionality (i.e. the density of the arrangement; often also referred to as re-orchestration that varies according to in-game progress) so as to provide an informative auditory commentary to the player that does not overshadow the immediate feedback of an action, such as offered by sound effects (cf. Gasselseder 2015a). More specifically, to support flow experience, the spectral and dynamic characteristics of a music cue must be ensured to leave sufficient space for sound effects to be clearly audible as to allow them to fulfill their potential to provide immediate feedback to the player. The sense of player skills and task difficulty is afforded by the moment-to-moment synchronization of the sense of control over an action and environment, which represent two fundamental components of the flow model (see Csikszentmihalyi 1991). Apart from making sound effects fit seamlessly into the master mix, it may sometimes also be necessary to keep music in the background and its arousing characteristics restrained. Following a paradigm of "less is more" prevents the music from overshadowing sound effects on an acoustic as much as connotative and semantic level (e.g. situational manifestations, as in the mismatch in spatiality and reverb between the mixes of non-diegetic and diegetic sounds). The beneficial effects of the "less is more" approach do not end with flow. The relationship between sound effect and game play experience also extends to the sense of agency. Depending on the perceived latency between an event and an incoming stimulus, players are more likely to ascribe an action to agents corresponding to avatar and extra-avatar characters (cf. Gasselseder 2015b; Jeannerod 1997; see "forward models" in the previous sections' discussion). This synchronization of premotor activation contributes to autotelic stages of flow, where the purpose of an action is fully intrinsically motivated and goals have been clearly defined and adopted by the player (the latter represents another component within the flow model of Csikszentmihalyi 1991).

5.5.3. Define the Goals of Game Play: Not the Destination, but the Path

As discussed earlier, autotelic experiences are closely related to absorption. But while the impact of sound effects is mainly constrained to that of action, in the case of absorption, the realm of music is only about to begin. When planning for the structural experience prompted by music alone, that is whether a particular cue is positively reinforcing (e.g. by fostering a sense of achievement), agitating (e.g. by using faster tempos and short rhythmical patterns) or overall heroic, we also touch issues of goal setting at various hierarchical stages within the dramaturgic arch. While working on the design of a game it often helps to define these goals as global and local. According to the flow model, in order to become deeply involved or absorbed in any activity, the task must be precisely understood moment by moment (cf. Csikszentmihalyi 1991). Yet, the player must first be introduced to goal-sets and by doing so will gradually engage with the situational context implied by the game world (cf. Gasselseder 2015a). This is initially accomplished by defining clear goals that act on a global level by outlining an opportunity. Here, the player is provided with an outlook towards challenges at several levels of complexity and is ideally made to believe that these lie within the player's capacity (i.e. balance of skill and difficulty). Usually these global goals are intertwined with the games' narrative and set up the motivational ties and purpose of the challenges that the player will be presented with (e.g. the journey of the hero; within Aristotelian drama theory also known as "desis" or "complication," exposition of which is followed by "peripeteia" or "turning point" and "lysis" or "resolution"). For music that aims to express global goals this follows a few aesthetic considerations as in the use of thematic material and instrumentation to convey the overall conflict and to situate the avatar and its world. Further down the situational context model, on the way to experiences of sensory-spatial immersion, the player's attentional focus towards an activity becomes narrower, which is mirrored by local goals that define the task at hand within the current scene (e.g. stealth game play to avoid being discovered by opponents or to distract them). As for aesthetic and motivational considerations, again, local goals operationalized in music aim to attain congruence with the current task (e.g. an atmospheric, slowly moving synth pad during periods of stealth or exploration, and an agitating, fast or accentuated orchestral arrangement during a combat scene). In this case a music cue may capture the immediacy of the situation by following several key actions or events as to underline their meaning within the progress of a subtask (i.e. a challenge). However, music often struggles to underscore action as well as other momentary changes within the accent structure of a task due to the nonlinear nature of game play as well as scoring conventions listeners have become accustomed to from previous exposure to music in the media.

5.5.4. Plan Your Actions: The Possible Comes First, the Impossible During the Implementation

Typical implementations in current generation games try to approximate the abovementioned key events by means of the vertical dimension within a music system, which in the case of re-orchestration involves the adding or subtracting of instrumental tracks from the arrangement in the progress of game play. While this gives players a hint towards the current state and likely evolution of the subtask, thus counting on musically induced effects of expectation and tension, it lacks the timing necessary to act as a pinpoint marker of success or failure (e.g. a critical strike followed by a short musical pause) and thus loses in emotional contagion. Despite the difficulty of its implementation, several solutions have been proposed for tackling the timing problem, one of which is to integrate event triggers within the game's logical structure that become active as soon as a certain threshold is passed in game play (e.g. a boss being finished in three out of ten strikes). This technique assumes that the relevant musical parameters of the various states are loaded and tracked by the music system. Under these constraints it may be easier to estimate the time available for a stinger or motive to transition between different states. However, the problem as for how to seamlessly connect to the previous music state remains. For that reason, it is advisable to set up a specific threshold range to allow the previous state to complete its phrase, beat or measure and switch into the stinger or transition (i.e. this technique is often also referred to as musical demarcation). Depending on whether we want to prepare the player for a key action or a culmination point of a challenge, different compositional requirements apply. A key action may be better accentuated by the sudden appearance of a short stinger, which, apart from moving onto the next sequence (i.e. horizontal dimension), can also be stacked on top of the currently active music state (i.e. vertical dimension) and act as a transitional device to the next state. Typically, stingers, when used as a transitional device, remain rather ambiguous with regard to their musical belonging (i.e. harmonic and metric ambiguity or even a short moment of silence/pause) and possess a point-by-point quality that reflects the time-sensitive nature of the action. In the case of the culmination point, a relatively short loopable phrase or motive may be more adequate to counteract the indeterminacy of game play. Good examples for this are musical effects, such as risers (e.g. the Shepard-Risset "infinite riser" (cf. Stevens 2018)), glissandi (i.e. a typical arrangement for orchestra would be a piccolo and flute section playing a fast glissando in octaves together with harp and a stopped crescendo in the cymbal), percussion hits or short ostinati.

5.5.5. *Keep It Coherent . . . and Your World From Collapsing*

Whereas both global and local goals should reference each other as to make them seem as originating from one world, they reflect distinctly different tasks. In the examples given above, it is apparent that prototypical characterizations of an aggressive hero (global) and the hiding from adversaries (local) do not go hand in hand. Thus, a music cue adjusted to local goals needs to convey the situational affordances of the task and action that is currently in demand, while keeping intact the integrity of the characterization of the avatar and its world. Several ways of maintaining integrity exist, whether that is by recalling or modulating themes/leitmotifs, instrumentation or effects in the mix chain, such as distortion, that appear in connection with a character or event. Being reminded of the overall purpose of one's actions by musical connotation lays out the corresponding intentions, plans and desires of the avatar in a way that centers attention towards the present, alters the sense of time and facilitates absorption as in being present in a different world. With this in mind, a congruent aesthetic of global and local goals ensures that personal growth from meeting challenges and developing skills is being attributed to the player, but also projected towards the avatar. Hence, we attain a stronger intro-relational link (i.e. the relational differential between avatar and player) if we follow along the later stages of the flow model from clear goals and opportunity, over immediate feedback, attention focus and being present, towards control, an altered sense of time and the transcendence of individuality that makes us engage in something larger than ourselves (cf. Csikszentmihalyi 1991). Contextualizing the motivational ties suggested by local goals within the global goal-set also avoids unwanted sensitivity towards inevitable weaknesses of the game design, such as in striking cases of gamification where a task demands actions that may not be thought to be representative of the avatars' characterization. In this connection, it is important to stress the potentially disruptive effect of sudden transitions between two local goals, as realized by switching between horizontal and vertical dimensions of the music system. For example, if the shift between two music cues, that are assigned to different local goals, is being performed too soon, the player is inclined to readjust the centering of attention, as empirical evidence suggests (Gasselseder 2014, 2015a). As outlined above, the process of goal setting requires the establishing of synchronization and absorption first before a strong sense of task focus can emerge. Sudden changes between these goal-sets break the fluid course of action that the player experiences while being in flow (e.g. to follow the examples given above, this could be an abrupt switch between the atmospheric pad assigned to the stealth scene and the epic orchestral arrangement assigned to the combat scene). Therefore, it is recommended to introduce separate transitions between music cues that move relatively seamlessly from one to another and, ideally, are linked to game events of high relevance (e.g. mastering a first step of the challenge or being close to finishing a challenge or completing a challenge). Depending on the task a player is charged with, the length of a transition can span from two to ten seconds. Accordingly, the transition should be adequately timed in order to avoid drawing attention to itself or cause confusion as for the continuation of game play.

5.6.   Final Remarks: A Game Over Is Just Another Beginning

Rest assured, this is not the end. At this point we barely touched the extent of interactive music. Apart from dynamic and adaptive forms, more elaborate methods, such as algorithmic and procedural implementations, will challenge your composition, sound design, sampling and programming skills in the near future. My advice is to stay current (or ahead) and to be prepared for big changes. After all, the techniques and methods will evolve, but the core principle of writing music for games will remain largely unaffected. Avoid drawing attention to the music and guide it towards the experience instead. Set the mood by positive or negative expression (emotional valence) and in this way offer players an easy path into absorption and suspension of disbelief. Connect them to the avatar and make them feel present by modulating the arousing characteristics of the music according to the player's actions. Remind the players why they are pressing a series of buttons while making them forget what they are actually pressing on. Motivate and reward your players as that will raise their sense of possible actions, especially after overcoming a difficult challenge. And use silence as a creative device and anchor of perceptual realism rather than as a void that appears when nothing worthy of musical commentary happens on screen. Contrasts such as these will make a score seem more dynamic and put your music back into the spotlight at its point of entrance. Yes, this somewhat contradicts the point that music is not supposed to attract attention to itself. This does not mean that game music should pass by your players without them even noticing. It rather means that whenever music gets a shining moment of exposition, it should refer to an idea that is bigger than itself, which is the game. Only a few musical devices are as motivating as presenting a grandiose theme at a moment of desperation. When the obstacles seem the steepest, the hero will rise. And so will you as a composer of interactive music as much as the immersive experience of the player. Now that you were warned, the rules are yours to be broken. Restart!

References

Allesch, C. G., and Krakauer, P. M., 2005. Understanding Our Experience of Music. *Musicae Scientiae*, 10, no. 1, pp. 41–63.

Audiokinetic, 2018. *Wwise*. Viewed Aug 6, 2018 www.audiokinetic.com/products/wwise

Boltz, M.G., Ebendorf, B. and Field, B., 2009. Audiovisual Interactions: The Impact of Visual Information on Music Perception and Memory. *Music Perception*, 27, no. 1, pp. 43–59.

Bruner, J.S., 1986. *Actual Minds, Possible Worlds*. Cambridge, MA: Harvard University Press.

Bruner, J.S. and Postman, L., 1949. On the Perception of Incongruity. *Journal of Personality*, 18, no. 2, pp. 206–223.

Cohen, A.J., 2001. Music as a Source of Emotion in Film. In: Juslin, P.N. and Sloboda, J.A. (Eds.), *Series in Affective Science: Music and Emotion: Theory and Research*. New York: Oxford University Press, pp. 249–272.

Crathorne, P.J., 2010. Video game genres and their music. Thesis at the Department of Music, University of Stellenbosch, Stellenbosch, RSA. Viewed Sept 3, 2018 http://scholar.sun.ac.za/handle/10019.1/4355

Csikszentmihalyi, M., 1991. *Flow: The Psychology of Optimal Experience*. New York: Harper and Row.

Dalton, P. and Fraenkel, N., 2012. Gorillas We Have Missed. Sustained Inattentional Deafness for Dynamic Events. *Cognition*, 124, no. 3, pp. 367–372.

Decety, J. and Grèzes, J., 2006. The Power of Simulation. Imagining One's Own and Other's Behavior. *Brain Research*, 1079, pp. 4–14.

Elias., 2018. *Elias 3* (software). Viewed Aug 6, 2018 https://eliassoftware.com

Epic Games, 2018. *Unreal Engine 4*. Viewed Aug 6, 2018 www.unrealengine.com/en-US/what-is-unreal-engine-4

Ermi, L. and Mäyrä, F., 2007. *Fundamental Components of the Gameplay Experience. Analyzing Immersion*. In de Castell, S. and Jenson, J. (Eds.), *Worlds in Play. International Perspectives on Digital Games Research*. New York: Peter Lang Publishing, pp. 37–54.

Fireflight, 2018. *FMOD* (software). Viewed Aug 6, 2018 www.fmod.com

Fitterer, D., 2015. *Audiosurf 2* (video game). Independent.

Franco, F. et al., 2014. Affect-Matching Music Improves Cognitive Performance in Adults and Young Children for Both Positive and Negative Emotions. *Psychology of Music*, 42, no. 6, pp. 869–887.

Gasselseder, H.P., 2014. Re-Scoring the Game's Score. Dynamic Music, Personality and Immersion in the Ludonarrative. *IADIS International Journal on WWW/Internet*, 12, no. 1, pp. 17–34.

Gasselseder, H.P., 2015a. Re-Sequencing the Ludic Orchestra. In: Koenitz, H. et al. (Eds.), *Interactive Storytelling: Lecture Notes in Computer Science*. Cham: Springer, pp. 458–469.

Gasselseder, H.P., 2015b. The Role of Agency in Ludoacoustic Immersion. Experiencing Recorded Sound and Music in Situational Context. In: Kalliris, G. and Charalampos, D. (Eds.), *AM '15. Proceedings of the Audio Mostly 2015 on Interaction with Sound*, Oct 7–10. Thessaloniki, GR and New York, NY: ACM.

Gasselseder, H.P., 2016. What You Hear Is Where You Are Is What I Hear. Optimising Immersive Experiences for Interindividual Differences. In: Bowen, J., Diprose, G. and Lambert, N. (Eds.), *EVA London 2016*. Electronic Visualisation and the Arts, Jul 11–13. London: BCS Learning and Development, pp. 61–69.

Gaver, W.W. and Mandler, G., 1987. Play It Again, Sam. On Liking Music. *Cognition and Emotion*, 1, no. 3, pp. 259–282.

Harrar, V., Harris, L.R. and Spence, C., 2016. Multisensory Integration Is Independent of Perceived Simultaneity. *Experimental Brain Research*, 235, no. 3, pp. 1–13.

Hoeckner, B. et al., 2011. Film Music Influences How Viewers Relate to Movie Characters. *Psychology of Aesthetics, Creativity, and the Arts*, 5, no. 2, pp. 146–153.

Irrational Games, 2013. *BioShock Infinite* (video game). 2K Games.

Jeannerod, M., 1997. *The Cognitive Neuroscience of Action*. Oxford, UK: Blackwell.

Jørgensen, K., 2007. On Transdiegetic Sounds in Computer Games. In: Fetveit, A. and Stald, G. (Eds.), *Northern Lights: Film and Media Studies Yearbook*, Vol. 5, No. 1. Bristol: Intellect Publications, pp. 105–117.

Koreimann, S., Gula, B. and Vitouch, O., 2014. Inattentional Deafness in Music. *Psychological Research*, 78, no. 3, pp. 304–312.

Lombard, M. and Ditton, T.B., 1997. At the Heart of It All. *Journal of Computer-Mediated Communication*, 3, no. 2. Viewed Sept 3, 2018 https://onlinelibrary.wiley.com/doi/full/10.1111/j.1083-6101.1997.tb00072.x

Maasø, A., 2000. Synchronisieren ist Unnorwegisch (Dubbing is not Norwegian). *Montage AV*, 9 pp. 147–171.

Mason, O.J. and Brady, F., 2009. The Psychotomimetic Effects of Short-Term Sensory Deprivation. *The Journal of Nervous and Mental Disease*, 197, no. 10, pp. 783–785.

McLeran, A., 2017. Interactive Music Systems for Games. In: Somberg, G. (Ed.), *Game Audio Programming*. Boca Raton, FL: CRC Press, pp. 209–222.

Melodrive, 2018. *Melodrive*. Adaptive Music Generation. Viewed Aug 6, 2018 http://melodrive.com

Örtqvist, D. and Liljedahl, M., 2010. Immersion and Gameplay Experience. *International Journal of Computer Games Technology*, no. 3, pp. 1–11.

Petrini, K. et al., 2011. The Music of Your Emotions. Neural Substrates Involved in Detection of Emotional Correspondence between Auditory and Visual Music Actions. *PLoS One*, 6, no. 4.

Popper, A.N. and Fay, R.R., 1997. Evolution of the Ear and Hearing. Issues and Questions. *Brain, Behavior and Evolution*, 50, no. 4, pp. 213–221.

Prechtl, A., 2016. *Adaptive Music Generation for Computer Games*. PhD Thesis. The Open University, Milton Keynes, UK.

Robillard, G. et al., 2003. Anxiety and Presence During VR Immersion. A Comparative Study of the Reactions of Phobic and Non-Phobic Participants in Therapeutic Virtual Environments Derived from Computer Games. *CyberPsychology and Behavior*, 6, no. 5, pp. 467–476.

Rocksteady, 2011. *Batman. Arkham City* (video game). Warner Bros. Interactive.

Schrader, C. et al., 2017. Rising to the Challenge. An Emotion-Driven Approach Toward Adaptive Serious Games. In: Ma, M. and Oikonomou, A. (Eds.), *Serious Games and Edutainment Applications Volume II*. Cham: Springer, pp. 3–28.

Simons, D.J. and Chabris, C.F., 1999. Gorillas in Our Midst. Sustained Inattentional Blindness for Dynamic Events. *Perception*, 28, no. 9, pp. 1059–1074.

Spence, C. and Driver, J., 1997. Audiovisual Links in Exogenous Covert Spatial Attention. *Perception and Psychophysics*, 59, pp. 1–22.

Stevens, R., 2018. Going Nowhere. Stasis, Ambiguity, and the Infinite Riser in Video Game Music. *Paper Presented at Ludomusicology 2018*, Apr 13–15. Leipzig, DE. Viewed Aug 6, 2018 www.ludomusicology.org/ludo2018/#citem_1121-940c

Stevens, R. and Raybould, D., 2011. *The Game Audio Tutorial*. Burlington, MA: Focal Press.

Stevens, R., Raybould, D. and Mcdermott, D., 2015. Extreme Ninjas Use Windows, Not Doors. In: *AES 56th International Conference*, Feb 11–13. London and New York: AES.

Sweet, M., 2014. *Writing Interactive Music for Video Games*. Upper Saddle River, NJ: Addison-Wesley.

Sweet, M., 2016. *Top 6 Adaptive Music Techniques in Games*. Pros and Cons. Viewed Aug 6, 2018 www.designingmusicnow.com/2016/06/13/advantages-disadvantages-common-interactive-music-techniques-used-video-games

Unity Technologies, 2018. *Unity 2018*. Viewed Aug 6, 2018 https://unity3d.com/unity

Visceral Games, 2011. *Dead Space 2* (video game). Electronic Arts.

Vitouch, O., 2001. When Your Ear Sets the Stage. Musical Context Effects in Film Perception. *Psychology of Music*, 29, no. 1, pp. 70–83.

Warner Bros. Games Montréal, 2013. *Batman. Arkham Origins* (video game). Warner Bros. Interactive.

Weibel, D. and Wissmath, B., 2011. Immersion in Computer Games. The Role of Spatial Presence and Flow. *International Journal of Computer Games Technology*, 2011, pp. 1–14, article ID 282345.

Weinel, J., 2018. *Inner Sound. Altered States of Consciousness in Electronic Music and Audio-Visual Media*. New York: Oxford University Press.

Williams, D. et al., 2016. A Perceptual and Affective Evaluation of an Affectively-Driven Engine for Video Game Soundtracking. *ACM Computers in Entertainment*, 14, no. 3. Viewed Sept 3, 2018 https://cie.acm.org/articles/perceptual-and-affective-evaluation-affectively-driven-engine-video-game-soundtracking/

Wirth, W. et al., 2007. A Process Model of the Formation of Spatial Presence Experiences. *Media Psychology*, 9, no. 3. Viewed Aug 6 2018 www.tandfonline.com/doi/abs/10.1080/15213260701283079

Zwaan, R.A., 1999. Situation Models. The Mental Leap into Imagined Worlds. *Current Directions in Psychological Science*, 8, no. 1, pp. 15–18.

Further Reading

Phillips, W. 2015. *A Composer's Guide to Game Music*. Cambridge, MA: The MIT Press.

Stevens, R. and D. Raybould 2015. *Game Audio Implementation*. Burlington, MA: Focal Press.

Summers, T. 2016. *Understanding Video Game Music*. Cambridge: Cambridge University Press.

Sweet, M. 2014. *Writing Interactive Music for Video Games*. Upper Saddle River, NJ: Addison-Wesley.

6
_____

Interfaces for Sound Installation

Greg Corness, Kristin Carlson and Prophecy Sun

6.1. Introduction

Our ears are what most people think of as how we experience sound and as the most basic interface with the audio world. The inner workings of the ear respond to different frequencies and volumes of sounds around us, while the outer ear, the pinna, manipulates the incoming pressure waves in a way that enables our brains to process the signals from our ears. In terms of an interface, this is still very passive. We consider an interface as a process for both perceiving and manipulating a system. We often think of an interface as a controller (keyboard, mouse, game controller), where the user's intention is transmitted to the system and the system provides pertinent feedback back to the user for their continued engagement with the system. However, the transmission of information is never complete or without errors. While a computer keyboard captures the letters I am typing, the sounds I make while typing are lost unless I design the keyboard to respond to this additional input. Similarly, a program may show only the information it has been programmed to provide and cannot give any further information. Our interactions with the physical world have similar limitations and expectations. Our experience with the world around us teaches us that we can hit a drum to make it sound—basic interface. We learn that hitting it in different ways can get different sounds but within limits—feedback on how we hit it. Our hands, eyes and skin all play a role as part of our interfacing with objects. We see two objects collide, perceive the crash and understand the connection between the action and the sound. We feel a vibration on our skin, hear a noise and make a connection—we get information about the system. Yet, if the drum sounded like a flute we would be surprised and reexamine how we engaged with the drum, which would mean reexamining the *interface*. Underlying these experiences are the laws of physics, which define how the objects around us respond to our actions to produce sound. Our experience teaches us about the physical world. This informs our expectations of how our actions engage a system, be it a drum or a computer, in producing sound. It also is the foundation of how the produced sound provides information on our actions (how hard we hit the drum) and the state of the system (how large the drum is, if the skin is tightened, etc.). Through our bodies we interface with sound objects having learned expectations on the relationship between sounds and objects, the actions (information) they respond to and the sound (feedback information) they respond with. Now, what if those expectations were changed? What if we could set up a world where we controlled how actions produce sounds? We could customize this basic human interface to sound, our bodies in a world where new relationships between sound and objects are artificially linked to create a new interface that provides an augmented experience for the visitor in an installation.

As art in the twentieth century moved away from representation through painted image to exploring means of commenting on other aspects of objects and human experience, the practice of *installation* became more prevalent. Duchamp described the modern aesthetic artistic choices as "not about 'the work' itself but the attention we bring to it" (Betts 2015: pg. 54). In our natural engagement with the world, sound provides a sonic map of our placement in it. Our perception of the constant changes in the sounds, their reverberation in space, the relative and changing volumes, changes in timbre, and other parameters of the sound all combine to produce a complex sonic environment. An audio installation sets up a space as a microcosm of the world where the parameters of sound draw attention to aspects of the space and the objects, including sounds themselves, to encourage an experience and reflection by the viewer. Such a space may be designed to be *precomposed*, such as a preset series of sounds as in an audio recording or video, or in an interactive format where the sounds are affected by the actions of the visitor.

Whether the installation is precomposed or interactive, the visitor's ears and brain remain the fundamental interface for their experience with the sounds. The artist/designer affects the visitor's experience by selecting and/or designing a set of sounds to create the sonic environment guiding the passive experience. When the interaction is included in the design, the visitor's body is included too. The artist/designer creates not only a subset of sounds for the environment but the relationships between the sounds and the actions of the visitor. As artist David Rokeby states:

> Most artworks start as a set of possibilities: the blank canvas, the empty page, the block of marble, etc. The act of realizing a work is a process of progressively narrowing the range of possibilities by a series of creative choices until one of the possible has been

manifested in the finished work. One might say that the interactive artist decides at some point in this process not to choose
from among the remaining possibilities but to create some sort of
audience-actuated choosing mechanism.

(Rokeby 1995, p. 136)

The audience activation of the choices is through the interface designed
into the system. It should be noted that in engaging with an interface, a
visitor constructs beliefs as to their level of control—what is part of the
installation, what they meant to do—and in this way adds the artificial
interface of the installation to their natural body interfacing in the world
(Dix et al. 2006). The visitor's natural body interface is augmented by a
system put in place artificially linking their actions to the sonic responses
of the installation. The artificial links are designed to encourage and support a path of reflection and experience.

    This chapter discusses interface design considerations for sound instal-
lations. To do this we highlight prominent artists and creatives that offer
interesting strategies and methods of production. While we will discuss
the technologies, we situate them within a larger context, and focus on
the concepts behind the interactions. We suggest that technologies offer
entry points into adopting new ways of learning and engaging with interfaces. However, while many specific technology options exist and are
ever evolving and constantly changing, the concepts and strategies behind
these technologies move much more slowly. Throughout we suggest a list
of design considerations that describe modes of interaction, which may
then be layered over specific types of interfaces.

6.2.   Engaging With Sound Installations

We may consider an interactive sound installation as any situation in
which a visitor's experience is evoked by the modulation of the sound
environment through some direct engagement of the visitor. This general
description covers a wide range of approaches. For example, a curated
soundwalk through a city allows visitors to hear different aspects of found
sounds that they would not notice otherwise on their own. The movement of a visitor in a room may be tracked by a camera and then be used
to change a soundscape or trigger different sound files. An installation
may require conspicuous viewers to place their hands on a sensor to capture their body temperature or pulse, using that data to change aspects of
the sound environment. Any number or types of sensors may be linked
to numerous processes that trigger, distort, modulate or even generate a
sound environment.

The interface associated with each individual interaction may appear to the visitor as inherent to the work. However, the link between an intended experience and the interface can be so intertwined that it is hard to separate the interface from the installation. This line can be even further blurred by the aesthetics of the installation. The design of the interface is so integral to the visitor's experience that the aesthetic choices, the metaphors of interaction, and other design parameters all contribute to the constructed belief system that creates the environment for a visitor to experience and reflect on (Dix et al. 2006). To illustrate the importance of interface design we present three example works: Janet Cardiff and George Bures Miller's *The Cabinet of Curiousness* (2010) and *Experiment in F# Minor* (2012), and Christine Sun Kim's *Game of Skill 2.0* (2015).

In Janet Cardiff and George Bures Miller's piece *The Cabinet of Curiousness*[1] (2010), audio files are triggered by the opening and closing of drawers in an antique wooden card catalog. The immersive experience of this work is directly influenced by Cardiff and Miller's unique interface design to playing audio files. The visual aesthetics of an antique wood cabinet reference the historical tradition of *Cabinets of Curiosity*, yet place the work as a modern adaptation through the use of sound objects rather than physical visual objects and the assumed technology for playing the sounds. Cardiff and Miller chose to use the opening and closing of the drawers as the interaction, further referencing curiosity and the traditional card catalog aesthetic. The overlapping of tangible, visual and sound aesthetics plays a large role in the experience of the work.

In contrast, Cardiff and Miller's piece *Experiment in F# Minor*[2] (2012) employs a very different visual and interactive aesthetic. The artwork consists of an arrangement of open speakers on a table, with a visual effect of a raw electronic experiment. The speakers play various sound and instrumental tracks, creating an abstract audio composition. Sensors around the edge of the table sense the presence and motion of viewers that fades the volume of various tracks up and down to produce a layered cacophony of sound. All aspects of the work accentuate an aesthetic of raw unstructured exploration that guides the experience of the visitor.

In Christine Sun Kim's *Game of Skill 2.0*[3] (2015), visitors engage with sound by pulling a device along a cable above their heads. The sound installation reflects the speed and direction of the visitor walking. The design of the interface accentuates two elements of the installation: (1) the temporal aspect of the sound is experienced spatially by the visitor, and (2) the artist has constructed the interface to not track easily, causing the visitor to concentrate on the mechanism for producing the sound. The forcing of perspective can change the visitor's experience from a passive experience into an encumbering active experience.

The above examples demonstrate the unique and symbiotic connection between the aesthetic of the interface, the aesthetics of the installation as a whole, and the intention of the artist. In each case the interface is part of the piece and not easily separable from the installation design as a whole. For example, the technology employed has been designed to elicit a particular interaction that the artist believes will support the reflection they want the visitor to experience. From this stance a general discussion of interfaces for sound installations would be a listing of example works and technologies. However, we need to realize that the technology is not the interface. The link between the drawers in *Cabinet of Curiousness* and the system triggering the sounds could be done in any number of ways. Similarly, *Experiment in F# Minor* and *Game of Skill 2.0* could utilize a different technology for their interface. The different technology may or may not result in a different experience depending on how the designer exploits the affordances of the technology, but not because of the technology itself. Fundamentally, the change is from how it links the visitor's actions to the system that is at the root of the installation. From this perspective we may now discuss general design considerations that align to interfaces for sound installations separate from the technology used to build them.

6.3. Design Considerations for Sound Interfaces

In analyzing a variety of sound interfaces for installation, we articulate five design categories to consider. These categories reflect how the actions of the visitors are mapped to actions of the installation (Agency), how obvious the interface is to the visitor (Explicit/Implicit), how functional the interface is (Reliability), how easy it is for the visitor to explore the options provided by the interface (Discoverability) and how the use of metaphors are integrated to reflect the actions of the visitor (Quality). These categories can be used to support design choices that will contribute to how a sound installation is experienced by the visitor.

These design considerations could be implemented a variety of ways, including through the acoustics of an object, the movement of the visitor in the space, or the way the visitor handles an object. One technical way to implement these design considerations are through mappings, which are the connections built between the visitor and the system. The four basic types of mappings include (See Figure 6.1):

- One-to-One: a single channel of sensing that affects a single parameter of the output, often in a control paradigm. The visitor presses a button that triggers a sound file.

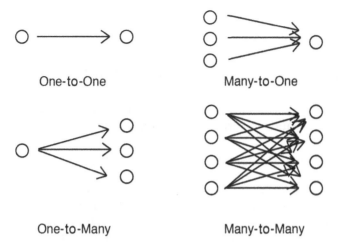

One-to-One   Many-to-One

One-to-Many   Many-to-Many

Figure 6.1   Mapping Strategies for Interface Design.

- One-to-Many: a single channel of sensing that effects two or more parameters of the response. This may be in a control paradigm or a more complex paradigm involving other channels of influence on the parameters. The visitor presses a button that changes both the pitch and volume of a sound file.
- Many-to-One: several channels of sensing that affect a single parameter. Often the channels of sensing are subjected to signal processing and/or other comparative processes to infer a final channel of information that is used to affect a single parameter of the response. The visitor waves their hand in the air, which changes the volume of a sound file.
- Many-to-Many: several channels of sensing all working together to affect several channels of response. As with Many-to-One, the channels of sensing are often processed to infer a single line of information. Similarly, the changes to the response parameters are inferred through an understood (or assumed) relationship between the parameters to present a single overall texture or quality from the response system as a whole. The visitor walks into the room. The direction they walk changes the pitch of the sound, the speed they walk changes the volume of the sound, the quadrant of the room they are in changes the tempo. While the visitor may be able to identify exactly which action caused the reaction, the work is experienced as a whole and a good design of the interface does not focus solely on the technical responses of the system.

Mappings will be used as a technical example throughout the following discussion of design considerations.

6.4. General Descriptions of Design Considerations

This section provides a more thorough description of the five design categories in application to various forms of interaction.

6.4.1. Agency: Responsive, Interactive and Generative

Agency refers to how a system responds when a visitor engages with the interface. While the term *interactive* is used ubiquitously, not every digital interaction is inter*active*. We separate the terms responsive, interactive and generative to indicate the *type* of agency a visitor has when engaging with a system. These designations label the extremes on multiple axis in an interaction. So, a system may be partway between responsive, interactive and generative while containing elements of each.

6.4.1.1. Responsive

Responsive refers to a single direction, the visitor does one thing and the system changes (see Figure 6.2). For example, a visitor pushes a button and a sound file is triggered. Responsive systems often use One-to-One mappings, or Many-to-One.

6.4.1.2. Interactive

An interactive system can move in two directions, meaning that the visitor does something and the system changes, but the system's response changes how the visitor can engage (see Figure 6.3). For example, a visitor

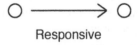

Responsive

Figure 6.2   A Responsive System Moves Information in a Single Direction.

Interactive

Figure 6.3   An Interactive System Moves Information in Two Directions.

pushes a button and a sound file is triggered but also the button changes. Then the visitor has to figure out how to push the button again for the interaction to progress. Interactive systems may use One-to-Many, Many-to-One or Many-to-Many mappings.

6.4.1.3. Generative

The third term is generative, where a visitor may or may not interact with the system, and the system creates its own content. A generative system can purely create content itself, or in *response* to a visitor, or in *interaction* with a visitor. An example of an interactive and generative seminal work is *Voyager* by George Lewis, a Columbia University trombone professor who designed a system to improvise with live performers (Lewis 1999). The system "listens" to the live performer and then generates its own improvised music based on what the live performer did. Generative systems may use any of the mapping options to control parameters that will generate sound and/or provide interaction.

6.4.2. Explicit/Implicit Interfaces

Explicit and Implicit refer to how visible the control functions are in an interface, which contributes to the visitor's understanding of how they are engaging with an installation.

6.4.2.1. Explicit Interfaces

Explicit interfaces are obvious to the visitor. There is often a clear presentation of the controls and the audience has feedback that they are directly affecting the installation. For example, a board full of knobs and dials, where the audience may not know what exactly each component will do, but it is clear that their actions should have a direct impact on the sound being played. Explicit interfaces may use a One-to-One mapping or One-to-Many that have clear feedback patterns to the visitor.

6.4.2.2. Implicit Interfaces

Implicit interfaces are not obvious to the visitor, and the visitor may not be aware that they are having any effect on the installation at all. An example would be a soundscape that changes subtly based on the color of clothing worn by the audience passing through the installation, but may not be recognized by the audience. Implicit interfaces may use a Many-to-One mapping or Many-to-Many that do not have clear, direct feedback patterns to the visitor.

6.4.3. Reliability

Reliability refers to *how* the interface is perceived to function. This includes technical aspects such as latency between the visitor's action and feedback, or misfires by the interface. It may also include design choices such as whether the interface does the same thing every time, or whether it does something different tomorrow. Reliability can illustrate how technical issues such as latency become important in the design, such as when a designed One-to-One mapping that should be clear is, for some reason, delayed or overlaps with another sound making it unclear as to the resulting response.

6.4.4. Discoverability

The design of an interface often aims to enable the visitor to engage in the installation *without* an instruction manual. Discoverability refers to the *affordances* of an interface that give clues about how to use it. Affordances are the obvious ways to grasp, move, trigger or engage with an interface that are apparent through its design. These affordances provide options that allow the visitor to discover on their own how to interact with an interface, play with a system or be inventive with their choices.

Maintaining discoverability over time can be challenging. Many visitors enter an installation looking for "the trick," wanting to figure out how to control it, test it and try to break it, and are then bored. Designing an interface for sound that allows for a variety of discovery options is one approach to keeping the visitor exploring, but what do you do if the experience of exploration becomes boring or overwhelming? For example, if a visitor walks into a room quickly and then stops, and hears sound mimicking a quick change and then a hold, they would assume that they control the sound through their movement in the room. While there are no instructions listed on the walls, the visitor might attempt a variety of movement options in order to "discover" what the system does and how they can innovate within the experience. Enabling layers or levels of experiences that become unlocked over time, or new environments that can emerge through certain interactions are methods for keeping visitors engaged and enchanted (McCarthy et al. 2006).

6.4.5. Quality

The quality of an interface refers to how design choices are able to support the visitor's perception of qualitative connection between events in the installation and the response of the installation. This could mean the decision to use a pressure sensor instead of a button to trigger a sound, because the way each device is pressed creates a different experience. Another example is if a visitor hits a button harder or faster, there is a similar qualitative response by the system. Quality can affect how the interface is presented to a visitor, how well the interface conveys the affordances or possible discoveries to the visitor, or can address the level of detail or nuance designed into the interface.

6.5.   Typology Description

We present a typology of interfaces to provide guidance for design opportunities when creating interfaces for sound installations. While this list is not exhaustive (it was challenging to find examples of crowdsourcing or data mining as an interface), it represents a categorization of current trends for interfaces. This section articulates interfaces as tangible objects, devices for signal processing and recording, motion tracking, position or spatial tracking, biosensors and machine control. Each interface type is described with examples, then uses the above design considerations to emphasize what is important about that type of interface when designing for interaction.

6.5.1.   Tangible Objects

Tangible interfaces are physical and material, with clear affordances. Consumer-level tangible interfaces can include items such as game controllers (i.e. the guitar controller for *Guitar Hero*, or a console controller) or cell phones (using an app or sending SMS data) that enable the audience to interact with an installation. There are many homemade interfaces that use buttons, pressure sensors, switches or strings to control sound. Historically, responsive tables such as *ReacTable* have been used to create an interface on a surface that can be manipulated with tangibles such as blocks with fiducial markers (a unique visual shape that can be recognized by a computer-vision system to see the tangible object's position or orientation) (Jordà et al. 2003). A useful online collection of tangible musical objects can be found on Martin Kaltenbrunner's website (2018).[4]

A current version of a tangible table includes the *SoundFORMS*[5] (2016) system that uses a shape-shifting display relying on synthesizer and drum machine triggers to physically visualize sound waves. These waves can be engaged with through the visitor's gestures to control aspects of the sound, providing a physical experience of music composition. *SoundFORMS* uses a combination of touch interaction and gestural interaction (hand shapes with a Kinect camera), though we included it in the tangible category because of its material qualities. Visually, the system is made of a grid of white plastic squares that create waves in the X, Y and Z axis.

SoundFORMS could be considered an interactive interface, with information moving in two directions (Colter et al. 2016). The interface uses a Kinect sensor to create a One-to-One mapping, in which the shape of the hand gesture controls the shape of the sound wave (i.e. timbre). The shape of the sound wave is reflected both sonically and visually in the tangible interface. There is discoverability in that the visitor can discover new sounds depending on how they explore different hand gestures and in that it uses an implicit interface because it does not indicate how you touch it. The quality of the interface's presentation is reflected in its physical shifts. The aesthetics of the installation are clean and beautiful, which encourages a visitor to engage and continue to engage.

Sonic Banana[6] is a work by Eric Singer of LEMUR (League of Electronic, Musical, Urban Robots). It is a flexible rod that uses bend sensors to detect the actions of the visitor (Singer 2010). The visitor can twist, bend and straighten the instrument to generate MIDI data that can be sent to any MIDI sound gear. Being a bendable rod, the visitor has control through MIDI over what sounds are played and how they are manipulated in reference to the amount of bend. The tactility of the interface affords an interesting link between effort used to bend the sensor, the resulting shape of the sensor, and the change in the sound being controlled. However, the physical act of bending also implies that "jumping" from extremely bent to straight would not be possible because the sensors would detect each intervening state of bentness. The visitor has full agency to the extent of the rod's affordances including its limitations.

The *Sonic Banana* (2013) interface is explicit because there is nothing else to do with the instrument besides manipulate it. The affordance of a flexible rod is to bend it. There is discoverability in that how it makes changes to the sounds will shift depending on how you manipulate the rod. While it is unclear how many sounds you can make and manipulate simultaneously, the physical affordances of the sonic banana encourage exploration and are reliable with immediate feedback.

The Handphone Table (1978)[7] is an installation by Laurie Anderson in which the visitors sit at a table with their hands over their ears, placing their elbows in a notch that transmits sound vibrations. Vibrations travel through the visitor's forearm bones and use their hands as resonating chambers. We use this as an example of a tangible system because it illustrates how we embody sound. This would be considered a responsive system because information only travels in one direction, and the visitor has agency in that they can control how and when their body is transmitting information to their ears. The interface is explicit because there is an image hung on the wall that shows two visitors using the system. There is minimal discoverability past the initial interaction, yet the system is simple and robust. The quality of the interaction is beautiful in that the gesture is simple and embracing while the table with notches looks hand carved and worn, creating a cozy and intimate experience.

6.5.2. Devices for Signal Processing and Recording

One approach to designing the interaction with sound in an installation is to design a direct interaction with the electronic signal representing the sound, be it analog or digital. This approach leverages techniques of digital or analog signal processing, often achieved using a variety of tools including audio software plug-ins, off-the-shelf audio effect devices or even custom designed circuits. By passing the electronic signal from devices such as microphones, recordings, synthesizers and even detectors capturing signals such as radio waves, cosmic waves, subsonic, ultrasonic, etc., new signals are created that can be used as audio sources or to further manipulate other source sounds. Such techniques are common in the context of performance from guitar effect boxes to Laptop Orchestras to electronic instruments themselves such as the theremin. More relevant to the topic of this chapter is how these techniques lay the foundation for installation interfaces, engaged with by uninitiated visitors.

An example of audio effect devices being used in installations can be experienced in several of the exhibits in the *Experience Music Project* in Seattle (Experience Music Project 2018). One exhibit in particular encourages the visitor to reflect on the signature sound of Jimi Hendrix. The visitor is provided with the raw (unprocessed) guitar tracks from an actual Hendrix recording session. This audio signal is then passed to a combination of effect pedals that duplicates the setup used by Hendrix. The visitor can then change the parameters on each pedal to try to reproduce the sound used for the final mix of the song. The visitor compares the sound they create with copy of the released version. The setup of the exhibit provides an explicit interface where the visitor is made directly aware of the signal path of the sound effect process by using knob controllers. The interaction is perceived as reliable since there is immediate feedback to the choices made by the visitor. There is high quality in the interaction because more twists in the knobs create louder levels (an obvious affordance), which also contributes to the quality of the resulting piece because the metaphor of interaction matches the resulting action. Also, the interface has high discoverability since the user is learning by trying. The success of the installation is largely due to the discoverability in the interface. The effort the visitor puts into this interaction encourages reflection on what Hendrix's experience of playing music would be like.

Another example of interface through signal processing can be seen in artist Victor Mazon Gardoqui's use of his custom hardware The SIG-NUM (2016).[8] This hardware enables an artist to use antennas and custom microphones to detect and capture a wide assortment of energy waves from vibrations in materials from ultrasound to radio waves. The device transforms these frequencies into control voltages that can then be used to manipulate sound, light, video or other media. The hardware provides the basis for interfaces that may be explicit or implicit. While signals/waves may be targeted explicitly, the hardware can also be used to exploit unexpected, uncontrolled, or unnoticed changes in these waves. It should be noted that working with these types of input materials raises the issue of visitors not necessarily being aware of changes in these spectrums. For example, they may not be aware of the fluctuations in the radio frequencies in the room or of subsonic vibrations of passing traffic. Hence, using these signals to change the sound in an installation may encourage more of a technical rather than visceral reflection. Such work can feel intriguing because the experience extends beyond the visitor's perceptual limits.

The use of signal processing can also be applied to microphones providing live signal of sounds in installations. Artists Olle Cornéer and Martin Lübcke's work *BACTERIAL ORCHESTRA* (2008)[9] serves as an example. In this installation the audience enters a sound ecology where individual sonic agents respond to the sounds of other agents and visitors. The piece consists of variously sized speakers and a variety of microphones each representing an agent in the ecology. Each agent listens to the sounds that are surrounding them and uses signal processing to affect the behavior of the agents. The agency of the visitor is quite low, but in some respects the quality is high in that the system is responding to the quality of the noise around it in an attempt to *fit in*. The complexity and indirect nature of signal processing makes the connection between events and responses by individual agents hard to track.

6.5.3. Motion Tracking

The connection between movement and sound has been explored in a number of different disciplines including dance performance, music performance, artwork and installation. Strategies, technology and aesthetics have been shared between artists working in these various disciplines, including the connection between sound and movement. Many of these interfaces employ cameras as sensors; however, other sensors such as infrared or ultrasonic proximity sensors, hall effect, and sensors that detect location sequentially (at a rate that affords the calculation of motion) are also used. Despite the technical differences, all these interfaces focus on movement as an interface by sensing the quality of how motion changes (not just the presence of motion). For example, at a basic level a security light turns on when any motion is detected in the area. Such lights can be *accidentally* triggered by animals or blowing objects. To avoid accidental triggering, a second level of analysis could be added in to use qualities of the object or motion as a parameter in the interface, such as size, density of an object, movement speed or direction.

Many qualities of motion have been explored as interfaces to installation. For example, the dance company Palindrome (based in Weimar, Germany) developed its own computer-vision software called Eyecon. The software measures the amount of movement rather than the presence or absence of body parts at given locations (Wechsler, Weiss and Dowling 2004). Using this information, dancers could trigger different sounds based on the quality of their movement. Slow movements would trigger a certain sound while fast movements would trigger another sound. By mapping the response based on quality, the visitor or observer could perceive a deeper connection with the movement. While movement as an interface is often implicit, it depends on how clear the mappings are. This makes movement very discoverable and transparent, unless the mappings are unclear and there is a lot of additional motion that obscures the mappings. Often the process of motion tracking can be dependent on light and other factors that can trip up the system into false responses.

Christine Sun Kim's *Game of Skill 2.0* (2015) (discussed earlier, where the visitor moves an object on a cable above their head) uses simple movement as part of the interface. The piece connects the visitor's experience of motion, including speed, with their experience of the sound through a handheld device that must be held properly on a track to enable listening. This interface makes the act of listening explicit through effort and yet the mapping of space to sound quickly becomes understandable so that effects such as "scrubbing" forward and backward, slow and fast, are easily discoverable and perceived to be reliable.

In Cardiff and Miller's piece *Experiment in F# Minor* (2013) the motions tracked are implicit as the visitor approaches and views a table of speakers. In this way, the agency of the visitor is quite discoverable as the interface focuses on actions that will commonly be performed. The agency remains limited, in part by the limited quality of motion detectable affecting the range of mapping options. The use of proximity sensors limits the quality of motion detectable by the system, but the simple sensor can be quite reliable in a gallery environment.

Beyond the mapping of quality of motion to quality of sound, the quality of the motion can be used to determine other factors in an installation. In the piece *Light Strings* the artists Seo and Corness used tracking of motion to determine the type of use the visitor was engaged in (2015). The work, a large area filled by vertically hung strands of fiber optics, allowed visitors three modes of interaction: to sit in the space and slowly touch light and sound objects, to walk around the space following the objects or to run through the space to create waves of light and sound. Switching between these modes was determined by the analysis of the quality of motion in the space.

6.5.4. *Position and Spatial Tracking*

Position is one of our most natural engagements with sound. We naturally use our ears to gauge our spatial relationship to objects in our environment. Recreating this in an installation has often been explored using a variety of technologies and approaches. At a basic level we can consider the practice of soundwalks as a form of navigational space or position-based sonic pieces as the visitor mediates their experience by moving to new positions. Cardiff and Bures Miller presents gallery goers with a similar experience in her piece *Forty Part Motet* (2001).[10] The work consists of 40 speakers placed in a circle, each with a single voice from a choir singing a motet. As with the soundwalk, the visitor controls their experience of the sound piece by moving to new positions. To hear a single voice the visitor could move close to a single speaker, or they could stand between two speakers closely to hear overlapping voices, or in the center to experience the whole work.

In other systems, the visitor's position can be used as the input to the interface. As demonstrated by the work of the dance company Palindrome, detecting the position of the body or even body parts can be used to influence a sonic system including selecting sound files, the signal processing (e.g. reverb, lowpass filters), or changing the mode of the system.

The concept of position can also apply to the relative positions of the visitor or objects on a surface. The controller *ReacTable*[11] (2003) is an intelligent touch table that detects the relative position of objects and their orientation on the table. This data can then be used to affect the sonic response of the system. Though this system is designed as a controller, it demonstrates the approach taken by numerous installations in which the relative position of objects, what they are pointing to, and other spatial information such as orientation can be used in a sonic installation.

6.5.5. *Biosensors*

Sound installations often use biosensors or other devices to trigger a sound or motion, or use the physiological data as an interface to control choices or navigation through the installation. Biosensors use physiological data as input such as EMG (muscular conductivity), EEG (brain conductivity), heart rate, breath or GSR (skin conductivity). Installations may be designed to be passively controlled, using physiological data as it occurs naturally, or may be designed to intervene with the audience and attempt to engage with their physiological response, such as by scaring or exciting the audience.

Atau Tanaka is a Japanese-born artist and researcher who has regularly worked with biosensors. In the piece *Sensors_Sonics_Sights*, performed live at FACT Liverpool in 2008, Tanaka along with Laurent Dailleau and Cecile Babiole performed a concert work using the *BioMuse* system, ultrasound OpenGL and a theremin to create an ephemeral and nuanced work. *Sensors_Sonics_Sights* had the three performers *playing* their physiological data in front of a screen with visualizations of their data changes as they created a soundscape using their three forms of biological data. Tanaka's work has primarily used EMG sensors such as the BioMuse system designed at Stanford University's Center for Computer Research in Music and Acoustics (CCRMA) for concert performance (Tanaka 2000). BioMuse used eight input channels to gather a variety of EMG data that was processed and analyzed by using algorithms including envelope following and spectral analysis. The resulting data controlled a Max/MSP patch that manipulated sound live as the performer moved and controlled their body as an interface.

The use of biosensors in *Sensors_Sonics_Sights* creates interactive systems because the playing of the "instruments" (the body) is continually changing the instrument/body. There is a two-way interaction of the body moving to create data that triggers and manipulates sound, which changes the body's reaction and creates a cycle of exploration and play. Biosensor interfaces seem like they should be explicit, however, body data is difficult to control and tends to be very noisy. The lack of intentional feedback from the system turns this into an implicit interface because the visitor is exploring many ways of moving and stressing the system to get a response. Discoverability of the *Sensors_Sonics_Sights* work is rich in that the multitude of options created by the three instruments, the unique performers and the mappings used to explore sound enable many layerings of sound textures, visuals and experiences. Tanaka states:

> . . . limitations can arise from the human body's own capabilities. This supports the argument to include the body as part of the instrument definition. A very basic limitation of the body is fatigue. The performer has a certain capacity for continued muscle exertion over time. Another limitation quite characteristic of biosignals is the difficulty of maintaining an absolutely stable tension level. The EMG signal is a truly living signal, and will invariably waver. This creates an output that is quite distinct from other types of continuous input devices.
>
> (Tanaka 2000: pg. 396–397)

Another contemporary artist who has explored technology as a way to extend the body's use as an instrument for installation includes Neil Harbisson[12] (Harbisson 2013). Harbisson is a New York City-based artist who works in sound and performance domains and had an antenna implanted in his skull to transmit visual data into sonic vibrations. Having extreme color blindness, Harbisson uses his antenna to create synesthesia to *hear* color (the process of simulating one sensory function to elicit another). In one work, *Piano Concerto No. 1*, Harbisson painted a Steinway piano while his antenna mapped the color data to sound, which was played as a sound performance piece. The antenna on Harbisson is an explicit interface that controls how color will be *heard* since he can choose to look at a given color to get the response from the system. While Harbisson knows how to *play* his instrument, the discoverability of how color is explored, mixed and applied and simulated audibly is unknown to the visitor yet appears discoverable and enchanting through many layers of development. While Harbisson controls the mappings between color and sound, the interface is designed to enable very specific types of agency.

6.5.6. Machine Controlling

Though sound installations often are based on the playing of recordings or digital sounds, installations may also focus on the production of acoustic sounds. In such cases the mapping is to a physical process that produces a sound. Rainlith 2[13] is an example of visitors manipulating an acoustic process (Gato 2011). Rainlith 2 is a large rainstick held on a stand by a rotational motor. The movements of the visitor are mapped to changes in the motor's rotation. As the motor rotates in response to the visitor's motion, the rainstick is moved creating the sound for the installation. This idea of mapping to the physical process for creating the sound is, at one level, no different than mapping to the digital process for creating the sound. And yet, by exposing the sound production as a physical process, the experience can often reference the notion of "player" and "instrument," encouraging the visitor to reflect on what it means to play sounds or music. The choice of a rainstick as the sound source means that the visitor does not believe they need to control the system accurately, so any delays or inaccuracies in the camera tracking system or the control of the motor are excused. Similarly, any problems encountered by the system in tracking multiple visitors are excused in the visitor's interest in exploring the range of sounds produced by an understood physical system made strange by the interface. In this way the artist has focused on implicit discoverability while, through the choices of sound, addressing issues of durability and agency.

Extending this concept of sound interaction through a physical process is sound interaction as a byproduct of the installation. For example, the work *Trash Mirror*[14] by Daniel Rozin attaches bits of trash to many motors, which are tilted to create a shadow image that matches the visitor (2001). The visitor then sees themselves in the trash, hence the reference to a mirror. However, a subtle and yet significant part of the experience is the sound produced by shifting the trash. The sound of crinkling bags, objects rubbing together, and shifting position all serve to remind the visitor of the materiality of the piece, that it is not a polished set of objects. The sound is experienced as intrinsic to the nature of the material even though it would be possible to create the piece without the noise, as done with his previous piece *Wooden Mirror*,[15] but the experience would not be the same (1999). The connection between the sound and the material leverages an embodied knowledge based on an experience of the sound of trash. It is at once irrelevant and redundant, yet integral to the understanding that the image is being produced by simple trash. In this way the sound that is the result of the interaction, part of the feedback mechanism for the interface, enhances the experience and pushes the reflection of the visitor.

6.6.   Conclusion

This chapter focuses on interfaces for sound installations, which controls how a visitor begins to engage with an interactive sound environment. While the visitor's body is the initial point of contact with an installation, there are a variety of additional devices designed to interact with and control sound. We articulate different mapping strategies and how they contribute to interactive experiences. This chapter presents a variety of design considerations for use in creating interfaces and, using a typology of interfaces, has provided examples and discussions of interfaces to illustrate the practical application of the presented design considerations.

To explore different approaches to interfaces for sound installations, we categorized the methods in which the visitor engages with the installation material. The categorization includes tangible objects, devices for signal processing, position and spatial tracking, motion tracking, biosensing and machine control. These categories are illustrated through examples of artist's works and offer a means to examine how the design considerations of agency, explicit/implicit, reliability, discoverability and quality can be used as a framework for designing interfaces that transcend the technical and focus on the experiential. These categories are not meant to be definitive or exhaustive but suggest opportunities for engagement through embodied ways. By focusing on installations as a whole-bodied and sensory experience we can see how interfaces offer affordances that shape the interaction. While one use of technology can be exchanged for another without much change in the experience, the exchange of one interface for another can completely change the overall experience for the viewer.

Notes

1. Janet Cardiff and George Bures Miller: *The Cabinet of Curiousness,* www.cardiff miller.com/artworks/inst/cabinet_of_curiousness.html.
2. Janet Cardiff and George Bures Miller: *Experiment in F# Minor,* www.cardiffmiller.com/artworks/inst/experiment_in_f.html.
3. Christine Sun Kim: *Game of Skill 2.0,* https://news.artnet.com/art-world/12-sound-artists-changing-perception-art-587054.
4. Martin Kaltenbrunner Tangible Database: http://modin.yuri.at/tangibles/.
5. MIT Media Lab *SoundFORM*: https://tangible.media.mit.edu/project/soundform/.
6. Eric Singer *Sonic Banana*: https://arts.mit.edu/musictech-eric-singer/.
7. Laurie Anderson *Handphone Table*: https://vimeo.com/19207943.
8. Victor Mazon Gardoqui's *SIGNUM*: http://victormazon.com/wrks/workshops/signum/.
9. Olle Cornéer and Martin Lübcke's work *BACTERIAL ORCHESTRA:* www.everyday listening.com/articles/2009/9/17/bacterial-orchestra.html. www.corneerlubcke.com/works/bacterial-orchestra/
10. Janet Cardiff and George Bures Miller, *Forty Part Motet*: https://youtu.be/ncWFLzVrwU4.
11. Sergi Jordà, Günter Geiger, Martin Kaltenbrunner, and Marcos Alonso, *Reactable*: http://reactable.com/.
12. Neil Harbisson, *Piano Concerto No. 1*: https://youtu.be/6P8O5JXlAJg.
13. Rui Gato, *Rainlith 2*: http://cdm.link/2011/06/rainlith-a-robotic-responsive-rainstick-powered-by-kinect/.
14. Daniel Rozin, *Trash Mirror*: www.smoothware.com/danny/newtrashmirror.html.
15. Daniel Rozin, *Wooden Mirror*: www.smoothware.com/danny/woodenmirror.html.

References

Anderson, L., 1978. *Handphone Table: Remembering Sound.* Mass MoCA.

Betts, J., 2015. The Visual Rhetorics of Liberty: Towards the Viewer as Producer. *The Poster*, 3, no. 1–2, pp. 49–58.

Cardiff, J. and Bures Miller, S., 2001. Forty Part Motet. Installation. Viewed Oct 9, 2018 www.cardiffmiller.com/artworks/inst/motet.html

Cardiff, J. and Bures Miller, S., 2010. The Cabinet of Curiousness. Installation. Viewed Oct 1, 2018 www.cardiffmiller.com/artworks/inst/cabinet_of_curiousness.html

Cardiff, J. and Bures Miller, S., 2012. Experiment in F# Minor. Installation. Viewed Oct 9, 2018 www.cardiffmiller.com/artworks/inst/experiment_in_f.html

Colter, A., Davivongsa, P., Haddad, D.D., Moore, H., Tice, B. and Ishii, H., 2016. SoundFORMS: Manipulating Sound Through Touch. In: *Proceedings of the 2016 CHI Conference Extended Abstracts on Human Factors in Computing Systems*, pp. 2425–2430.

Corneer, O. and Lubcke, M., 2008. Bacterial Orchestra. Viewed May 22, 2018 www.corneerlubcke.com/works/bacterial-orchestra/

Dix, A., Sheridan, J.G., Reeves, S., Benford, S. and O'Malley, C., 2006. Formalising Performative Interaction. In: Gilroy, S.W. and Harrison, M.D. (Eds.), *Interactive*

Systems. Design, Specification, and Verification. Berlin Heidelberg: Springer, pp. 15–25.

Experience Music Project, Jimi Hendrix Museum, Seattle, WA. Viewed Oct 9, 2018 www. voodoohendrix.com/jimi-hendrix-museum/

Gardoqui, V.M., 2016. SIGNUM. Viewed May 22, 2018 https://victormazon.com/wrks/ workshops/signum/

Gato, R., 2011. Rainlith 2. Viewed May 22, 2018 http://cdm.link/2011/06/rainlith-a-robo tic-responsive-rainstick-powered-by-kinect/

Harbisson, N., 2013. Piano Concerto No. 1. Viewed May 22, 2018 https://youtu.be/ 6P8O5JXlAJg

Jordà, S., Geiger, G., Kaltenbrunner, M. and Alonso, M., 2003. Viewed May 22, 2018 http://reactable.com/

Kaltenbrunner, M., 2018. Personal Website and Tangibles Database. Viewed May 22, 2018 http://modin.yuri.at/tangibles/

Lewis, G.E., 1999. Interacting with Latter-Day Musical Automata. *Contemporary Music Review*, 18, no. 3, pp. 99–112.

McCarthy, J., Wright, P., Wallace, J. and Dearden, A., 2006. The Experience of Enchantment in Human–Computer Interaction. *Personal and Ubiquitous Computing*, 10, no. 6, pp. 369–378.

Rokeby, D., 1995. Transforming Mirrors: Subjectivity and Control in Interactive Media. In: Penny, S. (Ed.), *Critical Issues in Electronic Media*. Albany, NY: State University of New York Press, pp. 133–158.

Rozin, D., 1999. Wooden Mirror. Viewed May 22, 2018 www.smoothware.com/danny/ woodenmirror.html

Rozin, D., 2001. Trash Mirror. Viewed May 22, 2018 www.smoothware.com/danny/ newtrashmirror.html

Seo, J.H. and Corness, G., 2015. Aesthetics of Immersion in Interactive Immersive Installation: Phenomenological Case Study. In: *The Proceedings of International Symposium of Electronic Arts* (ISEA). Vancouver, Canada.

Singer, E., 2010. Robots, Slime, Propane, and Sonic Bananas. Viewed Sept 12, 2017 https://youtu.be/_P2eh-ZdMx0?t=22m20s

Sun Kim, C., 2015. Game of Skill 2.0. MoMA PS1's Greater New York. http://christine sunkim.com/work/game-of-skill-2-0/

Tanaka, A., 2000. Musical Performance Practice on Sensor-Based Instruments. *Trends in Gestural Control of Music*, 13, no. 389–405, p. 284.

Wechsler, R., Weiss, F. and Dowling, P., 2004. EyeCon: A Motion Sensing Tool for Creating Interactive Dance, Music and Video Projections. In: *Proceedings of the Society for the Study of Artificial Intelligence and the Simulation of Behavior (SSAISB)'s Convention: Motion, Emotion and Cognition.* England.

Definitions

Affordances   Affordances are the obvious ways to grasp, move, trigger or engage with an interface that are apparent through its design. These affordances provide options that allow the visitor to discover on their own how to interact with an interface, play with a system or be inventive with their choices.

Biosensors   Biosensors use physiological data as input such as EMG (muscular conductivity), EEG (brain conductivity), heart rate, breath or GSR (skin conductivity).

Cabinets of Curiosity   Known also as wonder rooms, Curiosity Cabinets are collections of categorized objects and items.

Interactive Art   Interactive art is an art form that has gained popularity since the 1990s, which often includes audience or spectators in the artwork.

Theremin   Invented by Lev Sergeyevich Termen, aka Léon Theremin, this device consists of a main box-like structure with two antennas that control the volume and pitch of oscillating electromagnetic frequencies.

7

Sound Spatialization

Enda Bates, Brian Bridges and Adam Melvin

7.1. Introduction: Sound as Spatial

> . . . you are *in* the sound, the space is actually the sound,
> and you're this slug with a few discs that's quivering a bit.
>
> (Brophy 2014)

In a recent panel discussion on the future of sound and music, composer/
director Philip Brophy provided a humorous, yet eloquent, explanation of
his own attraction to sound over visual stimuli. For Brophy, it is the "total-
ity" of sound—that fact that it is "360 degree spherical in its diffusion, its
apparition, its aura [and] its presence"—that enables sound to overwhelm,
making it "impossibly social" and "impossibly engaging."

It is the parameter of space, and its fundamental importance to sound
installation, with which this chapter is concerned. Sound, as a perceptual
entity, has two dominant dimensions of expression: (1) the temporal development of actions, and (2) the *space* (physical and acoustic) those actions
"inhabit." The spatial aspect be can implied, as in the virtual "space" of pitch
relationships that music often articulates, or it may be physically present, as in
acoustic spaces and environments and the differentiating positions of objects
within them. As Labelle (2003) notes, "sound operates acoustically within
space, gaining definition in relation to the space in which it is heard and, in
turn, lending definition to spatial organization through sonic agitation."

7.1.1. Spatiality and Sound Installations

In sound-based installation practices, the parameter of space acquires further agency, both as a creative tool, and within a performative context: "the
place of sound"—one might also add, the place-*ment* of sound—"becomes
as much a part of auditory experience as the material of sound itself."

(Labelle 2006, p. 197). While they may have some element of temporal development, sound installations are predominantly spatial forms, through the combination of materials they incorporate and the framing of perspectives within a defined space.

In an installation context, sound harnesses this existing sense of spatiality, reinforcing the spatial impressions offered by the visual perception of the installation space, or (perhaps, more typically) subverting or expanding them. Beyond the choice and placement of speakers or other sounding mechanisms, the type of sound materials themselves (i.e. their frequency content, amplitude and how this evolves over time) may also change how audiences perceive the characteristics of a sonic space or the sound materials within its frame. In addition, the participant's own interaction, and the interposition of sounds from the wider environment, may also affect the auditory spatial perspective within an installation. This chapter will address some of these key attributes and parameters of sound and space in installation design.

7.1.2. Spatial Audio Cues: The Spatial Affordances of Sound

Spatial sound could be said to have two main dimensional attributes or cases: (1) the sense of spatial "presence" versus "absence" (or defined "location" versus indistinct "immersion") that derives from a sense of relative distance within a spatial field, and (2) its particular angular direction on a plane or within a 360-degree sphere surrounding a listener. Thus, we have localization *direction*, in terms of angles of *azimuth* and/or *elevation* (the horizontal and vertical planes), and a sense of *distance* (near/far), or perceptual clarity/distinctiveness standing as a proxy for this. The majority of what we can easily control about the sonic expression of space will relate to directional aspects of sound, at least in terms of a single source, moving with respect to a listener. However, when multiple sources are present, or when particular frequency ranges are emphasized/de-emphasized within materials, the second attribute may also become an accessible avenue for creative exploration.

Regarding direction, as a species, humans are able to compare the signals received at our two ears (*inter-aural cues*). Two aspects of the sound waves from an incoming auditory event are significant here: the differentiation between time of arrival (*interaural time difference*, or ITD), and level (*interaural level difference*, or ILD). The ITD is caused by the spacing between the two ears and a speed of sound that is relatively slow in comparison with other physical phenomena (often taken to be 340 meters per second at a given standard temperature at approximately 15 °C). ILD is caused by a *head shadowing* effect in sound, with higher frequencies (and, hence, shorter wavelengths). Any sound wave that contains frequency components which give rise to short wavelengths (and, hence, a small physical scale when propagating as waves) are reflected and diffracted by a bigger object such as the human head. As a result, a sound that comes from directly in front of the listener will arrive at each ear at the same time (ITD), and at the same level (ILD), whereas a sound coming from the right side will arrive earlier (ITD), and at a higher level in the right ear compared to the left (ILD); see Figure 7.1, below.

Certain listening conditions or certain types of materials may subvert these cues. The ITD for a given sound source will be same whether a sound source is in front or behind a listener. The *cone of confusion* is the region of a set of points that could produce the same interaural time differences (other cues are needed to resolve this confusion).

The frequency range of sound materials may also impact on the efficacy of the basic ITD and ILD cues themselves. The ILD head shadowing effect may be pronounced enough to lead to as much as a 10-dB reduction within ranges around 4 kHz (Hartmann 2013, p. 146). The process behind ITD cues will only operate up to approximately1.4 kHz for relatively steady state tones. Beyond this range, envelope or transient characteristics may facilitate the operation of this cue (ibid. p. 148); it may also be subverted through the presence of very low tones.

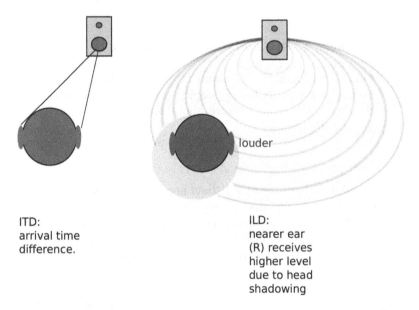

ITD:
arrival time
difference.

ILD:
nearer ear
(R) receives
higher level
due to head
shadowing

Figure 7.1 ITD: inter-aural time difference, and ILD: inter-aural level difference, compared.

Because of these limitations, sustained tones of between 1.4 and 4 kHz without significant harmonic content may be much more difficult to localize. Furthermore, sources whose frequencies coincide with the resonances of the ear canal (between approximately 2.5 and 4 kHz) may produce a perception of an extremely close, almost internal, source, evoking sonic intimacy or claustrophobia. To summarize, sound materials that contain a combination of significant transients or pronounced amplitude variation over the course of an envelope will facilitate ITD; ILD will be facilitated by the presence of higher frequency content; lower, sustained tones (e.g. drones) may be more difficult to localize.

The second component of spatial sound is the apparent distance of sources from the listener. The most obvious distance cue is amplitude, which gradually decreases as the distance to the listener increases. However, amplitude alone can only serve as a distance cue if the sound is highly familiar, such as a voice (e.g. we can easily distinguish between someone shouting far away, and someone whispering nearby, even if the amplitude of the two signals is quite similar). In interior locations, the relative levels of the direct signal, and the acoustic reverberation (commonly referred to as the *Direct-to-Reverberant ratio*, or DRR) also strongly contributes to the perception of auditory distance. In exterior locations, the high frequency content of very distant sounds will also be reduced due to air absorption.

7.1.3. Speakers, Spatial Audio and Installations

> One of the main things about my work is the physical aspect of sound. A lot of people think it's the narrative quality, but it's much more about how our bodies are affected by sound.
>
> (Cardiff, quoted in Tubridy 2007)

Broadly speaking, there are two distinct approaches to the use of audible space in installations. The first simply involves the allocation of a single loudspeaker for each sound source, positioned at every desired location in the performance space. In effect, each loudspeaker acts in isolation as a single, real source, which can be readily localized in terms of both direction and distance, regardless of the position or movement of the audience. Such an approach was adopted by Janet Cardiff's installation *Forty-Part Motet*, in which a recording of Thomas Tallis' 1575 choral work *Spem in Alium* is reproduced by 40 loudspeakers positioned in a circle, with one single singer reproduced by each individual loudspeaker (Tubridy 2007). Visitors to the installation therefore have the freedom to experience the entire spatial polyphony of the work from the center of the reproduction area or, alternatively, move closer to individual voices in the choir. While this approach is conceptually simple and offers some advantages in terms of robustness, it is also logistically demanding and may not be practical if large numbers of sounds, or the dynamic movement of sounds, is a requirement. In this instance, a second, more practical, solution may be the use of a spatialization technique, in which a smaller number of loudspeakers in a defined layout are used to virtually position sounds.

7.2. Spatialization Techniques Compared

A great many spatialization techniques have been developed over the past century; however, in general, the approach taken consists of one of the following:

1. The manipulation of level and/or time differences in pairs or multiple pairs of loudspeakers.
2. The reconstruction of a sound field over a listening area using a loudspeaker array.
3. The reconstruction of the ear signals using headphones.

The first approach, of manipulating either phase/time or, more usually, level differences between pairs of loudspeakers, is often referred to as *stereophony* of which two-channel stereophony (or just stereo) is by far the most common form. However, this exact same approach may also be extended to greater number of loudspeakers, as used in cinematic *surround sound*.

Ambisonics and *wave field synthesis* are two more esoteric techniques that attempt to reconstruct a sound field within a listening area using loudspeaker arrays. While these techniques are not quite as common as stereophony, their fundamentally different approach can be advantageous in certain contexts. For example, Ambisonics lends itself well to 360 video, and *virtual/augmented reality* (VR/AR) and is now widely used for the recording and production of spatial audio for such applications.

The third approach attempts to record or synthesize the two ear signals and directly reproduce the spatial auditory cues we perceive in normal hearing. This *binaural* approach (meaning having or relating to two ears) is highly applicable for a single listener, as it requires a strict separation of the two ear signals, such as when listening with headphones. While this technique also has a long history, much like Ambisonics, it was rarely used until the emergence of VR/AR in the twenty-first century, for which the delivery of spatial audio over headphones is highly common.

All of these different approaches generally feature both a recording technique, involving some type or arrangement of microphones, and a panning/encoding technique based in software or hardware. We will start however by looking at stereophony and see how these recording techniques and panners can be extended beyond simple two-channel stereo.

7.2.1. Stereophony

The word *stereophonic* derives from the Greek words *stereos*, meaning "firm, solid" and *phōnē*, meaning "sound, tone, voice." Two-channel stereo (or just stereo), which was first developed in the 1930s, has now become the standard audio format for most applications and is generally based on two loudspeakers positioned at a 60^0 angle in front of the listener. We can record stereo audio using two microphones that are routed to the left and right loudspeakers, or alternatively a monophonic audio file can be positioned between the loudspeakers using a panner. In the latter case, the position of the audio in the stereo field is controlled by simply adjusting the relative amplitude of the signal in each loudspeaker. This same principle can be extended to greater numbers of loudspeakers for sound spatialization and is the typical approach implemented in the spatial panners found in most digital audio workstations (DAWs), typically for common cinema surround sound formats such as 5.1 as shown in Figure 7.2.

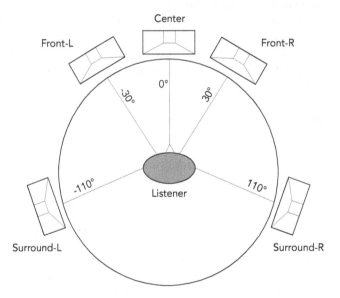

Figure 7.2 The 5.1 Loudspeaker Arrangement.

Source: Wikimedia commons, licensed under the Creative Commons Attribution-Share Alike 3.0 Unported licence

Likewise, different arrangements of multiple microphones can be used to record spatial audio for playback over such loudspeaker arrays (Theile 2000).

One of the challenges of multichannel stereophony is that these different microphone techniques and spatial panners are often closely tied to a specific, matching arrangement of loudspeakers (which is not the case with alternative techniques such as Ambisonics). So, while microphone arrangements such as the *Decca Tree* or the *Optimized Cardioid Triangle* (OCT) are well matched to the left, center and right loudspeakers found in cinema surround sound formats such as 5.1, they are less suited to other arrangements of loudspeakers such as a regular octagon or loudspeaker dome. In the case of spatial panners, simpler implementations are similarly tied to a specific loudspeaker arrangement while more sophisticated tools may be able to adapt to different arrangements as required. The simulation of distance is sometimes also included in these panners, although this is often implemented separately using reverberation, filtering and amplitude changes (Chowning 1971). It is important to note that while many stereophonic panners may visually suggest that sounds can be positioned at any point inside the array, this is often simply achieved by routing the audio signal to greater numbers of loudspeakers. In the case of a 5.1 arrangement, a sound at the center of the array would be simulated by simply routing audio to all 5 loudspeakers. However, this effect can be rather fragile and may not be effective if the listener is seated at an off-center position, or in the case of very widely spaced loudspeakers. For this reason, a number of new cinema surround formats such as Auro 3D and Dolby Atmos emerged in the 2000s which included elevated and overhead loudspeakers that can more reliably position sounds at different points inside the listening area.

In a stereophonic system, if a signal is panned to the position of a loudspeaker, then only that loudspeaker will be used, but a signal panned to any other position will be created as a *phantom image* by a pair of loudspeakers. This results in a slight yet perceptible change in timbre as a signal is panned from a loudspeaker position to a point in between a pair of loudspeakers. As a result, when amplitude panning is used to dynamically move a sound around the array, the number of contributing loudspeakers is constantly changing, producing a small yet clearly perceptible timbral shift that tends to emphasize the loudspeaker positions and distort the perceived trajectory (Pulkki 1999). For this reason, many panners include a control to spread the signal and ensure that multiple loudspeakers are always used regardless of the panned angle, thereby eliminating this panning artifact for moving sounds.

Just like two-channel stereo, multichannel stereophonic recording techniques can be based on coincident, near-coincident or spaced microphone arrays. As with stereo techniques, spaced microphone arrangements will produce an increased sense of "spaciousness" due to the capture of timing differences in the recording and the resulting decorrelation of the loud-speaker signals. However, this may also reduce localization accuracy. Coincident techniques will often produce more accurate localization, but with a less spacious and "open" sound. The choice of approach is as much an aesthetic issue as a technical concern, but in general, spaced microphone techniques will produce a more robust, spacious reproduction, particularly for listeners positioned away from the center of the loudspeaker array (the so-called *sweet spot*). In contrast, coincident techniques can potentially produce more accurate directionality, particularly for listeners in the sweet spot, but may be less optimal at off-centre positions. In either case, there is a close connection between the type of microphone array used and the specific layout of loudspeakers required to reproduce the recording. One novel, and highly practical, approach to spaced microphone recordings using portable, low-cost sound recorders has been developed by artist Augustine Leudar (Leudar 2014) and is potentially useful for site-specific installations.

7.2.2. *Ambisonics*

Ambisonics is an alternative spatialization technique that was first developed in the 1970s by Michael Gerzon, among others (Gerzon 1974). Although originally based on analog hardware, Ambisonics is now a complete system for the recording, panning and reproduction of spatial audio using specific microphones and software panners which, unlike stereophony, can readily be adapted to a variety of different loudspeaker arrangements (although regular symmetrical arrangements are preferred). Unlike stereophony, the different audio channels in an Ambisonic signal are not simply loudspeaker feeds, but instead comprise a three-dimensional description of the entire sound field that must be decoded for a particular loudspeaker configuration (or headphones). Most commonly this comprises just four audio channels labeled W, X, Y and Z, but higher-order formats are available that provide increased localization accuracy but require a greater number of audio channels.

Ambisonics can be recorded using a specific type of microphone containing multiple, coincident microphone capsules in a single housing. Such recordings can be smoothly rotated and processed in different ways, which is highly beneficial for VR applications, and usefully can also be decoded for different loudspeaker configurations. Existing sounds can also be positioned using an encoder that differs from a stereophonic panner in that its output does not comprise individual loudspeaker feeds but is instead an Ambisonics signal that must be decoded for playback (see Figure 7.3). Although this approach is somewhat more complex

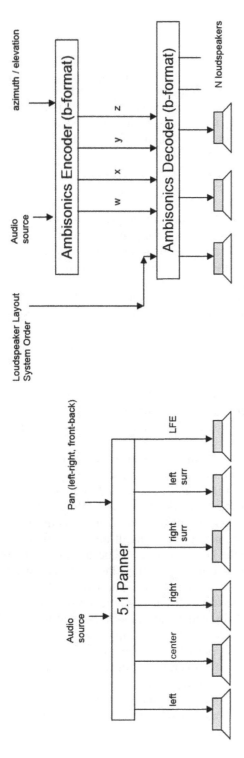

Figure 7.3 Stereophonic Panners Directly Output Loudspeaker Feeds while an Ambisonics Encoder's Output Must Be Decoded Before Playback.

than stereophony, the separation of encoding and decoding into separate stages allows for more flexibility in terms of the final reproduction system, whether that be a loudspeaker array or headphones. Ambisonics has been used in numerous art installations, with the work of composer and artist Natasha Barrett a prominent example (Barrett 2018). Much like coincident stereophonic recording techniques, Ambisonics performs optimally over a relatively small listening area, albeit with more flexibility in terms of the loudspeaker layout.

7.2.3. *Binaural*

Like stereo, *binaural recording* techniques were first developed in the 1930s and are generally intended for playback over headphones. Binaural techniques aim to recreate both binaural localization cues (ITD, ILD) and monaural cues due to the spectral filtering of the head pinna and torso, all of which are characterized by the *head-related transfer function* (HRTF). When these recordings are played back over headphones the effect is quite different from that of normal stereo. Instead of the highly unnatural (although by now quite familiar) effect of perceiving sounds as if inside the head, sounds are now externalized in much the same way as in normal hearing.

Binaural recordings can be made using either a dummy-head microphone or, alternatively, using a binaural headset in which two microphone capsules are inserted into the ears using an earbud-type arrangement. In the latter case, the recorded audio will be encoded with that particular individual's HRTF. As a result, the recording will closely mimic normal listening for that individual, but may be far less convincing for others. In the case of a dummy-head recording, a generic HRTF is used that is intended to work reasonably well for most listeners, but again the effectiveness of the spatial cues can vary considerably between individuals. In particular, reproducing sounds that appear to come from directly in front or behind or at different elevations can be difficult and can often be perceived as coming from inside the head. This occurs because the perception of these directions is largely determined by the filtering effect of HRTFs, which, much like fingerprints, are highly individualistic.

These recording techniques can also be used to capture a *binaural impulse response* (BIR), which can then be applied to an audio signal to artificially position that sound in the captured direction (much like convolution reverbs). Multiple BIRs can be used to dynamically position or move sounds in space, however, this comes with significant costs in terms of processing power and requires the capture of large numbers of BIR's for many different directions. An alternative approach is often used to render stereophonic or Ambisonic spatial audio over headphones by convolving each signal with a BIR for the angle corresponding to the position of the loudspeaker (generally referred to as virtual loudspeakers (McKeag and McGrath 1996)). Usefully, this only requires a limited number of BIRs (one for each loudspeaker feed) and is therefore a more efficient approach. The dynamic movement of sounds can then be created using standard panning tools to adjust the loudspeaker feeds prior to their convolution with BIRs.

This virtual loudspeaker approach is often used in combination with Ambisonics for sound spatialization in VR applications. When listening to stereo or binaural recordings on headphones, as we rotate our head, the entire sound field will also move, which is particularly problematic for VR. To avoid this, head-tracking can be used so that the perceived position of sounds does not change as the listener moves their head. This can potentially alleviate some of the problems with front-back reversals as the listener can rotate their head slightly to help resolve ambiguous source positions. To implement these systems, a head-tracker is attached to the headphones (or built into the VR headset) transmitting head rotation angles back to the reproduction system. This data is then used to smoothly rotate the entire sound field in the opposite direction (which is relatively straight-forward with Ambisonics) so that sounds hold their position regardless of the orientation of the listener.

Binaural audio has been widely used in art installations, particularly when headphone reproduction is required. The highly personal nature of this recording and reproduction technique (if recordings are made using two small microphones placed in the ears) is highly suitable for the production of mobile works, such as some of the audio works of Janet Cardiff, or the binaural sound art of Dallas Simpson.

7.3. Acoustic Environments, Installation Practices and Spatial Aesthetics

Sound installations can generally be considered as spatial territories and containers for a variety of sonic materials. These defined spaces may partly be perceived within the context of their visual aspect (from galleries to more specific spaces and sites) in which dimensions the borders of the space may be quite defined, halting at walls, doors or windows. However, the sonic borders of such territories and spaces may be more porous or even nonexistent. Indeed, while many authorities have sought to enumerate the differences between sonic space and visual reference points (Sterne 2012, p. 9; Blesser and Salter 2006, p. 21), the use of sound itself as a medium creates the potential to shrink and expand the perceived audiovisual boundaries of a given space: "sound necessarily exceeds itself, washing over borders" (Labelle 2003).

Rooms and buildings do provide some aspects of hard boundaries for sound, and the distinction between sound and vision within such spaces is not always so stark when considered on the basis of a more scientific model of such phenomena. The key distinction may be seen as related to the speed of the sound waves; echoes of sonic events are often heard due to reflections from various (hard) surfaces producing later arrivals of "copies" of the original vibrational impression, a temporal distinctiveness that is not a factor in our everyday visual experience. This can give rise to sonic boundaries that are analogous to visual borders. While a doorway into a room may be a particularly lax border-guard for external sonic events, the sonic environment within is truly a different land; the faint echoes of reflected sound heard within signal both its interiority and the particular dimensions (and materiality) of its construction. Installation artists who utilize sound will need to bear in mind the perceptual principles of what Blesser and Salter (2006, p. 1) term *aural architecture*: how the intersection between sound waves, objects and surfaces may give rise to a sense of a location's dimensions and structure.

7.3.1. Aural Architecture and Acoustic Territories

A generic room is a rectangular box, and this is the kind of spatial archetype that many performance and gallery spaces tend towards. However, though elegant in its visual simplicity, this type of space poses significant problems for the installation designer. Such rooms will resonate in various different ways based on their dimensions, the types of surface materials and the types of sound materials. These factors may contribute *frequency-based effects*, that is, the subtle "coloration" acoustic structures impose on sound materials (Blesser and Salter 2006, p. 42), and *temporal effects* (discrete echoes and diffuse reverberation).

The geometry of smaller rooms will affect the audibility of frequency content. Certain frequencies—those with wavelengths that are close to multiples of the room dimensions—will be amplified through reflections off the various surfaces. These resonance effects are known as *room modes*: the room will have a more pronounced response to certain frequencies (in effect, amplifying them) and will have a less active response in relation to others. Such responses are heavily dependent on location and may be particularly obvious in relation to certain lower frequencies; if certain bass materials sound quite "boomy," it may make sense to move a speaker or other sounding device to another part of the room. (Higher frequency materials are likely to pose less of a problem in terms of this differential response: modes within an octave increase significantly with frequency, making them hard to differentiate (Rumsey and McCormick 2006, p. 23)). Such active room acoustic responses can be creatively exploited, as in Alvin Lucier's *I am Sitting in a Room* (1969) which re-recorded the products of a room's audible response iteratively in order to process a recording of speech. In this case, the significant room responses are to be found in higher frequency region, including *comb filtering*: the effect of (very) short delays (produced by reflected sound), combining and constructively and destructively interfering with the original sound source (or other delayed materials). Small, especially narrow, spaces may provide distinctly colored responses that may be creatively exploited in terms of their ability to reinforce or subvert an audience's sense of aural perspective (Brant 1967).

In larger rooms, the temporal attributes of acoustic responses predominate. A large concert hall, church or even school gymnasium or swimming pool will have *reverberant* acoustic responses; multiple/repeated echoes from the hard surfaces, dispersed over time, will become blended together into a diffuse "afterimage" of the original sound event, which is known as *reverb*. The decay time of this reverberant tail can be related to the dimensions of the room and various figures (coefficients) for the materials of the room's surface. Reverb time is standardized as RT_{60}, or the amount of time it takes for the reverberant response to decay by 60 dB. Installations that occur within such large spaces will need to make use of strategies to relate to their sound materials to the space. Any rhythmic or transient details (including speech sounds) may become indistinct if the reverberant response predominates (beyond a certain distance from the source, termed the *critical distance*). In addition, there may be some discrete echoes that are audible in a given space and location; although it might be expected that these would interfere with localization, the *precedence effect*—sometimes termed the *Haas effect* (Haas 1951)—holds that localization direction is judged on the basis of the "first arriving wavefront" rather than the direction of any reflections (within approximately 50 msecs). As such, we are able to localize with a reasonable degree of accuracy within typical rooms.

Installation designers therefore need to bear in mind the role of rooms as active sonic environments. Thus, it may be helpful to consider a number of exemplar cases:

1. Anechoic or outdoor spaces: spaces that have no significant reflected sound created through the presence of reflecting walls. Such spaces may be created through the deployment of absorbent and diffusing surfaces (an anechoic chamber) or may be found in certain outdoor environments.
2. Small, colored spaces: smaller rooms in which modal characteristics or comb filtering effects (through reflections from hard walls) may impose changes upon the frequency content of sound materials.

3. Reverberant spaces: as noted above, larger spaces have less pronounced modal characteristics and more pronounced temporal responses. What is an appropriate reverb time will depend, to some extent, on the type of sound materials to be presented. Content with significant transient detail may benefit from a shorter reverb time; significant harmonic or slowly developing textural detail may benefit from a longer reverb time (many orchestral halls have reverb times of around two seconds; see (Beranek 2003)).

Typically, gallery spaces will tend towards the second case (small, colored spaces), but larger-scale or multi-room/multi-space installations may entail encounters with more reverberant spaces.

7.3.2. Soundscapes and Spatial Perspectives

Apart from the temporal and frequency characteristics, the presence and type of background sound may affect the suitability of a space for sound installations. Ambient sound materials may impose themselves upon a sound installation, either through their audibility and the "salience" of this audibility (through distractingly recognizable or repetitive sources) or, if of significant amplitude, through *masking*, in which relatively high amplitude sources may perceptually "cover" sources within adjacent frequency ranges.

More generally, such background, ambient sounds may help to articulate a spatial structure or opposition against which a sound installation may function. Emmerson (2007, pp. 97–101) proposes a spatial-territorial typology of musical sound events based on their inhabiting different *space frames* that are typified by two primary functional associations: *local* (a foreground or region of clearly perceived events) and *field* (a more diffuse ambient background context for local/foregrounded events). These frames may deal with the interrelationships between various composed sound materials and the wider context of the soundscape, and between ideas of spatial "presence" versus "absence" or "dispersion"; see Figure 7.4.

Emmerson suggests that listeners track sonic-performative space within concentric circles arising from gestures within a focused area, framing localized events within the frame of a *stage*, which is encapsulated within both an *arena* (performance/installation environment) and a wider background *landscape*. An important consideration for the sound designer here is that of audience position. Electroacoustic concerts, like private listening, typically employ a *fixed*, forward-facing audience orientation (Smalley 2007, p. 52; my emphasis) within the delineated "stage" of a

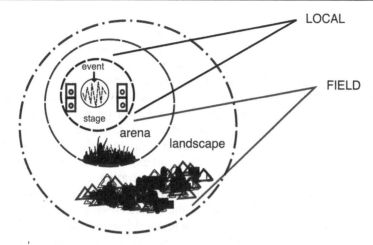

Figure 7.4 Local and Field Space Frames.

Source: Adapted from (Emmerson 2007)

stereo or circumspatial, multichannel speaker array; in turn, the spatial manipulation of sonic events serves to blur perceived boundaries between fields. Works may be mixed with an intended audio sweet spot (usually central) in mind, with the potential drawback that the optimal conditions for experiencing the work are confined to a relatively small zone. In contrast, *variable* listening positions may offer an alternative audio experience of a piece that, while arguably less ideal in a circumambient context, can nonetheless be considered equally favorable (Smalley 2007, p. 52). Arenas such as Stockholm's Audiorama embrace the contrasting experiences and, consequently, potential multiple readings of a sonic work that result from different listening perspectives by placing random seating positions facing in various directions within an extensive, circumambient speaker array.

The common practice of *sonic diffusion* furthers the potential for fluidifying the boundaries of sonic space frames. Here, a designated individual (often the composer) manually controls the amplitude and spatial distribution of an electroacoustic piece (Harrison 1998) to various speakers around the performance space; in its most basic form, this may involve mixing a stereo composition to a multichannel array. Sound is adjusted "live" in response to the particular speaker system as well as to the acoustic and spatial properties of the arena; the interface (mixer) and the room become "instruments" (Gibbs 2007, p. 132). (Some speakers may be positioned specifically to produce the impression of greater distance via the off-axis response of being turned away from the audience, producing the attenuation of high frequencies (Harrison 1998).) Diffusion practices are not uncontroversial; they may introduce *phase cancellations* through the sonic interactions of multiple copies of audio with different amounts of delay, based on the different positions of multiple speakers (Piché, cited in Harrison 1998). As Harrison responds (ibid.), the acoustics of a performance space are, in any case, likely to cause phase cancelations for audience members further away from any fixed "stage" position.

In the installation context, multiple "stages" may be provided via speakers or sounding mechanisms, across several acoustic arenas, with local or field assignment, depending on the relative distance or nature of sound materials. Works may allow and indeed encourage further movement from the listener(s) in order to facilitate their full experience. Multiple, conflicting, complementary, cumulative or interpenetrating audio zones may be articulated, within which the listener can discover their own favored listening position. Indeed, the speakers themselves may constitute their own spatial "corners," creating quasi-architectural half-box havens of immobility (Bachelard 1964, p. 137) that entice the listener. Bruce Nauman's *Raw Materials* (2005) depends on this very dynamic; amid the complex of samples emanating from 22 speaker channels placed within the vast turbine hall of London's Tate Modern, audience members move towards individual speakers to discern the piece's separate audio threads. Ultrasonic technology and the use of *audio spotlights*, extends this concept further, enabling the sound designer to isolate entirely egocentric zones of listening, whose material is otherwise excluded from the rest of a given audio space resulting in a more aggressive, almost *anti*-sweet spot. Elsewhere, multiple and mobile listening positions may be used as a means of prompting a greater sense of integration between audio content and its architectural vessel as well as interaction within mixed-media contexts, resulting in what Smalley terms *peripatetic* listening (2007, p. 52). Like the broader definition of installation art proposed by de Oliveira et al., here, audio scrutiny is quelled "in favour of a consideration of the relationships between a number of elements or of the interaction between things and their contexts" (1994, p. 8).

7.3.3. Immersion Versus Localization in Spatial Audio Aesthetics: Space, Place and Scale

In many ways, both Emmerson's space-frame model and Smalley's three vantage points of listening depend on an overall dynamic of *immersion versus localization*. Although it is a commonplace, almost trite term today, the notion of immersion does, at least, resonate with the ideas implied in this chapter's epigraph, namely, the acknowledgement of sound and sound installation as a physical, dimensional entity in which we are *immersed*, "a spacious thing that you can inhabit" (Brewster 1998). "Walking through [sound] in its resonant state provides an experience similar to perusing a landscape but from the inside with all of your body instead of from the outside with just your eyes" (ibid.). It is sonic immersion that arguably provides the conditions—the backdrop—for experiencing more localized convergences of sound or isolated sonic gestures as found in the Cardiff and Naumann examples mentioned earlier. This relationship between immersive sound *en masse* and its component strata—"a universe dotted with little stars of sound, moving in compact nebulae or isolated" (Xenakis 1992, p. 237)—forms the basis for the employment of space as a compositional parameter in the musical works of composers such as Xenakis and Stockhausen. For both composers, the spatial dispersal of instrumentalists around a concert space allows the musical material of a given piece to become "purer" (Xenakis in Varga 1996, p. 97), more "transparent" (Stockhausen in Cott 1974, p. 187) than if emanated from a single, fixed (ensemble) source, thanks to the increased dynamic of spatial mobility between gestures that the greater distance between players (and reduced effects of frequency-based masking) facilitates. Immersion is thus the somewhat disorderly state that enables the composer to "tame" space (Varga 1996, p. 97). (Presumably silence may also be considered immersive, since it functions in a similar manner.) Grimshaw interrogates the concept of immersion further, citing Slater's distinction between immersion and "presence." Here, "presence is the feeling of being within a space in which there is potential to act. It is subjective whereas immersion is objective and relates to the potential of the . . . environment to facilitate presence" (Grimshaw 2015, p. 280). A fundamental aspect of what both writers distinguish as presence is the sense of a "coherent 'place' that you are 'in'" (Slater 2003, p. 2; cited in Grimshaw 2015, p. 287). It may therefore be more beneficial to consider sonic installations as architectural constructs or environmental sites, rather than draw the more obvious comparisons that tend to be made with musical composition. Thierry du Duve's notion of *site*, as discussed by Mavash (2007, p. 55) and Mittosian (1997, p. 64), offers a useful point of reference in this respect. De Duve theorizes site in the context of architectural design as a harmonization of three fundamental concepts: (the cultural associations of) *place, space* (as a perceived grid of references) and (human/embodied measurements of) *scale*, noting the propensity for many examples of contemporary art, sculpture and architecture to spotlight two of these parameters while compromising the third. Unlike "fixed," visual entities, installed sound might be considered as encompassing a shifting dialogue between all three. As Labelle (2003) notes, when used as a creative medium, "sound is both subject *and* object." Sonic material, its spatial diffusion and guise as a spatial form in itself can be used to evoke *place*, map *space* and both articulate

and manipulate *scale* to achieve what is, for all intents and purposes, a variant of Du Duve's site, albeit in flux.

7.4. Conclusion

Sound installation practices entail a number of creative affordances through the presented audio's relationship with the acoustics of the space, its speaker system and the audience's perspective. The manipulation of the audio via various spatialization/diffusion techniques activates, reinforces and subverts the relationship with the presentation space. Whether through the more *ad hoc* approaches of point source, multiple audio-channel-per-speaker approach (as in Cardiff's work), diffusion of identical materials through multiple loudspeakers or more technically considered approaches such as Ambisonics or binaural techniques, contemporary sound artists and sound designers have a wide variety of approaches at their disposal through which to activate various types of auditory-spatial experience. Aesthetic dynamics of local and field frames, and diffusion as activating aspects of a presentational space or moderating a fixed auditory perspective, may challenge the audience's relationship with a given space. Installation art's concern for the redefinition of space finds in sound a means to reconfigure apparent dimensions and, indeed, relationships with neighboring spaces and soundscapes. The intimacy and presence of the localized sound event, as against the immersion entailed by diffusion, provides listener-participants in any sound installation with spatialized experiences ranging from the proximal (or even internal) to the expansive and enveloping. That such experiences can sometimes operate in tension with the audience's preconceptions of a given space via fairly simple technological means is one of the reasons why sound is a powerful modality in the hands of an installation artist. Sound's ripples activate and challenge our perceptions of, and egocentric perspective within, space and place.

References

Bachelard, G., 1964. *The Poetics of Space*. Translated from the French by Jolas, M. Boston: Beacon Press.

Barrett, N., 2018. Installations. Viewed June 28, 2018 www.natashabarrett.org/installations.html

Beranek, L., 2003. *Concert Halls and Opera Houses: Music, Acoustics, and Architecture*. New York: Springer.

Blesser, B. and Salter, L., 2006. *Spaces Speak: Are You Listening?* Cambridge, MA: The MIT Press.

Brant, H., 1967. Space as an Essential Aspect of Musical Composition. In: Schwarz, E. and Childs, B. (Eds.), *Contemporary Composers on Contemporary Music*. New York: Holt, Rinehart and Winston, pp. 221–242.

Brewster, M., 1998. Where, There or Here? An Essay About Sound as Sculpture. Viewed June 28, 2017 http://michaelbrewsterart.com/news/where-there-or-here-sound-as-sculpture-michael-brewster

Brophy, P., 2014. Melbourne Conversations: What's the Future of Sound and Music? (panel discussion—Fox. R. chair). Knowledge Melbourne. Viewed Aug 5, 2017 www.youtube.com/watch?v=w7eO5Aj2IGA

Chowning, J., 1971. The Simulation of Moving Sound Sources. *Journal of the Audio Engineering Society*, 19, pp. 2–6.

Cott, J., 1974. *Stockhausen: Conversations with the Composer*. London: Picador.

De Oliveira, N., Oxley, N. and Petry, M., 1994. *Installation Art*. London: Thames & Hudson.

Emmerson, S., 2007. *Living Electronic Music*. Aldershot: Ashgate.

Gerzon, M.A., 1974. Sound Reproduction Systems, US Patent No. 1494751.

Gibbs, T., 2007. *The Fundamentals of Sonic Art & Sound Design*. Lausanne, Switzerland: AVA.

Grimshaw, M., 2015. Presence Through Sound. In: Wöllner, C. (Ed.), *Body, Sound and Space in Music and Beyond: Multimodal Explorations*. Abingdon, Oxon and New York: Routledge, pp. 279–297.

Haas, H., 1951. On the Influence of a Single Echo on the Intelligibility of Speech. *Acustica*, 1, pp. 49–58.

Harrison, J., 1998. Sound, Space, Sculpture: Some Thoughts on the 'What', 'How'and 'Why'of Sound Diffusion. *Organised Sound*, 3, no. 2, pp. 117–127.

Hartmann, W., 2013 *Principles of Musical Acoustics*. New York: Springer.

LaBelle, B., 2003/2007. Short Circuit: Sound Art and the Museum. *Journal BOL*, no. 6, Insa Art Space, Korea, pp. 155–175.

LaBelle, B., 2006. *Background Noise: Perspectives on Sound Art*. New York and London: Continuum.

Leudar, A., 2014. An Alternative Approach to 3D Audio Recording and Reproduction. *Divergence Press*, Dec, 3, no. 1.

Mavash, K., 2007. Site + Sound: Space. In: Muecke, M.W. and Zach, M.S. (Eds.), *Resonance: Essays on the Intersection of Music and Architecture*. Ames, Indiana: Culcidae Architectural Press, pp. 53–75.

McKeag, A. and McGrath, D., 1996. Sound Field Format to Binaural Decoder with Head Tracking. *Preprint of the Audio Engineering Society for the 6th Australian Regional Convention*, no. 4302.

Mittosian, C., 1997. Public Art/Public Space. In: Ramesar, A. (Ed.), *Urban Regeneration: A Challenge for Public Art*. Barcelona: Universitat de Barcelona, pp. 61–70.

Pulkki, V., 1999. Uniform Spreading of Amplitude Panned Virtual Sources. In: *Proceedings of the 1999 IEEE Workshop on Applications of Signal Processing to Audio and Acoustics*. Mohonk Mountain House.

Rumsey, F. and McCormick, T., 2006. *Sound and Recording*. Oxford: Focal Press.

Smalley, D., 2007. Space—Form and the Acousmatic Image. *Organised Sound*, 12, no. 1, pp. 35–58.

Sterne, J., 2012. Sonic Imaginations. In: Sterne, J. (Ed.), *The Sound Studies Reader*. London: Routledge, pp. 1–18.

Theile, G., 2000. Multichannel Natural Recording Based on Psychoacoustic Principles. *Preprint of the Audio Engineering Society for the 108th Convention*, no. 5156.Tubridy, D., 2007. Sounding Spaces Aurality in Samuel Beckett, Janet Cardiff and Bruce Nauman. *Performance Research*, 12, no. 1, pp. 5–11. doi:10.1080/13528160701397984

Varga, B.A., 1996. *Conversations with Iannis Xenakis*. London: Faber & Faber.

Xenakis, I., 1992. *Formalized Music: Thought and Mathematics in Music*, Revised edition. Hillsdale, NY: Pendragon Press.

8

Thinking Sound-Motion Objects

Rolf Inge Godøy

8.1. Introduction

The focus of this chapter is on short fragments of sound, typically in the 0.3- to 3-second duration range, such as that of a single trumpet tone, a rapid harp glissando, a bottle tumbling down a staircase or a quick filter sweep on a synthesizer. The idea is that sound fragments in this duration range are perceived and conceived holistically as coherent units, and notably so, as units combining sensations of sound and body motion. We shall call these units *sound-motion objects*, units that we believe have a special role in sound design because they fit well with the predisposition of both our perceptual and motor apparatus to break up any continuous and complex sensory stream into delimited and readily tractable chunks.

Sound is here regarded as closely linked with sensations of sound-producing body motion such as hitting, plucking, stroking, shaking, scraping or blowing, in short, linked with motion that transfers energy from our body to an instrument (or our vocal apparatus). Massive and lifelong experiences of links between sound and sound-producing body motion have given us robust mental schemas for sound perception, making links of sound and motion automatic in listening situations, even making us project these schemas onto sound generated by electronic means.

The preference for sound-motion objects in the 0.3- to 3-second duration range stems from the combined constraints of sound-producing body motion (need for rests, for pre-programing, for hierarchical control) and of our perceptual apparatus (limits of short-term memory and of attention spans). This duration range is also what is needed for perceiving salient features such as the dynamic, pitch and timbre contours of the music, as well as the overall sense of motion, affect and style, making the duration range of sound-motion objects crucial for the perception and generation of music.

To elaborate on our ideas of sound-motion objects, we shall first give
an overview of some important challenges and perspectives in the field of
sound design, as well as broaden our scope to include considerations of
multimodality and timescales that are at the base of sound-motion objects.
This will be followed by an overview of the main features of sound-motion
objects and an overview of practical tools for an analysis-by-synthesis
approach to sound-motion objects. Also, there will be some ideas on montage for using these objects in musical composition and/or production,
followed lastly by a summary of the main issues for further theoretical and
practical work on sound-motion objects.

8.2. Challenges

One of the main challenges of sound design may be defined as enabling
the creator to fulfill his or her musical intentions. This is assuming that
there are obstacles between an image of sound to be made and the actual
sonic realization of this image, obstacles that may at least be partially
overcome given good diagnostic tools as well as good control tools in the
generation of sound. Sound design should here be understood to include
any means of sound generation, be that of electroacoustic sound based on
recorded or processed environmental sound, or on analog electronic, on
digital electronic or on any kind of unprocessed acoustic (vocal or instrumental) sound, ranging from that of solo performances to that of any kind
of ensemble, up to whole symphony orchestras.

As for the sought-after sound, it may be "unheard," or it may be more a
matter of recycling familiar sound, or sound with the readily recognizable
output "fingerprint" of some often-used synthesis methods (e.g. the typical
FM or granular synthesis sounds). Or, the challenge may be to work from
an initial vague idea to a concrete result. Musicians and producers may
tend to use verbal metaphors for sound features, for instance, "fat," "slim,"
"rough," "smooth," and so on—metaphors that may work well enough in
some circumstances (Porcello 2004) and may also suggest that these people have a mental image of the wished-for sounds.

Assuming that the musicians/producers have some clear idea of the
wished-for sounds, there are needless to say a number of obstacles in
making these sounds. In the realm of "classical" vocal or instrumental
sound design, voices or instruments come with both constraints and possibilities, what we may refer to as *idioms*, encouraging an adaptation of
the aimed for sound to these idioms. The orchestration and writings on
orchestration of Nikolay Rimsky-Korsakov are particularly instructive
in this respect, demonstrating a strategy of first analyzing musical intentions, then analyzing the constraints/idioms of the instruments to be used

and finally combining these two domains for the maximal ergonomic and acoustic results, thus maximizing the musical-aesthetic outcomes (Rimsky-Korsakov 1965).

With electroacoustic sound generation, both analog electronic and digital, there are also challenges of mastering the generative constraints (see, e.g. Farnell 2010). Although digital sound synthesis in principle can generate any thinkable sound, there are difficulties in controlling features of the resultant sound because of the large number of required control dimensions and because of the required temporal resolution. This has resulted in a number of synthesis models (e.g. additive, subtractive, source-filter, physical models, diphone, FM), which all tend to have some characteristic features ("fingerprints") in their output. We have also seen an intense interface development within the field of New Interfaces for Musical Expression (NIME), with the aim of enabling more spontaneous control of features, in line with that of traditional instruments. And there are also combinations here, such as in the use of environmental sound in different guises of editing, processing and montage, as well as in the use of acoustic-electronic combinations, such as in the real-time processing of acoustic sound and in the use of so-called extended instruments.

All through these different approaches to sound generation, there remains the challenge of coming as close as possible to the intended sound. One possible strategy here could be that of firstly developing suitable diagnostic tools for detecting salient perceptual features, and then secondly extending these diagnostic tools to become generative control tools. As proposed by Pierre Schaeffer in the middle of the twentieth century, the most effective strategy to realize this could be in focusing on short sound fragments, typically in the 0.3- to 3-seconds duration range, hence on what came to be known as *sound objects* in the theory of Schaeffer (Schaeffer 1966; Chion 1983).

This more "local" focus enables systematic and in-depth focus on the perceptually salient features of musical sound. In particular, this object focus makes a non-symbolic approach to sound features possible, not bound by Western notation symbols, and instead representing overall dynamic, pitch-related, and timbre-related sound features as shapes, thus effectively erasing the difference between notated and non-notated music. This thinking of all sound features as shapes is then a universal diagnostic tool and may be extended to become a tool for creation, implemented in the strategy of *analysis-by-synthesis* (cf. section 8.10 of this chapter).

It is worth noting that Schaeffer mentioned three main challenges of music theory in the 1950s and 1960s—new technologies, new aesthetics and the music of non-Western cultures—and of these three, the last one was considered the most significant because experiences of non-Western music forces us to think of musical sound in novel ways. In particular, this meant shifting the focus of music theory away from the generative processes, regardless whatever generative scheme may have been used (instrumental, electronic, various synthesis models), towards the subjective perceptual features of the sound objects.

8.3. Perspectives

The idea of sound objects had a pragmatic origin in the early days of the *musique concrète* in the use of looped fragments of sound on phonograph disks called *closed groove*. Before the advent of the tape recorder, mixing of sounds was done by starting and stopping the playback of these looped fragments on the phonographs. This had the side effect of making composers listen to these looped sounds many times, and also make them experience shifts of attention away from the immediate and more everyday significance of the sound (such as the event-centered listening of ecological acoustics, cf. Gaver 1993) towards the sonic features, such as the overall dynamic, pitch-related and timbral envelopes of the sounds, as well as the various internal features of spectral and temporal distribution. This shift of perspective was crucial for the subsequent notion of sonic objects and was later labeled as "reduced listening" in what came to be identified as a "phenomenological reduction," linked with more general introspective methods for studying human perception (Schaeffer 1966; Chion 1983).

This shift of attention necessitated a differentiation of what a sound object actually was, what we could call ontological differentiations. Schaeffer was careful to point out that a sound object is *not* a physical entity or a recording or a score but, rather, an image in our minds held together by our intentional focus on a set of features. This shifting away from source features to sound features was called "acousmatic listening," meaning that we do not see the source of the sound, and more generally, that we disregard most of other generative elements that may have been associated with the sound objects.

This distancing from the immediate causes of the sound object was taken one step further with the idea of *anamorphosis*, or what could be rendered as *warping*, signifying that there are a number of nonlinearities between features of the physical signal and our subjective experiences of the sound object (see Schaeffer 1998 for various illustrations). This meant that subjective perception was considered to be the most important element of music theory, but this did not mean that the acoustic features were irrelevant for the music. Rather, this meant that exploring the *correlations* between the physical signal and these subjective perceptions was seen as a long-term goal of music theory.

The main strategy of Schaeffer's music theory was then to focus on our subjective images of sound objects, starting with a differentiation of the overall global sensations, and gradually moving downward into differentiating progressively finer details of the sound objects. The method by which to do this was to ask questions, starting with the seemingly simple question of "what are we hearing now?" This Socratic method of seemingly naive questioning was retained all through the subsequent, quite elaborate so-called *typology* and *morphology* of sound objects (see below), leading Schaeffer to state that this was a questionnaire and not a balance sheet.

The common principle all through Schaeffer's theory is to characterize features as shapes. Shapes are intrinsically holistic in the sense that they are distributed as figures and are not just abstract symbols. Thinking of features as shapes is the general method for classifying sound object features and can also be used as a practical tool in sound design. This means that temporally unfolding features are transformed into "instantaneous" shape images, that is, into *piecewise* cumulative memory images. When working in sound design, one of the principal strategies could then be that of shifting our attention between the real-time unfolding of sound and the instantaneous overview images of the various concurrent sound features as shapes.

8.4. Production-Perception

The so-called "reduced listening" and its dissociation of perceptual features from everyday significations, as well as the dissociation of perception from production, has engendered some misunderstandings and disagreements in the reading of Schaeffer's music theory texts, and it could be useful to differentiate somewhat the various components here. Adapting a general model from so-called morphodynamical theory, a remarkable endeavor focused on shapes in human experience (Thom 1983; Petitot 1985; Godøy 1997), we can establish a distinction between a *control sphere* and a *morphology sphere* in studying perception (Petitot 1990). The control sphere includes the input, or control parameters, to a generative process, for instance, in digital sound synthesis or in motor control, whereas the morphology sphere includes the perceived output. The relationship between these two spheres may vary between being quite linear and being highly nonlinear, but the main point is not to confuse the generation of features with the perception of features.

We have had a number of problematic issues of non-valid couplings of features in musical composition, and Schaeffer's theory of sound objects encourages us to disregard most generative schemes in favor of the perceptually salient output features. Well-known examples of this include the use of generative schemes that may have little perceptual effects for most listeners, such as in various elaborate serial generative models, but other schemes may have more immediate and salient outcomes, such as the statistical distributions of sound events in complex textures in the generative models of Iannis Xenakis (1992). We may see analogous discrepancies between control input and output features in, for instance, FM synthesis, where linear changes in the control input may result in nonlinear changes in the perceived output. And we may come across similar discrepancies in cases of sonification where there is a mapping from one domain (e.g. medical data) to another domain (e.g. melodic profiles) where this mapping may or may not make sense. Similar problems of mappings may also be found in NIME, where there may be unclear relationships between what performers do and the sound that we hear.

Interestingly, traditional musical instruments often have multidimensional changes in the sound output from what may be conceived as unidimensional input changes (e.g. changes in the combined dynamic and spectral sound output on a piano resulting from unidimensional change in the velocity of the key motion). But in these cases, it seems fair to assume that we have learned to associate acoustically quite different sounds (e.g. a bowed and a pizzicato violin sound) with the same instrument through massive exposure to these relationships.

What these questions boil down to is the relationships between the generative features and the emergent perceptual features as suggested by the morphodynamical model. If we dig deeper into the meaning of various sound features, we see that they are about patterns of change over time, and hence, are basically all shape-related, as was indeed the main point of the morphodynamical theory. If we take this one step further, we can also see that notions of shape are intimately linked with notions of body motion, and this is the basis for what has been called motor theories of perception (Liberman and Mattingly 1985; Galantucci, Fowler and Turvey 2006).

8.5. Multimodality

Although we may think of music as primarily a sonic phenomenon, it is not controversial to claim that music also is an eminently multimodal art form. This is obvious from the fact that music has traditionally been produced by body motion and in settings where we see both the musicians and the instruments being used, hence we have both motion and vision saliently present in our musical experience. These modalities will in turn entail other sensations, for instance, the sense of touch of the effectors (fingers, hands, arms, lips, tongue, etc.) in contact with the instruments and the sense of effort of moving the effectors, as well as the derivatives of effector motion, such as the sense of velocity and acceleration. In addition, we have the widespread use of various shape-, motion-, and touch-related metaphors in sound-related discourse (cf. section 8.2 above), as well as a multitude of graphical representations of musical features, so there can be little doubt about the multimodal nature of musical experience.

This multimodal nature of musical experience fits well with the notion of sound features as shapes in Schaeffer's and related theories of sound objects (see, e.g. Delalande et al. 1996). The basic issue here is that musical sound, and music-related body motion, are both ephemeral, present here-and-now and gone in the next moment, yet fortunately leaving behind memory images that we can (variably so) recall and inspect at will. The nature of such memory images is not well explored in cognitive psychology, but there can be no doubt about their existence. Otherwise, human communication could hardly be possible.

What has emerged from several decades of research on auditory and motion perception is that there are strong links between these modalities and that schemas from one modality may be projected onto the other modality. This means that there is a top-down, and in parallel, a bottom-up influence of motion images on sound images and, conversely, of sound images on motion images (Griffiths and Warren 2004; Bizley and Cohen 2013).

And there are two elements of sound and motion links of particular importance in our context of thinking sound-motion objects:

- *Motormimetic cognition*, meaning the ongoing mental simulation of the assumed sound-producing body motion generating the sounds we are hearing, as has been claimed by so-called motor theories of perception (Liberman and Mattingly 1985; Galantucci, Fowler and Turvey 2006). It has also been suggested that in seeing, there is an analogous mental simulation of manually tracing visual patterns (Berthoz 1997). We have in the last decades seen a number of publications that support this view of motor sensations as integral also to music perception (see, e.g. Godøy 2001, 2003, 2006; Godøy and Leman 2010).
- *Posture-based cognition*, meaning that motion planning and control is hierarchical (Grafton and Hamilton 2007) and focused on a series of postures at particularly salient moments in time (Rosenbaum et al. 2007). The shape and position of the effectors (fingers, hands, arms, feet, lips, tongue, etc.) at salient moments in music such as at downbeats and other accents, we have chosen to call *key-postures*, and furthermore, we believe there are continuous motion trajectories between these key-postures (Godøy 2014, 2017).

In addition to these two main sound-motion links, there are also other significant links such as of the overall sense of motion (also referred to as *quantity of motion*), with the overall sense of energy in the music, as well as derivatives of motion such as velocity, acceleration and jerk, which may correspond to the motion sensations of the sound such as relative smoothness or abruptness.

Also, the sense of motion spans different timescales, from the very rapid to the rather slow, and may be linked to sensations of sound texture with rapid motion, such as in tremolos and trills, or to sensations of slower motion, such as that of the gait of various metrical patterns. At even slower timescales, such as of the entire sound object, sensations of motion may be linked with the overall shapes of the sound object, such as the overall dynamic contour of a long crescendo, or the overall pitch contour of a prolonged melodic curve.

From a motormimetic point of view, motor images are special in that they span so much of human cognition (i.e. that our sense of body motion is so fundamental that it transcends modal divides). In this sense, there should then be no conflict between the so-called reduced listening of Schaeffer and motor images of sound objects, because these motor images are general and at the very root of thinking as such, in fact becoming what could be called an *amodal* feature of our cognitive apparatus.

8.6. Timescales

We have different timescales at work in music, ranging from the small in the sub-millisecond range of single vibrations to the large of several minutes or even hours for whole works. But the different timescales are usually superimposed so that we have the very fast motion of audible pitch in parallel with the slower "sub-audio" features (<20 hz) of various tone events, and slower still of extended features like melodic and rhythmic fragments. To get a better understanding of the different timescales and their corresponding sound and motion features, we have used the following three main timescale categories in our research:

- *Micro*, meaning continuous sound and motion, with sound features such as loudness, pitch and overall (or quasi-stationary) timbral features.
- *Meso*, which is typically the sound-motion object timescale, meaning motion chunks and corresponding sound chunks with overall shapes of salient dynamic, pitch-related and timbral features, as well as salient modes of motion, sensations of affect and style.
- *Macro*, which is typically the concatenation of meso timescale chunks, and can extend to several minutes and even hours, and may variably so convey overall features of genre and style.

In addition to these timescale categories, we may also have combinations of different layers with different speeds, for instance in sound-motion objects at the meso timescale with a protracted or slowly moving or sustained background combined with a fast-moving layer of short and densely spaced sound events, such as combinations of "wet" and "dry" sound in effects processing (cf. so-called morphology features in section 8.9 below).

The point of differentiating these timescales is to enhance our understanding of how the various features are contributing to the total perceptual experience of the music, and to see why the meso timescale has a privileged status as the basis for the sound-motion objects. The following factors converge in making the meso timescale particularly significant:

- What we can call the *quantal nature* of sound due to basic envelopes, for instance, a plucked or a percussive sound will have distinct attack and decay shapes, making this timescale crucial for the very nature of the sound. This could also be seen as conditioned by the energy envelope of instruments (i.e. the timescale for energy dissipation after a pluck or an impact) (Godøy 2013).
- *Biomechanical constraints* of body motion will typically dictate a need for rests or shifting of posture, but most of all, the *constraints of motor control* will tend to favor motion chunks at the meso timescale as action gestalts (Klapp and Jagacinski 2011), something that seems to also fit quite well with recent ideas on intermittency in motor control (Karniel 2013; Loram et al. 2014; Sakaguchi, Tanaka and Inoue 2015).
- *Preferred durations in perception and cognition*, that is, what seems to be the optimal durations for attention spans (Michon 1978; Pöppel 1997).
- *Duration threshold for salient features* are typically in the meso timescale, such as for perceiving rhythmic, textural, melodic and harmonic patterns, or for sensing mode of motion, affect and style (Godøy 2008; Gjerdingen and Perrott 2008).

In sum, there is a convergence of elements here that privileges the meso timescale chunk, which, hence, will be the basis for further discussions of sound-motion objects here.

8.7. Objects

It may seem paradoxical that music, consisting of continuous streams of sound and motion, is also crucially dependent on discontinuous and *piecewise* cumulative perceptual images, that is, on what we think of as objects. This has confounded philosophers and psychologists since the end of the nineteenth century (Husserl 1991; Godøy 2010) and is still quite enigmatic, but we have seen the proposal of some innovative models to account for this duality in the form of having continuous perceptual sensations kept in some kind of cumulative buffer while being processed together with previous schematic knowledge in our minds (e.g. as suggested in Grossberg 2003). In our case, there is the clear utility of focusing on object images, both because the meso timescale captures so many of the salient musical features and because the piecewise preprogramming of motion chunks fits well with intermittency in motor control.

The epistemological power of object images has been well argued by the mentioned morphodynamical theory. The main advantages for sound design are the following:

- *Holistic representation* of the temporally distributed substrate of sound and motion, reflecting the entire sound-motion fragment as an entity that owes its salient features to the whole extent of the fragment.
- *Non-symbolic representation*, which does not collapse the sound-motion into some more abstract entity, or set of entities, but preserves the temporal and spectral unfolding of the sound and motion.
- *Reflecting the quantal nature of sound and motion*, that is, reflecting the entire sound-motion fragment cumulatively as an "instantaneous" entity in our minds, both in the recall and in the preprogramming for motion control (i.e. the envisaging of the entire sound-motion object "in a now," to use Husserl's expression) (Husserl 1991; Godøy 2010).

The crucial point is that we have the quantal elements of sound and motion in a subjectively instantaneous, or "all-at-once," type of shape image in our minds. Furthermore, shapes are, as argued above, a universal and amodal representation of sound and motion features, enabling a mapping of features across the domains of sound and motion.

As we move on to differentiate the features of sound-motion objects, we shall use Schaeffer's terms:

- *Typology* denotes the object timescale shapes such as the overall dynamic, pitch-related and timbre-related envelopes.
- *Morphology* includes all the within-object shapes, hence the temporal and spectral details from the overall features and a progressive differentiation of sub-features, sub-sub-features, and so forth, as long as such detail differentiation may seem to be useful.

The first step in the method of sound-motion objects is then to select a fragment of sound and corresponding motion, be that from a "naturally" occurring sound event (i.e. something that happens because of causal relationship in our environment, such as a hammer hitting a metal plate, a harp glissando, or an ornament by fingers on a keyboard), or from an edited event (i.e. something cut out and/or processed from another context, such as the middle of a protracted organ sound, a reversed cymbal sound, or a torrent of granulated speech sounds). Schaeffer suggested some criteria for this with the idea of what he called "the suitable object," mainly in view of how such an object could be used further in a compositional context.

The main point here though is that any such object, provided it is within the approximate duration range mentioned above (0.3 to 3 seconds) will, regardless of whether it stems from a real event or from some editing/processing, have an overall shape as well as an internal fabric, and these two elements will then be the topic of the ensuing typology and morphology of the sound object, but here notably extended to also include the corresponding body motion image (Godøy 2006).

8.8. Typology

Schaeffer called the typology a "first sorting" of sound objects according to overall dynamic and pitch-related content. As for the dynamic content, there are three main categories:

- *Sustained*, meaning prolonged and relatively stable sound.
- *Impulsive*, an abrupt onset as in a percussive or plucked sound, followed by a decay.
- *Iterative*, a rapidly fluctuating sound, such as in a tremolo or a trill.

A crucial feature in our perspective is that these three sound categories correspond to three categories of body motion, categories that are qualitatively different because of their use of muscles: that which is sustained with continuous effort, by impulse with discontinuous effort (so-called *ballistic* motion), and iteratively with back-and-forth or shaking/trembling effort.

It should also be noted that there are so-called *phase-transitions* (Haken, Kelso and Bunz 1985) between these categories, meaning that with changes in the duration and/or rate of sound-motion events, there may be a change from one category to another. For instance, if an impulsive sound is repeated more and more often, it may turn into an iterative sound and, conversely, if an iterative sound is slowed down below a certain threshold, it may turn into a series of impulsive sounds.

As for the pitch-related features, there are also three main categories:

- *Pitched*, meaning sound with a fairly clear and stable pitch or set of pitches.
- *Variably pitched*, a clearly pitched sound, but with the pitch varying, for example, by glissando, undulations or series of different pitches.
- *Non-pitched*, meaning any sound with an inharmonic or noise-dominated spectrum.

These categories were combined into a 3 x 3 matrix and nicely illustrated with sound examples in Schaeffer 1998.

It should also be kept in mind that the typology is very general, applicable to all kinds of music, also various kinds of Western classical music, but is independent of notation concepts. Typological categories may also be applied to not-so-well-researched areas of Western music and to orchestration, which is of particular interest in our context of sound design. The typology provides a rudimentary but useful framework for differentiating components of orchestral textures according to whether they have roles of sustained, impulsive or iterative textural content. The sustained role will typically be background sound, the "wet" component of a texture, the impulsive will typically have the attack reinforcement function, and the iterative will often have the rapid textural flutter role, such as in tremolo or trills, bordering on the morphological feature of *grain* (see the next section). Similarly, the three very coarse pitch-related categories can tell us something about the pitch-related motion within an orchestral texture (i.e. whether it is more or less stationary, moving or complex).

8.9. Morphology

The morphology of sound objects concerns internal sound-motion features and is a very extensive scheme with detailed differentiations. The main principle is always that of a top-down qualification, starting out with a seemingly simple question followed by progressively more detailed questions about sub-dimensions, sub-sub-dimensions and so forth. Moreover, like the typology, the morphology scheme may be applied to all kinds of music, also Western classical music, and several of the feature dimensions are quite useful for qualifying until now ill-discerned elements of orchestral scores such as the following:

- *Grain*, meaning rapid fluctuations in the sound, such as with various tremolos and trills, or with the natural texture of various instruments, such as maracas, washboards or deep tones on the double bass or on the bassoon.

- *Gait*, meaning the slower motion within the sound, such as slow fluctuations of the timbre by a wah-wah mute or the opening and closing of a filter.

There is much more to the morphology as it concerns the internal fabric of sound-motion objects and ranges from continuous quasi-stationary sound features of tone color, formants, harmonicity, inharmonicity and the corresponding stationary posture shapes to various fluctuations and micro-motion features such as of noise, shimmer, rumbling and the mentioned grain, also here with corresponding motion features.

Lastly, a common crucial feature of both the typology and the morphology is that of being concrete and non-symbolic. Schaeffer was very clear in opposing the notion of the concrete to that of the abstract, meaning that the concrete is distributed and has shapes, and the abstract is symbolic and does not have distributed shapes.

8.10. Analysis-by-Synthesis

The conceptual apparatus of feature differentiations presented with the typology and morphology can be used in the exploration and design of sound-motion objects by:

1. Detecting a feature (e.g. the grain of a sound-motion object).
2. Placing it in a multidimensional matrix (e.g. for specifying its rate and amplitude).
3. Making incremental variations (e.g. from minimum to maximum).
4. Evaluating the results and choosing the right rate and amplitude grain of the wished-for sound-motion object.

This approach is called analysis-by-synthesis (Risset 1991) and is, in practice, well known to most people who have worked with sound synthesis and/or production—and, in fact, also to musicians and conductors—as the incremental adjustment of the different features of a sound until it becomes what is wanted. There are innumerable well-known experiences of this in sound synthesis, for instance, in the regulation of the attack time for a sound from close to instantaneous (a couple of milliseconds) to very long (up to one second or even more), resulting in output sounds varying from percussive to very soft/gradual in its attack feature. Another example might be in introducing the "right" amount of inharmonicity in a wished-for "metallic" or "bell-like" sound-motion object by incrementally mistuning the partials in an additive synthesis model until the subjective experience of the sound is right, together with adjusting the attack rate to what makes the sound seem to originate from a hammer hitting metal, as well as individuating the decay rate of the partials so that the overall decay of the sound is as wished for (e.g. higher partials decaying fast, lower partials more slowly).

In short, analysis-by-synthesis means making multidimensional and incremental variants. The challenge is to implement this in whatever practical work scheme is at hand, be it in digital synthesis or in scoring for traditional instruments.

8.11. Montage

The focus on sound-motion objects by no means prevents us from thinking more large-scale musical compositions. Sound-motion objects may be combined in various ways, and this we can refer to as *montage*. Montage is well known from other contexts, in particular from film theory, and has also been widely used in musical composition, eminently so in the mentioned musique concrète, but also in a number of other epochs and styles. Needless to say, it has been quite often used in various contemporary popular music cultures by what is referred to as "sampling." The basic feature of montage is that by a concatenation of elements, there will emerge a new entity that has new features not present at the level of its constituent parts. Another view of montage could thus be that there is a change of resolution where there is a fusion of smaller-scale elements into new units with their own timescale-specific features. There are a number of ways to do this, and Schaeffer singled out two cases of object combinations:

- *Composed objects*: the superposition of different objects, typically a sharp attack object (e.g. a percussive sound) on top of a sustained sound, with a resultant sound having a sharp attack and a prolonged sustain.
- *Composite objects*: the concatenation (and partial overlap) of successive objects so as to produce a new larger object (e.g. a trumpet tone cross-faded into an oboe tone so as to create a new object that evolves from trumpet-like to oboe-like). This is similar to the so-called phenomenon of coarticulation, meaning the contextual overlap and smearing of two or more consecutive sounds into a qualitatively new sound (see Godøy 2018 for details).

On a more large-scale level, there are of course countless possible combinations, however, the challenge from a sound design perspective is that the perceptual effects of montage is not a well-researched area. The traditional Western music theory view is that grand-scale formal designs are crucial for the experience of the music, however, this has not been well studied, and the few studies we know of seem to suggest that we should be skeptical of some inherited claims in this area (Eitan and Granot 2008).

Montage could actually be a kind of analysis-by-synthesis procedure in the sense of variant recombinations of sound-motion chunks. Indeed, with readily available means for digital music editing, it is fairly easy to generate a number of reedited variants of large-scale musical works and then have some kind of perceptual assessments of the effects of such reediting.

8.12. Summary

This chapter is based on the fusion of two main ideas:

1. Focusing on delimited fragments of sound, on what Schaeffer called sound objects, as the prime strategy for feature exploration in sound design, and;
2. Extending this sound-object focus to also include the associated sensations of body motion with the concept of sound-motion objects.

The main reason for the object perspective of point 1 was that it represents the convergence of important features of sound production (combined instrumental/vocal constraints and body motion constraints) and sound perception (duration includes the majority of salient features, and also seems optimal for attention and memory). The main reason for including the body's motion sensations (real and/or imagined) in these sound-motion objects (point 2) was that sound objects are fundamentally multimodal and seem to be influenced in their constitution by motor schemas (i.e. by projecting motor images onto sound).

The concept introduced for this in Godøy 2001 was *motormimetic cognition*, and this concept may then be exploited as a source for the salient presence of otherwise ephemeral sound and motion as shapes, as both quasi-stationary postures and as unfolding trajectory shapes, that is as salient shape images in our minds.

In practical terms, this means drawing sound-motion features, on paper, screen, or just in our minds, and then enacting the shapes as sound-motion objects, then once again drawing their features, once again enacting the shapes as sound-motion objects, and so on; it essentially means continuing an iterative process, resulting in the gradual accumulation of sound-motion features knowledge. Using available means for motion capture, animation and immediate playback, the detail scrutiny of sound-motion links in real musical contexts is now becoming easier to do (e.g. as reported in Godøy, Song and Dahl 2017). Such detail scrutiny of sound and motion links may help us realize that the distinction between sound and body motion in music is less obvious than we may have previously thought.

References

Berthoz, A., 1997. *Le sense du mouvement*. Paris: Odile Jacob.

Bizley, J.K. and Cohen, Y.E., 2013. The What, Where and How of Auditory-Object Perception. *Nature Reviews Neuroscience*, 14, pp. 693–707.

Chion, M., 1983. *Guide des objets sonores*. Paris: INA/GRM Buchet/Chastel.

Delalande, F., Formosa, M., Frémiot, M., Gobin, P., Malbosc, P., Mandelbrojt, J. and Pedler, E., 1996. *Les unités sémio- tiques temporelles: Éléments nouveaux d'analyse musicale*. Marseille: ÉditionsMIM-DocumentsMusurgia.

Eitan, Z. and Granot, R.Y., 2008. Growing Oranges on Mozart's Apple Tree: 'Inner Form' and Aesthetic Judgment. *Music Perception*, 25, no. 5, pp. 397–417.

Farnell, A., 2010. *Designing Sound*. Cambridge, MA: The MIT Press.

Galantucci, B., Fowler, C.A. and Turvey, M.T., 2006. The Motor Theory of Speech Perception Reviewed. *Psychonomic Bulletin & Review*, 13, no. 3, pp. 361–377.

Gaver, W.W., 1993. What in the World Do We Hear? An Ecological Approach to Auditory Event Perception. *Ecological Psychology*, 5, no. 1, pp. 1–29.

Gjerdingen, R. and Perrott, D., 2008. Scanning the Dial: The Rapid Recognition of Music Genres. *Journal of New Music Research*, 37, no. 2, pp. 93–100.

Godøy, R.I., 1997. *Formalization and Epistemology*. Oslo: Scandinavian University Press.

Godøy, R.I., 2001. Imagined Action, Excitation, and Resonance. In: Godoy, R.I. and Jorgensen, H. (Eds.), *Musical Imagery*. Lisse: Swets and Zeitlinger, pp. 237–250.

Godøy, R.I., 2003. Motor-Mimetic Music Cognition. *Leonardo*, 36, no. 4, pp. 317–319.

Godøy, R.I., 2006. Gestural-Sonorous Objects: Embodied Extensions of Schaeffer's Conceptual Apparatus. *Organised Sound*, 11, no. 2, pp. 149–157.

Godøy, R.I., 2008. Reflections on Chunking in Music. In: Schneider, A. (Ed.), *Systematic and Comparative Musicology: Concepts, Methods, Findings*. Frankfurt: Peter Lang, pp. 117–132.

Godøy, R.I., 2010. Thinking Now-Points in Music-Related Movement. In: Bader, R., Neuhaus, C. and Morgenstern, U. (Eds.), *Concepts, Experiments and Fieldwork. Studies in Systematic Musicology and Ethnomusicology*. Frankfurt: Peter Lang, pp. 245–260.

Godøy, R.I., 2013. Quantal Elements in Musical Experience. In: Bader, R. (Ed.), *Sound—Perception—Performance. Current Research in Systematic Musicology*, Vol. 1. Berlin and Heidelberg: Springer, pp. 113–128.

Godøy, R.I., 2014. Understanding Coarticulation in Musical Experience. In: Aramaki, M., Derrien, M., Kronland-Martinet, R. and Ystad, S. (Eds.), *Sound, Music, and Motion: Lecture Notes in Computer Science*. Berlin: Springer, pp. 535–547.

Godøy, R.I., 2017. Key-Postures, Trajectories and Sonic Shapes. In: Leech-Wilkinson, D. and Prior, H. (Eds.), *Music & Shape*. Oxford: Oxford University Press, pp. 4–29.

Godøy, R.I., 2018. Sonic Object Cognition. In: Bader, R. (Ed.), *Springer Handbook of Systematic Musicology*. Berlin: Springer Nature, pp. 761–777.

Godøy, R.I. and Leman, M. (Eds.), 2010. *Musical Gestures: Sound, Movement, and Meaning*. New York: Routledge.

Godøy, R.I., Song, M.H. and Dahl, S., 2017. Exploring Sound-Motion Textures in Drum Set Performance. In: *Proceedings of the SMC Conferences*. ISSN 2518–3672, Espoo: Aalto University, pp. 145–152.

Grafton, S.T. and Hamilton, A.F., 2007. Evidence for a Distributed Hierarchy of Action Representation in the Brain. *Human Movement Science*, 26, pp. 590–616.

Griffiths, T.D. and Warren, J.D., 2004. What Is an Auditory Object? *Nature Reviews Neuroscience*, 5, no. 11, pp. 887–892.

Grossberg, S., 2003. Resonant Neural Dynamics of Speech Perception. *Journal of Phonetics* 31, pp. 423–445.

Haken, H., Kelso, J. and Bunz, H., 1985. A Theoretical Model of Phase Transitions in Human Hand Movements. *Biological Cybernetics*, 51, no. 5, pp. 347–356.

Husserl, E., 1991. *On the Phenomenology of the Consciousness of Internal Time, 1893–1917*. English translation by Brough, J.B. Doredrecht, Boston and London: Kluwer Academic Publishers.

Karniel, A., 2013. The Minimum Transition Hypothesis for Intermittent Hierarchical Motor Control. *Frontiers in Computational Neuroscience*, 7. doi:10.3389/fncom.2013.00012

Klapp, S.T. and Jagacinski, R.J., 2011. Gestalt Principles in the Control of Motor Action. *Psychological Bulletin*, 137, no. 3, pp. 443–462.

Liberman, A.M. and Mattingly, I.G., 1985. The Motor Theory of Speech Perception Revised. *Cognition*, 21, pp. 1–36.

Loram, I.D., van De Kamp, C., Lakie, M., Gollee, H. and Gawthrop, P.J., 2014. Does the Motor System Need Intermittent Control? *Exercise and Sport Science Review*, 42, no. 3, pp. 117–125.

Michon, J., 1978. The Making of the Present: A Tutorial Review. In: Requin, J. (Ed.), *Attention and Performance VII*. Hillsdale, NJ: Erlbaum, pp. 89–111.

Petitot, J., 1985. *Morphogenèse du Sens I*. Paris: Presses Universitaires de France.

Petitot, J., 1990. Forme. In: *Encyclopædia Universalis*. Paris: Encyclopædia Universalis.

Pöppel, E., 1997. A Hierarchical Model of Time Perception. *Trends in Cognitive Science*, 1, no. 2, pp. 56–61.

Porcello, T., 2004. Speaking of Sound: Language and the Professionalization of Sound-Recording Engineers. *Social Studies of Science*, 34, no. 5, pp. 733–758.

Rimsky-Korsakov, N., 1965. *Principles of Orchestration*. New York: Dover Publications Inc.

Risset, J.C., 1991. Timbre Analysis by Synthesis: Representations, Imitations and Variants for Musical Composition. In: De Poli, G., Piccialli, A. and Roads, C. (Eds.), *Representations of Musical Signals*. Cambridge, MA, and London: The MIT Press, pp. 7–43.

Rosenbaum, D., Cohen, R.G., Jax, S.A., Weiss, D.J. and van der Wel, R., 2007. The Problem of Serial Order in Behavior: Lashley's Legacy. *Human Movement Science*, 26, no. 4, pp. 525–554.

Sakaguchi, Y., Tanaka, M. and Inoue, Y., 2015. Adaptive Intermittent Control: A Computational Model Explaining Motor Intermittency Observed in Human Behavior. *Neural Networks*, 67, pp. 92–109. http://dx.doi.org/10.1016/j.neunet.2015.03.012.

Schaeffer, P., 1966. *Traité des objets musicaux*. Paris: Éditions du Seuil.

Schaeffer, P. (with sound examples by Reibel, G. and Ferreyra, B.), 1998. (first published in 1967) *Solfège de l'objet sonore*. Paris: INA/GRM.

Thom, R., 1983. *Paraboles et catastrophes*. Paris: Flammarion.

Xenakis, I., 1992. *Formalized Music (Revised edition)*. Stuyvesant: Pendragon Press.

Climb!—A Composition Case Study

Actualizing and Replicating Virtual Spaces in Classical Music Composition and Performance

Maria Kallionpää and Hans-Peter Gasselseder

9.1. Introduction

In one way or another, the concept of space is present in any classical music performance. The musicians are aware of the importance of spatiality and physicality in the context of every performance and they know how to adapt their movements, articulation, interpretation, tempi and so forth to the prevailing acoustic qualities of the room and the characteristics of their instrument, either intuitively or consciously. Seasoned musicians who frequently perform in "*acoustically diverse spaces*" learn how to change their playing style accordingly to attain the desired artistic results (Nonken 2014, p. 77). They are trained to "play the room" and to integrate themselves into it in a real-time situation (McNutt 2003, p. 298). It is the job of the performers to constantly deal with the temporality, varying acoustic conditions and issues related to the sound production principles of the musical instruments. Nonken explains that,

> [. . .] in its sonic conception and graphic representation on the page, the music elegantly confronted the temporal issues of resonance with which pianists grapple every moment at the keyboard. All pianists deal with the uncertainties and disorders that come from the nature of our instrument: its legendary decay, its unstable and idiosyncratic tuning, how different registers speak in various ways, and how every instrument seems to resonate somewhat curiously, amplifying certain frequencies and muting others, in relation to its own inherently awed construct and the peculiarities of the space in which it resides.
> [. . .] I was heartened to find a group of composers who considered these aspects of instrumental reality—the instrumental body, and the physical reality of how sound works—as defining aesthetic

concerns. They engaged with the phenomena of the piano in real time, demanding from the performer spontaneity and the ability to respond with alacrity to the instrument in time and space.

(Nonken 2014, pp. 9–10)

While all this also applies to the traditional classical music repertoires, the importance of being able to navigate within different acoustic qualities and spaces (for example, a large, reverberant concert hall versus an auditorium with dry acoustic character) is even more prominent in the context of contemporary electroacoustic repertoires. This chapter reflects the concept of space from the viewpoints of physical, acoustic and virtual environments. The following discussion will also regard "space" as an abstraction related to the compositional structure and form, as certain musical events happen in different passages of the score or composition. Moreover, to deliver a coherent performance of the piece, the musician needs to time their actions within a certain timeline that must also be interpreted as a spatial entity or phenomenon. Although a musical composition is usually perceived as a linear entity, "its versatile conceptual structure can be viewed as hyperstructures which can encode branching compositions and incorporate other multimedia elements and annotations" (cf. Kallionpää and Gasselseder 2016, p. 465; Byrd and Crawford 2002, pp. 249–272). Such versatile dimensions of spatiality and form will be further discussed in the presented case study *Climb!*, in which the pianist navigates through a nonlinear musical adventure within the virtual space with the help of an interactive system.

9.2. Spatiality as Part of the Musical Structure

The pursuit of exploiting the performing space as part of the musical aesthetics has been going on for centuries. The composers representing the Venetian polychoral style (generally known as "*cori spezzati*") of the Renaissance Era already knew how to extend the capabilities of the choirs by placing their singers on different sides of the cathedral and letting them interact with each other in responsive musical dialogues. Giovanni Gabrieli (1557–1612), a composer and organist of the San Marco Cathedral, was a master of this style. Alongside such composers as Hans Leo Hassler (1564–1612) and Michael Praetorius (1571–1621) (among the others), he composed works that paved the way for the later development of the concertato style (predecessor of the sonata form, concerto grosso and cantata). Although there is no proof that Praetorius ever visited Italy, he was well familiarized with the style. However, he leaned more towards the German style of polychoral writing, which focused more on the role of the orchestra (and in general on the instrumentation of the compositions). Carver argues that the polychoral music of the latter composers can be equated with the best Venetian music of the time (Carver 1988, pp. 210–217).

As discussed above, spatiality in music can also be taken into account as more of an abstract phenomenon related to the structure of a musical composition. A musician who performs a large-scale work, such as a sonata or concerto, needs to be able to handle it as a kind of an object with its own characteristics, identity and proportions. For a musical composition to be viewed as a spatial entity, its temporal nature must be accepted. As argued by Taylor, "in order to see music as an object, its temporal dimension with its continually disappearing parts has to be frozen so that the work can be expressed as a simultaneity—in other words, become spatialized (Taylor 2016, p. 93)." Furthermore, according to the same author, the "objective basis for this spatialization in music is created by the principle of repetition" (ibid.). Conversely, with regard to contemporary music, Chadabe states that one of the most important steps of its development was to start seeing musical compositions as living "processes" rather than "fixed objects" (Chadabe 1996, p. 43).

The twentieth- and twenty-first-century composers have shown strong interest in including spatiality as part of their musical expression and technical parameters. Henri Dutilleux's *Timbre, Espace, Mouvement* (1978) is an illustrative example of such thinking. The work is subtitled "La Nuit Etoilée," which refers to the painting of Vincent van Gogh, of which the composer created his own musical interpretation. The two-dimensional events of the painting get translated into music via musical gestures, instrumentation and the special positioning of the instruments (the cellos) within the concert hall. Customizing the seating of the orchestral musicians has a direct impact on how the music sounds, as in this way the composer can control not only how the listeners perceive the sounds (or the composition as a whole) but also how the musicians perform the music in the concert situation. Letting the performers or listeners wander around in the concert space for this purpose is relatively popular among the contemporary music composers. Moreover, electronic or electroacoustic compositions frequently make use of the possibility of moving the sound from one speaker to another while the listeners are placed in the middle of the concert space, surrounded by four or more sound sources. The sound quality, room and space, instrumental techniques and sound source itself form the essence of the performer's and listener's experience. The audience's reception of an acoustic solo piano performance is different than that of a laptop orchestra's multichannel performance, in which physical elements are minimized and thus make it harder for the audience to distinguish the sound source (Kallionpää 2014, p. 13). Furthermore, the space itself can function as an instrument. For example, Kuusisaari discusses a performance of Répons by Pierre Boulez at the innovatively structured Pierre Boulez Hall in Berlin, stating that the concert hall worked as "a gigantic instrument" (Kuusisaari 2018, p. 25).

Spatial solutions are used as part of composers' tool kits for a variety of reasons, ranging from dramatic, visual, harmonic or choreographic purposes, to name a few. Kallionpää has previously been involved in several performances of this kind of work, the most recent being the opera *ON/OFF* (2017) by Sara Caneva. In this work, the musicians are first placed at the very back of the concert hall, from where they then gradually move towards the orchestra pit.

As discussed by Kwon, several different terminologies refer to artworks or artistic practice tied to a certain space or site. Whether site-determined, site-oriented, site-referenced, site-conscious, site-responsive or site-related and so on, all originate from the idea of site-specific practices of the 1960s and 1970s "[. . .] which incorporated the physical conditions of a particular location as integral to the production, presentation, and reception of art (Kwon 2002, p. 1)." Certain musical compositions are site specific, too. In such works the venue (or the objects located in a particular space) forms an important part of the structure, aesthetics, and expressive contents of the composition. Thus, the role of the specific room and its qualities must be taken into account from the very beginning. Whatever it is that the space represents to the composer, it will be "written in" to the music. Kallionpää has composed two site-specific pieces. *Dialogo* (2008) was composed for the organ at the Dutch Church in London (and premiered at the Spitalfields Festival 2008), and *Le Safari de la Mort* (2013) for solo flute (premiered by Jeff Brown in June 2013). The latter title refers to the sculpture by Jean Tinguely that is located at the Tinguely Museum in Basel. The high, fragile, squeaky noises produced by the sculpture are treated as a part of the musical language and expression (Kallionpää 2013, p. 1). Because of this, the composition cannot be played in its original form in any other space or location, which of course limits the amount of its future performances. However, composers often deal with this by producing more "portable" versions of their pieces. Perhaps one of the most famous examples of site-specific compositions is *Organ²/ASLSP* (*As Slow as Possible*) by John Cage. It was composed in 1987 for the St. Burchardi church in Halberstadt, Germany, and the planned duration of the work is 639 years. The work shall be played there (or rather stay ringing there) until the year 2640. Although different versions of the work exist (it was originally based on Cage's piano composition *ASLSP* 1985 and an earlier organ version also preceded it), *Organ²/ASLSP* (*As Slow as Possible*) should be regarded as site specific because its performance could not take place anywhere else. The organ on which the piece is currently being played was specifically manufactured for the purposes of the project and special modifications were made to the venue to reduce the volume of the instrument.

Many spectral music composers regard spatiality as an essential part of their works. For example, Edgard Varèse who invented the "ionization" composition technique, stated that the future music needs to be spatial, meaning that there should be a dynamic acoustic space onto which different elements of sounds are to be reflected (Moscovich 1997, p. 22). Also, Giacinto Scelsi discussed spatial position as part of a spectral composition, alongside density, dynamics and smooth or rough particles (ibid.). James Tenney's work *Form 1* (1993) makes use of space by arranging the players in four groups: "[. . .] *in front of, behind, to the left of, and to the right of, the audience*" (Tenney 1993, p. 1). The score is divided into 57 different time segments or bars, each with the duration of 20 seconds. The player chooses the played pitches from the selection provided by the composer. The listeners perceive the work differently depending on how they are located (or how they move around) in the concert venue. Whereas various composers might position the musicians in a specific way in the concert space to attain dramatic purposes, in spectral music the sound itself is the focal point. Rather than concentrating on the pitch material, compositions are created around the *sound as a whole*. For example, composers split the sounds into harmonic partials and analyze them mathematically. Moreover, they also deal with aspects of psychoacoustics, the choice of sound sources, and review the sound as an acoustic phenomenon as a vital part of their composition processes (Moscovich, p. 21). Regarding the concepts of time and space this way applies to Grisey's works (for example, *Epilogue (Les espaces acoustiques VI)* (1985)), Tristan Murail's compositions (for example *Territoires de l'oubli* (1977)) and to the works of various other spectral composers. As discussed by Nonken, Murail "takes care to recognize, in his notation, the variables at play: the instrument, the space, and the body of the performer" (Nonken 2014, p. 79).

9.3. Virtual or Artificially Generated Environment as a Performing Space

As discussed earlier in this chapter, space plays an important role within a musical performance, and it frequently finds its way into the structure and dramatic content of a composition. But what if the entire space, or part of it, were virtual or otherwise artificially created? The general availability of fast and efficient computers and software has enabled composers and sound designers to construct such artificial environments and, thus, almost any acoustic, musical, instrumental or technical parameter can be altered or extended (Kallionpää 2014, p. 6). Moreover, the existing instrumental and

expressive skills and aptitudes of a performer can be extended or multiplied
(ibid.). By having an opportunity to create new acoustic environments or
altering the existing ones according to their artistic needs, composers are
in full control of the acoustic space. This applies especially to electronic
or acousmatic music because, on top of the sounds themselves, the space
in which the music "takes place" also needs to be generated. The intended
technical and acoustic setup has to be considered as early as during the
composition process. For example, it makes a difference whether an electronic music piece is designed for a set of stereo, quadraphonic or eight
speakers. In his *1900: Concerto for Piano and Genelec Orchestra* (2013)
composer Uljas Pulkkis (together with Professor Tapio Lokki) explored
the potential of creating an entire acoustic space with the help of Genelec
studio monitor speakers. The unique feature of this concerto is that both
the orchestra and the conductor are omitted and each orchestral part was
created digitally with the help of virtual orchestration tools, such as (semi-
step) multi-sampled instruments produced by the Vienna Symphonic
Library. The role of the conductor gets handed over to the soloist who,
together with the sound technician, controls the volume, tempi and tuning.
Despite not engaging any live musicians, the "virtual orchestra" follows
the soloist by reacting to the triggers she supplies. Pulkkis stated that the
benefit of this kind of a setup is that altering the tuning in the middle of a
performance becomes easier and the orchestra will always play the same
way regardless of the acoustic circumstances (Pulkkis 2013, p. 13). This
way the entire "performing space" and all its qualities get carved and customized by the composer, and the limitations of acoustic rules and natural
sound production qualities of the concert instruments are circumvented.

Composer and film director Michel van der Aa challenges the concept
of space in his interdisciplinary music theater works. His aesthetics seamlessly encompass the media of music, sound, video art and film, for which
van der Aa has been described as *a full-out virtuoso of mixed media* (Oestreich 2017, p. 1). In his works, new kinds of musical and visual spaces are
created with the help of innovative technologies. For example, his opera
Blank Out (2016) engages three-dimensional projection techniques that are
combined with live performance, prerecorded musical material and sound
processing. In this way van der Aa provides the audience varying visual
perspectives of what is happening onstage. Among other technical and aesthetic methods and ideas, the work exploits scale models that the singers
handle onstage in order to communicate the musical drama and story line.
The models are sometimes shown to the audience in their natural sizes,
sometimes projected as a three-dimensional format in a gigantic scale.
Rather than putting together a traditional opera, van der Aa's approach
provides the listeners an immersive spatial experience. In addition to his
work on *Blank Out*, the composer has also exploited three-dimensional

techniques in his film opera *Sunken Garden* (2011–2012). Furthermore, his previous music dramas *The Book of Disquiet* (2008), *After Life* (2005–2006), *One* (2002), *Vuur* (2001) and *Writing to Vermeer* (1999) likewise evince unique interactions between the live musicians, prerecorded material, video and soundtrack.

As mentioned above, a musical composition can be partly or completely constructed in a virtual environment, in which case its performance also takes place within an abstract, artificially generated space. For a better definition of interactive scenarios such as the one described above, one may draw on a few analogies found in computer games, thanks to a common ground in the overall concept of play and virtuality. While the discipline of game studies also may struggle to find consensus about central terminologies such as game audio and spatiality, a few notable contributions have been made towards embracing the complexities and potential of the virtual (cf. Collins 2008, p. 2).

Grimshaw and Gardner present the term "sonic virtuality" to describe the perceptual nature of sonic contents that may appear in the realm of the physical as much as in the context of computer games. However, the terms "virtual world" and "virtual environment" usually refer to the computer-generated environments of computer games (Grimshaw and Garner 2015, p. 160). All chances of provoking an Uncanny Valley effect aside, one might assert a similar rational for designing virtual spatiality and physical acoustics. Gasselseder (2016) finds further parallels between physical and virtual spaces by discussing the innate relationship between auditory perception and sense of agency in his situational context model (see Figure 9.1 for an excerpt of its functional schematics, which is relevant to the following discussion).

Drawing from seminal studies on directing attention and mental simulation, it is argued that real-life interactions in the acoustic realm typically require the ability to automatically synchronize incoming multimodal stimuli and identify their underlying actions at a lower, subpersonal level of cognitive processing (see Bruner and Postman 1949; Popper and Fay 1997; Blakemore and Decety 2001). Within premotor activation (i.e., the planning, spatial guidance and understanding of abstract rules and movement executed by the self and others), the gathered information feeds into "forward models" or hierarchical nodes that compare the desired and current state so as to encode global specifications of an action that are controlled for and adapted to their goals and underlying motivations (cf. Jeannerod 1997). Ultimately, these abilities allow musicians to both separate their own actions from those made by others and to expect and adapt their interaction to future developments within a performance. However, most importantly, forward models pave the way to understanding and subsequently contextualizing why a particular action (such as a strongly

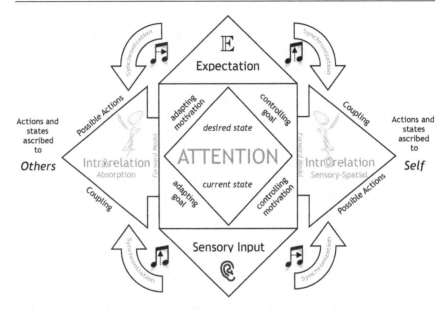

Figure 9.1 Excerpt of a Functional Schematic of "Situational Context" (Gasselseder 2016): Music score symbols with arrows pointing sideways indicate structures of horizontal expression in music; music score symbols with arrows pointing upwards indicate structures of vertical expression in music. Relational differentials refer to different hierarchical processing modes with regards to agency interactions that are externally (intra) or internally (intro) stimulated/projected.

accentuated chord progression) was performed at a specific point in time and how to respond.

A basic sense of spatiality is imposed onto a situational context by assigning the membership of an action and its occurrence in time. Not surprisingly, this sequence may be reversed in the situation of media reception. Consequently, a preexisting conscious sense of agency is followed by covert automatic processing that couples premotor action with a virtual avatar (whether that is a game character or an audio effect applied onto one's own electroacoustic performance). Thanks to the basal processes of synchronization, this sequence offers the highest yield in lowering uncertainty, gaining an understanding and focusing on the most information-rich characteristics within the modalities encountered in the virtual. For music, the arousal characteristic, that is, an emotional state ascribed to the intensity of an experience, conveys information that offers the highest value for synchronizing incoming sensory streams in real-life environments. Telling whether a listener finds pleasure or valence in a particular piece of music may take some time, whereas arousal can typically be assessed almost immediately. That is why the first few moments of a real-life performance situation are generally governed by arousal conveyed in "vertical" expression (a term commonly referring to the structural features of music taking place simultaneously at a given moment in time, such as arrangement/orchestration or sonic colors), followed by an appreciation of the temporal unfolding of the "horizontal" hierarchical structure (similarly here, horizontal refers to features unfolding over a certain period of time, such as formal or metrical structure). Here the focus lies on the performed actions and their immediate synchronization to the constant stream of multimodal inputs that may arise within a well-defined environment (i.e. a performance stage). Within the virtual, during the first moments of contact, less value is attributed to vertical expression due to synchronization yet to be achieved on the premotor stage. No couplings, let alone expectations, can be derived from a virtual avatar, such as an artificially intelligent piano, or a virtual environment without first knowing about its possible actions. If one were to rely on vertical expression in music, one would find oneself overwhelmed with explaining why a virtual object accentuated a musical passage at a higher dynamic velocity.

Rather than guessing at a repertoire of actions from individual observations, a more efficient way to determine the possible actions or capabilities of a virtual object is by inferring its coordination and limits by virtue of physical or virtual constraints/conditions. Generating these higher-order representations takes more time and, thus, finds higher information value in characteristics of music that are less inclined to change spontaneously. By setting the tone and overall situational context, the horizontal-hierarchical structure in music (whose valence, with its varying degrees between pleasure-displeasure, may contribute) offers the best match for the temporal resolution of higher-order representations of contextual constraints. In the case of valence, depending on whether a piece of music indicates a positive or negative quality, a corresponding set of forward models will have performers and listeners prepared to synchronize and expect the most likely action repertoire (cf. Gasselseder 2016). To give another example for electroacoustic music, an audio effect applied to an acoustic piano may be presented via channel-based (i.e. signal balanced between multiple speakers) or object-based panning (i.e. wave-field synthesis, binaural playback) methods. Depending on which one of these two preconditions apply, a musician may expect that the audio effect will move between speakers and interact with the preexisting acoustic environment (channel-based panning) or within the speaker array and span an acoustic environment that allows it to radiate in all directions (three-dimensional object-based panning). Hence, it is mandatory to provide a sufficient number of acoustic cues to give the performer(s) a sense of the capabilities and characteristics of electroacoustic objects that call for interaction.

Whether these capabilities are restricted by physical (as in a channel-based speaker setup or a piano played by artificial intelligence) or virtual means (such as in a three-dimensional audio setup or a sound generated by a synthesizer), their constraints and thus interactional behavior must be communicated to the performer(s) beforehand. These factors will have to be taken into account when designing virtual sonic spaces as part of an electroacoustic composition as they may impact the style and interaction mind-set of the performer(s), as will be shown in later sections of this chapter. Failing to consider these principles will have performers struggle to play along with the electronic elements of the score and further prompt a sense of incoherence among the audience. Thus, in electroacoustic works, the disciplines of composition and sound design are compelled to tap into the format of high-performance human-computer interaction, of which the most prominent are computer games.

Another design element common to both electroacoustic performance and computer games is music per se. Similar to film music soundtracks, game music scores have raised interest among concert organizers and listeners. As reported by Collins, game music concerts can help to increase the popularity of symphonic orchestras especially by attracting younger audiences (Collins 2008, p. 1). Concerts and festivals focusing on computer game music have been organized in various cities and countries (e.g. the All Your Bass festival in January 2018, where Kallionpää's interactive game composition was also staged, and which was hosted by the National Videogame Arcade in Nottingham). Although game music scores are regularly performed by orchestras, their performances often lack interactive content. Performing computer music scores the same way as any classical music repertoire strips them from their responsiveness and interactivity, which is the key component of their originality. However, there have been attempts to take this into account. For example, *Dear Esther*, a first-person exploration video game that contains the music score composed by Jessica Curry and which was developed by The Chinese Room, was played through as part of the concert program (on January 19, 2018) at the All Your Bass festival in Nottingham. The play-through was accompanied by live narration and a real-time orchestral performance.

Interactivity plays an increasingly large role in the context of modern computer games, which is why game developers have been interested in making their computer music scores procedural, too. In general, procedural content generation (PCG) refers to dynamic, algorithmic generation of game content. It also enables music composition that "adapts more granularly to player experience, avoiding needless repetition and providing an evolving, more emotionally intelligent sound-track" (Plans and Morelli 2012, p. 192) As demonstrated in a series of empirical studies by Gasselseder, implementations devising dynamic music improve the gaming experience by means of increased immersive experiences ranging from suspension of disbelief, flow and spatial presence to the realm of interaction as much as the plausibility of game characters (Gasselseder 2015, p. 464ff). Various game composers and researchers have successfully imported classical composition techniques into the game music context, but there has generally been less research on how to apply interactive game scoring systems into classical music composition. Nonetheless, interactive scoring systems have inspired some exploration in the field of contemporary music. For example, Huw Davies has composed works in which he has combined the methodologies used in computer game music scoring and classical electroacoustic techniques. His works *Deus Est Machina* (2014) and *Starfields* (2014) are computer game-inspired, purely electronic works that exploit algorithmic composition techniques to generate music. The composer uses a preprogrammed synthetic voice to communicate the narrative and to create accompanying musical material. Instead of experiencing the composition in a regular concert room, the "musical performance" takes place in a virtual space that has its own sets of rules, mechanisms and modes of communication. Similar to the abovementioned works by Davies, sonic virtual environments of interactive electronic or electroacoustic compositions are usually replicated for the gamer/listener via headphones or speakers. Audio games make use of kindred techniques. Instead of receiving visual cues, the gamer needs to understand the game environment by means of listening only and, thus, to be able to navigate on the basis of that aural image. Such games were originally developed for the needs of the visually impaired and have since appealed to a wider audience of gamers.

The principles of procedural music are related to algorithmic composition techniques. The latter are frequently used both in dynamic computer game soundtracks and in contemporary classical repertoires. Essl argues that such techniques may develop into "an inspirational machine" by which to expand the limits of our musical creativity, experience and expression (Essl 2006, p. 1). Essl's list of works largely consists of compositions applying algorithmic techniques, such as the entire *Sequitur* series (a collection of compositions for a variety of different solo instruments) as well as the "infinite real-time composition" Lexikon-Sonate (1992–2016). The latter uses a specially designed computer program, which "is an infinite music installation that can run on a computer for years without repeating itself. It can also be used as a computer instrument for live performances" (Essl 1992, p. 1).

9.4. Case Study: Composing and Performing *Climb!* (2017)

In 2008, Collins argued that "[. . .] at this stage, games are so new to academic study that we are not yet able to develop truly useful theories without basic, substantial empirical research into their practice, production and consumption" (p. 2). Since then, game studies and associated disciplines such as musicology and cognitive psychology have come a long way towards developing reliable tools for the design and evaluation of audio content in computer games. However, the empirical data gathered has mostly focused on well-established practices, thus limiting the potential of testing and cross-validating advanced implementations of interactive music within other forms of high performance real-time interactive media. Alongside its obvious artistic goals, *Climb!* as a research project was the authors' response to the desideratum of combining two of the most challenging art forms in interaction design. The research team attempted to contribute to the development of new theories by using a combination of methods applied in computer game- and arts research (Kallionpää and Gasselseder 2016, p. 45). The authors intended to provide theoretical knowledge on how the use of interactive music systems affects the mechanisms of composing and how the procedural or dynamic music systems used in computer games could be used as part of the composer's artistic vocabulary.

Conversely, embracing procedural aesthetics from the perspective of an electroacoustic piano performance promised insights for future implementations of dynamic music that would encompass an immersive and highly optimized feedback system with coherent transitions allowing one to move through a narrative-dramaturgically structured musical scene. By writing music for a system that would simultaneously be a classical music virtuoso composition and a computer game, the composer, Maria Kallionpää, expected to access more versatile sonic environments with real-time tempo, rhythm and harmonic variation. Rather than composing a piece with a fixed form, *a fixed object* (Chadabe 1996, p. 43), one of the aims of creating *Climb!* was to challenge the concept of form in a classical music composition. Spatiality was also present throughout the creative process, as one of the research goals was to explore how to replicate and actualize virtual spaces in a concert hall setting. Furthermore, one of the aims was to distinguish how the role and working process of a composer using real-time electronic systems differs from the traditional means of composing. The objective was to produce new compositions (or musical games) that make use of the extended possibilities of automatization of musical parameters, as well as the widened virtuoso qualities of the performers, and finally to develop a new instrument (procedural music engine) and new performance techniques. To reflect the research object (*Climb!*), the results were also compared to other existing examples of interactive compositions, sonic artworks and computer game scores exploiting procedural or dynamic music systems.

9.4.1. What Is Climb!?

As discussed above, *Climb!* is a nonlinear composition that combines the concepts of a classical virtuoso piece and a computer game. The overall concept and narrative structure were originally presented by Kallionpää and Gasselseder in 2016 (pp. 42–48), and the work was then realized as a collaboration with the Mixed Reality Laboratory of the University of Nottingham. *Climb!* consists of three different "macro compositions" and a number of "micro compositions" (events) that reflect the narrative of the game, which tells the story of an avatar who climbs to the top of a mountain. In the end of the very first movement (*Basecamp*), the performer chooses between three different endings (musical motifs) that all function as codes that trigger the interactive system to take the performer to a distinctive location on the score (or to enter one of the "macro compositions"). Every performance starts with the *Basecamp* excerpt and ends at the *Summit*, but everything that happens in between varies every time. The "macro compositions" symbolize the paths that one can take and the "micro compositions" signify the smaller-scale happenings that take place within the journey. Musical challenges or choices are presented: they determine, depending on how the performer plays, whether they continue to progress along their current path or are diverted onto another one. Written for a Disklavier grand piano, the work engages the pianist in a musical dialogue where there is no fixed musical form. As in the game "snakes and ladders," or in the story books where the reader can decide the next development of the story line, the performer's/gamer's actions define how the story continues.

As the performer plays key passages, both the music and the instrument respond, jumping to new points in the score, transforming the sound or independently playing along with them—physically in the case of the Disklavier. This has been enabled by the bespoke software engine "Muzicodes" that enables the musical fragments embedded in the score to work as codes that allow for branching the form and triggering various functions and effects, such as activating the automatized parts of the Disklavier or simulating weather conditions by applying effect filters onto the audio stream. These weather simulations involve further partially randomized methods of sound processing (changing the sonic color) that are being constantly applied. Under some circumstances these colorations may pose an additional challenge in playing the codes correctly and may thus have an impact on the player's success and landing in different passages on the score. This, as well as the intricate nature of its branching system, is why every performance of *Climb!* is always different. The Disklavier is played by both a human performer and the automatized system alongside the pianist, which is driven by a MIDI input (Kallionpää and Gasselseder 2016, p. 465). The technical rider of the piece incorporates a Yamaha Disklavier Grand, interactive system (including the digital score system MEI and the Muzicodes software developed at the Mixed Reality Lab of the University of Nottingham), a smartphone app and an online archive where all the performances are stored.

Interactive visualizations projected behind the performer add to the immersive experience of the audience and support proper synchronization of perceptual stimuli stemming from the acoustic and electronic elements of the performance. The smartphone application enables the listeners and performers to revisit the previous performances of *Climb!*. By using the app the concert audience can follow the development of the musical structure and the narrative of the game in real time. All the routes ever taken in the game are stored in the online archive so that listeners can compare the decisions taken by the performers. Furthermore, instead of just a collection

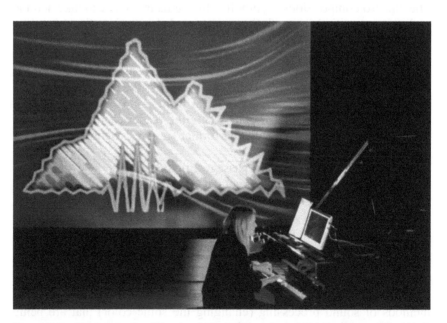

Figure 9.2 The Interactive-Visuals of *Climb!* Captured at the Premiere of the Work on June 9, 2017. Pictured here: Maria Kallionpää, piano and composition.

of the recordings of the piece, the composer regards the online archive to be an artistic entity on its own right, in which all the performances of the pianists are simultaneously present.

9.4.2. Climb! *A Process From the Perspective of a Composer-Performer*

Computer programs are being created to address the needs of specific composition projects. Such technological explorations can often be seen as an integral part of the artistic expression that cannot be distinguished from the rest of composing the work (Kallionpää 2014, p. 39). In the case of *Climb!*, the composition and its interactive system were created simultaneously. On top of the artistic needs affecting to the development of the technological solutions, new musical ideas inspired by the technology also enriched the composition itself (Kallionpää and Gasselseder 2016, p. 468). An interactive system like this resembles "an invisible chamber music partner" leading some musicians to claim that electroacoustic works can never be treated as solo compositions, but should instead be seen as "concertos in a modern guise" (Schwartz 1989, p. 102). This statement was also confirmed in some of the responses of the anonymized questionnaires that were used in the premiere to collect information from the audience members. Some listeners de facto reported that the piece seemed like a concerto for a human pianist and an electronic system (according to data collected during the premiere of the piece on June 9, 2017, at the Djanogly Recital Hall, University of Nottingham).

The role of the performer was present throughout the composition process, and the roles of the composer and performer constantly overlapped. Rather than completing a piece that would then be premiered, the composing process of *Climb!* was rather different. The work was performed in various stages of its development and improvements were made throughout the process, both in terms of the composition, its structure and applied software. The first drafts and a prototype were already presented in December 2016, when only the very first musical fragments were completed and not all the technological functions were yet defined. This kind of an approach enabled the composer and technology development team to obtain better results by having a chance to regularly test the system. For example, more branching options of the musical material were added later, and other kind of improvements, such as disabling the possibility of getting stuck in a loop, were also included in the system. To let the audience listen to two versions of *Climb!*, the piece was usually played twice in each concert. During the premiere, it became apparent that more branching options were required, as the performer ended up playing two very similar performances. This was counter to the conceptual goal, which was to create an interactive work that would not repeat itself. Moreover, the questionnaires discussed above were distributed to the audience between the two performances and the concept of the piece was explained to them. In this way, the research team could gather information about the audience's perception with and without prior knowledge of the composition and its mechanisms.

The need for system improvements was discovered in the circumstances of the concert. Such problems do not necessarily come up in the laboratory setting because one does not need to deal with the changing acoustic conditions, varying concert instruments and time pressure. Thus, each concert performance was analyzed and problems fixed accordingly. To optimize the functionality of such an interactive system in a real-time situation, it is important to design it for maximum stability. The work has now reached its final form as a composition but it still requires a more portable system that can be run by any sound technician outside of the immediate research team. This will allow for wider distribution of the piece among individual performers, concert organizers and festival programmers.

9.4.3.   Dealing With Space in Climb! Performances

As discussed above, in electronic music the acoustic space and qualities have to be defined by the composer. She or he decides how the sound sources transmit the sonic material (for example, one needs to know how many speakers will be used and how the sound material is divided or moved between them). *Climb!* is a work that should be performed in a concert hall setting, but one in which the electronic music characteristics apply. Some aspects of the piece are virtual. For example, it contains the previously discussed story line reminiscent of that of a computer game, and sound effects that support the narrative and that were designed with the help of the computer. The randomized sound processing that symbolizes the weather effects, directly affects the sonic outcome, and the performer should adapt their articulation, style and even tempi accordingly. Artificial creation of acoustic circumstances brings its own challenges. The pianist cannot just work in the usual dialogue with the actual space in which she or he is performing, but needs to her or his cues from the interactive game system. The composer/performer of *Climb!* has been subjected to these challenges within its several performances. For example, whereas the sound effects were quite audible for the performer in the premiere (and thus the virtual space was more easily navigable), hearing the effects in another performance taking place at the Reflective Conservatoire Conference in London (January 21, 2018) was more difficult because of the positioning of the speakers. Thus, the pianist could not really take the sound processing into account during the latter performance, which is why some incoherence appeared between the piano texture and the sound effects. As reported in the questionnaires collected during that performance, some listeners found this somewhat disturbing. Similar problems with interaction with a live electronic system are also discussed by McNutt (2003, p. 298), who refers to a situation in which she was to improvise with a live system but could not actually hear it. When considering *Climb!*, the fine balance between the live piano performance and the virtual environment is crucial. According to the audience questionnaires, it is essential that on top of the pianists' performance, the interactive visual contents and sound processing work seamlessly. Any delays or discrepancies between different parts are conspicuous.

In order to counteract the reported shortcomings of hall acoustics, monitoring and speaker positions, the authors set out to develop a solution that would warrant a stabile as well as easily replicable balance between acoustic and electronic elements in order to ensure optimal conditions for expressive performances. Apart from increasing spatial awareness, the aims were twofold: (1) to empower the performer by providing an authentic virtual acoustic space that mixes seamlessly with physical acoustic stimuli, and (2) to enable the comparative analysis of the interaction between players, instruments and room acoustics. The reasoning behind these aims stemmed from two main desiderata—absorption and hybrid listening—that had come to our attention during the design and testing phase of the project.

The first desideratum, absorption, directly related to the initial aim to integrate principles of gamification not only in structural terms of the composition but also as part of the sonic realization during the performance. Gamification requires the immersion of the player rather than the demanding that the player make a conscious effort to mesmerize an audience. Games achieve the former by imposing a different context on performed actions as compared to non-game conditions. "Playing" music as part of a game means to aim for success in game play—and not to play for the sake of creating pleasing music or to tailor it to an audience's specific tastes. Whether the audience will be taken on this figurative journey depends on the state of involvement of the player/performer. Such a rationale follows the logic of "absorption"; a term coined by French philosopher Denis Diderot (1757) who contended that an audiences' appeal of expression and transportation lies in actors not acknowledging the stage or the presence of an audience. Only by noticing its own absence in the authentic (i.e. unaffected) appearance of the actor/performer, can an audience transcend and empathize with the expressive accent structure underlying the situational context. If a performer can easily become absorbed/immersed in the expressive quality intended, she or he is more likely to succeed in the execution of such rendering. Yet, difficulties in reaching this state may arise when dealing with different situational demands ranging from acoustics, audience noise, and unpleasant tactile feedback to bad lighting, to name only a few. Contextual information gained through gamification helps to maintain strategic focus by way of increased attention in response to challenge. Hence, a benefiting factor of gamification may be seen in the shifting of focus from "that one is doing" to "what one is doing"; essentially alleviating one's focus on the actual performance and replacing it with the feedback gained from a larger system/ecology. By not having to think about its execution but only its situational affordance, the performance gains in processing fluency (as stipulated by the dual processing model) and expression by virtue of potentially augmenting the performer while simultaneously supporting a performative augmentation in the process (see Kallionpää and Gasselseder 2016; Gasselseder 2016). The second desideratum, hybrid listening, directly connects to the concept of immersability in that it intends to tackle the merging of acoustic and electronic sound sources into one super instrument (see Kallionpää 2014 for further discussions on the concept of super instruments). Being presented with an acoustic as well as an electronic source at different spatial locations and with unknown translational accuracy at the audience perspective, the performer finds it difficult to make judgements and adjustments in tonal quality. This becomes even more apparent in performance scenarios of electroacoustic music where the performer is not provided with appropriate monitoring solutions—often being left alone with the acoustic instrument while the only set of speakers is pointing towards the audience.

Our bespoke binaural room simulation platform, named "Virtualizer," aims to solve these desiderata by accurately reproducing any acoustic space and speaker configuration via headphones at zero latency. Thanks to a personalization process that incorporates a short one-time measurement of binaural impulse responses capturing the head-related transfer function (HRTF) for each performer, it is possible to create an exact rendition of an existing room and speaker setup that can be further interpolated into any speaker combination (e.g. 64 channels) and head movements by way of modeling a virtual array of coincident signals. Using this method while listening on headphones, a multichannel audio recording sounds perceptually indistinguishable from listening to speakers in a physical room. Preliminary testing indeed suggests elevated levels of absorption and subsequent involvement of the performer. Furthermore, the method also considers the aforementioned desideratum of hybrid listening by enabling the placement of virtual sound sources inside the acoustic instrument. This is realized by applying a measured frequency response compensation onto the audio signal in order to allow a close to linear transmission via bone-conduction (BC) headphones. While these types of headphones typically lack in fidelity, they offer good comfort levels and do not hinder acoustic sounds from entering the ear canal due to the positioning of drivers behind the ear.

Thanks to the acquired compensation curves, virtually all drawbacks of BC transmission can be eliminated so that we feel confident in proposing BC headphones as an optimal monitoring tool for electroacoustic music performance. In the case of *Climb!* we opted to play back the electronic mix elements over BC headphones while leaving the acoustic elements to the natural hearing of the performer. This approach requires that the captured impulse responses directly mirror the acoustic space in which the performance is planned to take place. Ultimately, rather than trying to find a sonic balance between acoustic and electronic elements, the performer is confronted with a mix that makes it seem as if both elements were being emitted from a single super instrument.

9.5.   Conclusions

Whether or not one pays attention to it, the concept of space forms an important part of musical culture. It can be used as part of the musical dramaturgy, compositional structure or even to represent an independent musical instrument. The space can be electronic, naturally acoustic, a virtual environment or a mixture of artificially created sonic entities. The presence of space and spatialization applies to both the composers and performers, who will need to define their relation to it and learn how to navigate within it. The performers need to find a way to adapt their performing techniques to the prevailing acoustic circumstances, regardless of whether they are naturally acoustic or artificially generated. Moreover, the space can be used to formulate a musical structure, similar, for example, to the polychoral style of the Renaissance Era, or the way that the contemporary spectral composers have engaged it to emphasize the harmonic structures of their works. Furthermore, with the growing interest in interactive computer games (and audio games), the mechanisms of interactive computer game scoring systems, as well as the possibility of generating virtual environments, will provide potential for extending the technical means and modes of expression in classical music. Thanks to developments in software design, the qualities of space can be modified or challenged. The acoustic characteristics of a concert hall do not limit the musicians of the twenty-first century, but can be replaced, replicated, optimized or used to enhance the expressive potential of a musical composition and the performing experience of musicians today.

References

Blakemore, S.J. and Decety, J., 2001. From the Perception of Action to the Understanding of Intention. *Nature Reviews Neuroscience*, 2, pp. 561–567.

Bruner, J.S. and Postman, L., 1949. On the Perception of Incongruity. A Paradigm. *Journal of Personality*, 18, pp. 206–223.

Byrd, D. and Crawford, T., 2002. Problems of Music Information Retrieval in the Real World. *Information Processing and Management*, 38, no. 2, pp. 249–272.

Carver, A.F., 1988. *Volume 1. The Development of Sacred Polychoral Music to the Time of Schütz*. Cambridge: Cambridge University Press.

Chadabe, J., 1996. The History of Electronic Music as a Reflection of Structural Paradigms. *The Leonardo Music Journal*, 6. Cambridge, MA: The MIT Press.

Collins, K., 2008. *Game Sound. An Introduction to the History, Theory, and Practice of Video Game Music and Sound Design*. Cambridge, MA: The MIT Press.

Diderot, D., 1757. *Le fils naturel ou les epreuves de la vertu*. Comedie en cinq actes, et en prose. Viewed Jan 16, 2018 http://books.google.co.uk/books?id=ansHAAAAQAAJ

Essl, K., 1992. *Lexikon-Sonate. Algorithmic Music Generator*. Infinite Realtime Composition for Computer-Controlled Piano.

Essl, K., 2006. Algorithmic Composition. In: Collins, N. and d'Escrivan, J. (Eds.), *The Cambridge Companion to Electronic Music*. Cambridge: Cambridge University Press.

Gasselseder, H.P., 2015. Re-Sequencing the Ludic Orchestra: Evaluating the Immersive Effects of Dynamic Music and Situational Context in Video Games. In: Marcus, A. (Ed.), *Design, User Experience, and Usability: Design Discourse: 4th International Conference, DUXU 2015, Held as Part of HCI International 2015*, Aug 2–7, Proceedings Part I. Los Angeles, CA: Springer. Lecture Notes in Computer Science (LNCS), No. 9186, pp. 458–469.

Gasselseder, H.P., 2016. What You Hear Is Where You Are Is What I Hear. Optimising Immersive Experiences for Interindividual Differences. In: Bowen, J., Diprose, G. and Lambert, N. (Eds.), *EVA London 2016: Electronic Visualisation and the Arts*. London: BCS Learning and Development Ltd. Electronic Workshops in Computing, Electronic Workshops in Computing, pp. 61–69.

Grimshaw, M. and Garner, T., 2015. *Sonic Virtuality: Sound as Emergent Perception*. New York: Oxford University Press.

Jeannerod, M., 1997. *The Cognitive Neuroscience of Action*. Oxford, UK: Blackwell.

Kallionpää, M., 2013. *Le Safari de la Mort: Musical Score*. Helsinki: Music Finland.

Kallionpää, M., 2014. *Beyond the Piano: The Super Instrument. Widening the Instrumental Capacities in the Context of the Piano Music of the 21st Century*. DPhil thesis, University of Oxford.

Kallionpää, M. and Gasselseder, H.P., 2016. The Imaginary Friend: Crossing Over Computer Scoring Techniques and Musical Expression. In: Bowen, J.P., Diprose, G. and Lambert, N. (Eds.), *Electronic Visualisation and the Arts (EVA 2016)*, Jul 12–14. London: BCS Learning and Development Ltd.

Kuusisaari, H., 2018. Sali Soi Humanismia. *Rondo Classic*, Mar, no. 3, pp. 24–25.

Kwon, M., 2002. *One Place After Another: Site-Specific Art and Locational Identity*. Cambridge, MA: The MIT Press.

McNutt, E., 2003. Performing Electroacoustic Music: A Wider View of Interactivity. *Organised Sound*, 8, no. 3, pp. 297–304.

Moscovich, V., 1997. French Spectral Music: An Introduction. *Tempo: A Quarterly Review of Modern Music*, no. 200, pp. 21–27.

Nonken, M., 2014. *The Spectral Piano: From Liszt, Scriabin, and Debussy to the Digital Age/with a Contributory Chapter by Hugues Dufort*. Cambridge, UK: Cambridge University Press.

Oestreich, J.R., 2017. Review: 3-D Enhances an Enigmatic New Opera. *The New York Times*. Viewed Jan 17, 2018 www.nytimes.com/2017/09/22/arts/music/blank-out-opera-park-avenue-armory.html

Plans, D. and Morelli, D., 2012. Experience-Driven Procedural Music Generation for Games. In: *192 IEE Transactions on Computational Intelligence and AI in Games*, 4, no. 3. Viewed Jan 8, 2018 http://axon.cs.byu.edu/Dan/673/papers/plans.pdf

Popper, A.N. and Fay, R.R., 1997. Evolution of the Ear and Hearing. Issues and Questions. *Brain, Behavior, and Evolution*, 50, no. 4, pp. 213–221.

Pulkkis, U., 2013. *1900-Konsertto Pianolle ja Genelec-orkesterille, Teoksen Tekninen Toteutus. Kompositio* 3/2013. Helsinki: Kompositio.

Schwartz, E., 1989. *Electronic Music: A Listener's Guide*. New York: Da Capo Press.

Taylor, B., 2016. *The Melody of Time: Music and Temporality in the Romantic Era*. New York: Oxford University Press.

Tenney, J., 1993. *Form 1: Musical Score*. Toronto, ON: Canadian Music Centre.

Further Reading

Antescofo, 2007. Viewed Jan 27, 2018 http://repmus.ircam.fr/antescofo Live score following system developed at IRCAM.

Benford, S., Greenhalgh, C., Hazzard, A., Chamberlain, A., Kallionpää, M., Weigl, D. and Page, K.R., 2018. Designing the Audience Journey Through Repeated Experiences. In: *ACM CHI Conference on Human Factors in Computing Systems*, Apr 21–26. Montreal, Canada: Proceedings.

Davies, H., 2014a. *Towards a More Versatile Dynamic-Music for Video games: Approaches to Compositional Considerations and Techniques for Continuous Music*. DPhil Thesis, University of Oxford.

Davies, H., 2014b. *DPhil Composition Portfolio*. Oxford: University of Oxford.

Greenhalgh, C., Benford, S. and Hazzard, A., 2016. Muzicode$: Composing and Performing Musical Codes. *Audio Mostly 2016*. Norrköping, Sweden, Oct 4–6, Proceedings. New York: ACM Press.

Kallionpää, M., Greenhalgh, C., Hazzard, A., Weigl, D., Page, K. and Benford, S., 2017. Composing and Realising a Game-Like Performance for Disklavier and Electronics. In: *New Interfaces for Musical Expression (NIME'17)*. Copenhagen, Denmark, May 15–18. New York: ACM, pp. 464–469. http://www.nime.org/proceedings/2017/nime2017_paper0088.pdf

Sweet, M., 2015. *Writing Interactive Music for Video Games. A Composer's Guide*. Upper-Saddle River, NJ: Pearson Education.

10

Interactive Music on the Web

Sang Won Lee, Benjamin Taylor and Georg Essl

10.1.   Introduction

In this chapter we will be discussing Web Audio. Web Audio is a Java-Script API that enables audio synthesis and analysis inside a web browser. This chapter aims to present both introductory material and a discussion of a range of completed artistic projects realized with Web Audio. Our goal is to give you enough of a foundation that you can begin making your own creative projects with Web Audio. The range of uses for Web Audio is vast. It can be used as a core technology to provide audio synthesis and analysis capabilities in such applications as games, art installations, live performances, mobile music and others.

The history of audio on the web stretches back to before the advent of the Web Audio API. You can learn more about these historical developments from William Duckworth's 2005 book *Virtual Music: How the Web Got Wired for Sound* (Duckworth 2005). The Web Audio API arises also out of a long history of audio programming languages, which you can read about in detail in Ge Wang's chapter on the subject (Wang 2008). The main predecessor of Web Audio was Adobe's Flash Player, which could render interactive, media-rich content. However, Flash is disappearing rapidly, since its functionality is increasingly implemented directly in web browsers through such technologies as HTML5, CSS and JavaScript.

Before we dive into technical discussions, we will explore where Web Audio shines, and where it is probably not the best choice. This will set the stage for the rest of the chapter, which is divided into two parts: an introduction to Web Audio programming is presented in section 2, and a discussion of extensive applications and artistic projects realized in Web Audio follows in section 3. Understanding the basics of sound synthesis will be helpful for understanding our discussion here. Furthermore, an introduction to JavaScript and browser-based programming is beyond the scope of this chapter. There is a wealth of online resources and books available that provide great introductions to JavaScript (Haverbeke 2014).

10.1.1. Why Web Audio?

Web Audio offers many advantages for writing digital audio applications and interactive music projects. The web is a highly accessible, cross-platform distribution channel. It opens doors for social and collaborative music applications that connect users from around the world. Developers can take advantage of a wealth of web development resources in the form of tutorials and open-source toolkits. Through all of these assets, Web Audio lowers barriers in development, distribution and participation in comparison to alternatives available to developers and artists.

Foremost among Web Audio's advantages is the accessibility of web-based projects. Applications built on the web are hosted on a server and can be accessed at any time over the internet. This shifts the paradigm from installing an application to visiting a website that hosts an application. For example, one can search for "online guitar tuner" using a search engine on a smartphone and find a web app to tune the strings of one's guitar. This allows users to access audio technology that was previously available only as physical hardware or downloadable software. A working example is available at (Lewis 2015). Users no longer need to own audio gear to accomplish certain tasks, since they can find and use suitable applications online.

The web lends itself naturally to collaborative and social applications. Users already interact with other people in their browsers through email, social media, blogs and other online communities. The social usage of web browsers over decades has created interaction norms that users have come to expect of web-based applications, such as the ability to share information via social media, invite friends to web apps, meet new people with similar interests and collaborate to create new content. To that end, web development offers a number of methods for transmitting, sharing and synchronizing data to support these types of connections, making the web immediately conducive to collaborative music making.

The advantages of Web Audio can be seen also from a developer's perspective. Web-based applications are *cross-platform*, meaning they depend not on the technologies of a particular underlying platform (e.g. an operating system), but on common technologies like HTML, CSS and JavaScript, which are implemented by web browsers. Because web browsers are available on a range of platforms, web-based applications transcend hardware and operating system borders, the only requirement being that a (standards-compliant) web browser exist for a given device. Web Audio can therefore function on an assortment of hardware, including smartphones, tablets, laptops, traditional desktop computers and—at least in theory—any future hardware that can run a web browser. This greatly benefits developers, who can reach broader audiences across various contexts, from a user on a laptop in their living room to a pedestrian using their smartphone on the street. This makes web applications potentially *cross-context*, as well as cross-platform.

More broadly, web developers can take advantage of the abundant resources available for developing web-based applications. The inherent openness of Web 2.0 naturally encourages users and developers to generate technical resources that help their peers create online content and advanced interactive applications. These resources include the source code of existing web pages, online tutorials that incorporate interactive demos and open-source software projects. In particular, the number of available programming libraries that can expedite the development of web-based interactive media is constantly growing. Third-party programming libraries and corresponding application programming interfaces (APIs) help developers to quickly scaffold interactive music and audio applications. These include libraries for sound synthesis (Roberts, Wakefield and Wright 2013; Mann 2015), musical interfaces (Choi and Berger 2013; Taylor et al. 2014) and audio analysis (Rawlinson, Segal and Fiala 2015). Researchers and practitioners have additionally made efforts to export programs written in other programming languages (such as Csound, Pure Data and Faust) to modules that can be embedded in a web application (Lazzarini et al. 2015; Letz, Orlarey and Fober 2017; Piquemal 2017). The ease of developing software for the web has even encouraged developers to use Web Audio in building desktop applications (McKegg 2016).

Web Audio's integration with the internet makes it the go-to audio engine for a variety of forms of new media art, including online sound artworks, internet art installations and other projects. For example, if an artist wants to sonify real-time Wikipedia edits, Web Audio offers an audio engine for that project right in the browser. Building these projects on the web also enables artists to benefit from an abundance of online audio-visual media databases, which offer a virtually infinite number of sonic materials that artists can access.

In summary, Web Audio may be a good option for an audio application if easy access, broad distribution, social aspects, interactivity or media-rich content are priorities. Web Audio is also the primary option for developing web-based sound art projects with the internet, internet data or internet culture as subject matter. Finally, Web Audio might be a good option for those seeking to develop cross-platform applications, or those looking for a simpler development process and want to take advantage of the many resources and toolkits available to web developers.

10.1.2. Why Not Web Audio?

Developers should understand the technical and practical implications of creating web-based music and audio applications before choosing Web Audio as a development platform. As a web-based application runs within a web browser, its computational performance is limited compared to that of native applications. In particular, JavaScript code execution is slower than that of programs written in compiled programming languages, such as C or C++. Therefore, for an application that runs computationally intensive algorithms, such as signal processing techniques for real-time audio input or real-time machine learning techniques, the performance required may exceed the capabilities of a web-based environment. In these situations, building a native application may be preferable or even necessary. In particular, web browsers are designed to prioritize a responsive user experience, so all computations run in a single thread that also responds to user interaction.[1] Therefore, any program code that a developer writes in addition to the basic Web Audio pipeline (e.g. a spectrogram visualization running in the main thread and retrieving frequency-domain data from an AnalyserNode) may be interrupted by a user interacting with the user interface, such as by resizing the browser window. These interruptions can result in delays and glitches, which can be significant in music. However, as algorithms become more efficient and increased computing power becomes more widely available, this issue will be less significant in the long term. A potential approach to address this issue in a web-based application is to use cloud computing, which may require additional development to distribute the computing process and to coordinate distributed processes (Hindle 2014, 2015).

Because an internet connection is essential for accessing a web-based application, the use of web-based applications creates the additional technical requirement of connectivity. This requirement may be a barrier in situations where a wireless connection is not available, or to a population without access to affordable internet connectivity. In particular, relying on web-based music applications in performance contexts introduces a number of risks in practice. In such a scenario, having a separate, preferably wired, connection is desirable but can be costly or inaccessible to certain devices. Furthermore, traditional music performance venues may not have an internet connection because performers have historically had no need for such connectivity. Rather, the expectation is that using electronic devices in a concert hall is prohibited, particularly during a performance. Even with a public internet connection available, musicians may not want to rely on a wireless connection that may also be used simultaneously by a large audience, especially if an application requires a continuous, high-speed connection. For example, a performer may wish to stream real-time audio and video from a remote performance site.

While platform independence can be a major benefit of web-based application programming, the uncontrollable variability of the environment in which an application is used can present a challenge. For instance, as opposed to a sound artwork exhibited in a gallery, in which the artist can control the environment and design the trajectory of interaction in order to create a desired experience, web-based media may be consumed using various devices and in various contexts. Relevant factors include the devices that users choose and the contexts in which the devices are used. Differences in form factor, input method (touchscreen versus keyboard and pointing device) and potentially computational power from one device to the next can be a challenge for sound designers. Additional development may be necessary to adapt to these differences, a practice known to user interface designers as "responsive design." The cross-context property in the usage of web browsers may account for the richness and open-endedness of web-based music and audio applications.

Additionally, web-based interactive media in general shares challenges in its maintenance with digital information and software (Hedstrom 1997). Platform independence may not be guaranteed over a long period of time as platforms develop and change. The additional efforts needed to preserve both backward and forward compatibility can be a threat to the longevity of an application. For example, browser updates, changes to web standards—especially to the Web Audio API—security policies and changes to any third-party libraries used in an application may break functionality or render the application unusable until it is updated. If not maintained continuously, web-based interactive media lacks a way to archive the work in its functional form; code text can be archived, but without also preserving the environment in which the code is executed, the operability of the archived code cannot be guaranteed. For example, many forms of artwork or information exist as physical artifacts (paintings and sculptures) or have canonical protocols for archiving (sheet music, academic papers, films, music recordings). However, it is yet unknown how we can best archive or preserve web-based media or software in general, and so developers must make continuous efforts to maintain the usability of media they create. If they do not, their applications may cease to function as intended; in the worst-case scenario, users will no longer be able to access these applications at all without outdated browsers and devices. In addition, it can be costly to maintain a server to host a web application, depending on its scale (database, hard disk space, computational power required) and one cannot guarantee the availability or cost of third-party services (such as Github, which is free of charge at the time of writing, but is limited to hosting a static website and may charge hosting fees in the future). In general, the ephemeral nature of web-based interactive media can be problematic in the long term, and may be a threat to the platform. Finally, a developer should consider the cross-browser compatibility of a web application for different types of web browsers.

We do not want to discourage readers universally from using web-based interactive applications. Rather, we wish to familiarize developers with the advantages and disadvantages they need to consider in deciding whether or not a web application is appropriate for the presentation of their sound-based interactive media. For example, researchers and musicians have frequently used web-based music applications for audience participation in concerts, as this makes the application easier to distribute and more accessible, regardless of hardware specifics. In this case, the immediate benefits of Web Audio lie in its ubiquity and accessibility, which have motivated artists to discover novel performance practices. Developing a native application is always an option that can easily substitute for web-based media while affording developers more technical opportunities.

10.2. Fundamentals of Web Audio

10.2.1. Web Audio API: Basic Use

The general structure of the Web Audio API mimics that of virtually all audio processing systems. Audio data flows from signal sources (microphones, signal generators) through signal processing units (filters, compressors, analyzers) to a destination (speakers, headphones). Within this basic framework there is great flexibility, and many aspects of the sound processing setup can be manipulated and controlled by connecting the script to user interaction. For example, the frequency of an oscillator may be made subject to mouse interactions.

The *audio context* is the basic canvas on which a specific Web Audio structure is realized. Given that multiple web pages may have different audio needs—and that even within a web page, one may want to switch between audio behaviors—these contexts provide a means of specifying the active structure and associated behaviors. The first step for virtually all Web Audio applications, then, is to create a new context.

Once an audio context is established, it can be populated with content. The flow of audio signals is controlled by a connected graph that links audio nodes. There are three types of audio nodes: *sources*, *processing units* and *destinations*. Sources generate new audio streams (e.g. according to a waveform algorithm or by reading a buffer). Processing units modify incoming streams.

10.2.1.1.   Hello Web Audio API

To understand this concretely, consider this simple Web Audio program:

```
1 const AudioContext = window.AudioContext
                       || window.webkitAudio
                       Context;
2
3 const actx = new AudioContext();
4
  const oscillator = actx.createOscillator();
5
  oscillator.connect(actx.destination);
6
  oscillator.frequency.value = 440;
7
  oscillator.start(0);
8
9
```

The first line grabs the audio context of the browser. Then, it creates a new instance of the context for this audio setup.[2] In this simple example, we want to create a sine tone. An oscillator (line 6) provides the source of sinusoidal oscillation. In order for it to make sound, it needs to be connected to the speaker output. This is represented by the destination of the audio context (line 7). We change the frequency of the oscillator, then get going by telling the oscillator to start oscillating. The default oscillator type outputs a sine wave, but other types are also available ("sine," "square," "sawtooth," "triangle"). Every Web Audio structure has this basic nature of connecting audio nodes. More complex Web Audio programs simply use more nodes and connect them in more complex ways. They may trigger different nodes dynamically over time and based on user interaction.

10.2.1.2.   Creating a Signal Flow

Let's take a look at how we can build a slightly more complex audio graph by adding some processing units. This example assumes the creation of an audio context, as demonstrated in the previous example.

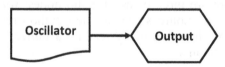

Figure 10.1   Routing of Audio Nodes in the First Example.

```
 1 const oscillator = actx.createOscillator();
 2 oscillator.type = "triangle";
 3 oscillator.frequency.value = 440;
 4 oscillator.start(0);
 5
 6 const gain = actx.createGain();
 7 gain.gain.value = 0.1;
 8
 9 const filter = actx.createBiquadFilter();
10 filter.type = "lowpass";
11 filter.frequency.value = 700;
12 filter.Q.value = 5;
13
14 oscillator.connect(gain);
15 gain.connect(filter);
16 filter.connect(actx.destination);
```

Here, we add a volume node and a low-pass filter in between our oscillator source and our destination speaker. A gain node is created on line 6, which scales the values of the audio stream. A filter is then created on line 9, and its type, frequency and Q are set. Finally, we create our signal flow by connecting the audio nodes together (lines 14–16). At this point, we have a versatile Web Audio synthesizer; through changing the parameters of these nodes, we can create volume envelopes and timbre envelopes and play different pitches. Note that line 4, which starts the sound synthesis, could have been mapped to a user interaction, such as a mouse click event.

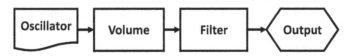

Figure 10.2   Routing of Audio Nodes in the Second Example.

10.2.2. Integrating Interaction

The core input devices through which we, as users, interact with a web browser are some combination of a keyboard, a mouse and a touchscreen, just as in native software applications. In addition, web applications can access a device's integrated sensors, such as microphones, cameras, accelerometers and gyroscopes, as well as other types of input devices that may be connected, such as MIDI controllers and audio interfaces. The majority of applications focus on using the basic controllers: keyboard and mouse or touch surface.

We can make our synthesizer interactive by using the mouse to control its frequency. To do this, we must create an *event listener* to monitor mouse movement (line 3). We give that event listener a *callback function*, which is the second parameter of addEventListener (lines 3–9), and which will be executed every time a user moves the mouse on the web page. Inside the callback function, we first find the mouse's current distance from the top of the viewport of the client page (line 5), then find the current height of the browser window (line 6), and finally use a little math to map the mouse's *y* position to a suitable frequency value (in this case, one within a range of 20–1000 Hz), and set our oscillator's frequency to that number (line 8).

```
1 var y, height;

2

3 document.addEventListener('mousemove',
  function(e) {

4

5    y =  e.clientY;

6    height = window.innerHeight;

7

8    oscillator.frequency.value = (y/height)  *
     980.0 + 20.0;

9 });
```

10.2.3. Beyond the Basics

Here, we will review techniques for sequencing audio events, loading audio files and receiving audio input from a user's microphone.

10.2.3.1. Timing

Timing Web Audio events can be a challenge. JavaScript is by nature asynchronous, meaning code events are added to a queue and executed as soon as resources are available. Therefore, the timing of events is imprecise. There is no guarantee that a code block will be executed at any particular point in time. This presents a substantial predicament in writing Web Audio programs, which may require precise rhythmic timing of sound events. To remedy this, the Web Audio API provides its own internal clock. All Web Audio events can be scheduled relative to this clock, available in the API as currentTime. The currentTime begins at 0 when the audio context is created, and increases incrementally, returning the number of seconds since the context was created. Audio parameters can be changed in relation to this time.

Let us now look at how we could use the Web Audio clock to create a rhythmically precise melody with our synthesizer by changing the frequency parameter of an oscillator.

```
1 let now = actx.currentTime;

2

3 oscillator.frequency.setValueAtTime(400, now);

4 oscillator.frequency.setValueAtTime(500, now + 1);

5 oscillator.frequency.setValueAtTime(600, now + 2);
```

On the first line, we learn the value of the current time (in seconds) according to the Web Audio clock and store it in a variable called now. Then, we schedule three events to happen: an event to change our synthesizer's frequency to 400 Hz immediately, an event to change its frequency to 500 Hz one second later (now+1) and an event to change its frequency to 600 Hz two seconds later (now+2). There are other audio parameter API functions that relate to time. An important one to note here is cancelScheduledValues(startTime), which removes all scheduled parameter changes after the given time. Thus, timing information can be dynamically added and removed from any audio parameter, resulting in a rather flexible framework for managing time-related events. In practice, developers must often schedule just a few events at a time rather than scheduling them all in advance, as the program may need to change audio events in response to a user's interaction. For this case, Wilson offers a programming trick that combines the JavaScript clock with the Web Audio clock, as an alternative that can look ahead only for a shorter time window and still be precise in timing (see Wilson 2013 for more detail).

10.2.3.2. Playing an Audio File

There may come a time when you wish to incorporate audio files into your application. This is possible using Web Audio's *buffer source* node, which acts as an audio file player. The buffer source node contains a buffer property, and can output that buffer's values as an audio stream. Playing an audio file requires a bit more code than our previous examples, because we must first request the audio file from a server using an asynchronous request. This is the complete code you will need to play an audio file:

```
1 var player = actx.createBufferSource();
2 player.loop = true;
3 player.connect(actx.destination);
4
5 function loadBuffer(node, url, play) {
6
7   var request = new XMLHttpRequest();
8   request.open("GET", url, true);
9   request.responseType = "arraybuffer";
10   request.addEventListener("load", function(){
11
12     if (request.status === 200){
13       actx.decodeAudioData(request.response,
       function(buffer) {
14         node.buffer = buffer;
15         // in case you want to play the
         sample immediately:
16         if(play) node.start();
17       })
18     } else {
19       console.log("Could not load file");
20     }
21   });
22   request.send();
23 }
24
25 loadBuffer(player, "./sound.mp3", true);
```

In this example, we create a buffer source node and connect it to our speaker destination (lines 1–3). However, our buffer source node's buffer is currently empty, so we need to grab a file from the server and place that file's contents into our buffer. We do this using an asynchronous XMLHttpRequest (line 7), which will get our buffer from a specific URL on our server (line 8). We tell our program to let us know when the file is finished loading (line 10), and if the audio file was found and loaded successfully. We then decode the audio file into data that our buffer source node will understand (line 13). Finally, when the audio file is decoded, we tell our buffer source node to use the resulting buffer (line 14) and play right away, if instructed to do so. As you can see, this is a fairly complex process, so it is a good idea to encapsulate this code in a function that you can call for any file you wish to load. We do this here by wrapping this code in the function loadBuffer. Note that the audio is playable not immediately after the function call (line 25) but after the audio is loaded and decoded (line 14).

10.2.3.3. Microphone Input

A final audio source that we will demonstrate here is the user's microphone. We can open an audio stream from the user's microphone using the browser's getUserMedia API. Note that this is not part of the Web Audio API; rather, it is a separate browser API for accessing the video and audio hardware of the user's device. However, we can connect this audio stream to our Web Audio graph using Web Audio's media stream source node (line 3, below). The media stream source node can then be connected to our output destination (line 4) or any other processing unit. *N.B.: sending microphone input directly to speaker output may create feedback.*

```
1 navigator.getUserMedia({audio: true},
  function(stream) {

2

3   var microphone = context.
    createMediaStreamSource(stream);

4   microphone.connect(actx.destination);

5

6 }, console.log);
```

As you have seen in these examples, the Web Audio API provides a variety of methods to generate audio streams: signal generators, audio buffer players and microphone inputs. These audio streams can each be connected to processing units and sent through fully customizable audio graphs before they reach the user's speakers. All of these audio nodes have parameters that can be scheduled with precision using the Web Audio API's internal clock. Furthermore, these parameters—and the audio graph itself—are dynamic, and can be modified at any time in response to user interaction.

Now that we have a basic understanding of how to use the Web Audio API, let's review some of its most salient applications and use cases. If you want to learn more about programming aspects, we recommend digging into the documentation for the Web Audio API and studying the code of the many examples that can be found on the web.

10.3. Web Audio Applications

10.3.1. *Musical Interaction on the Web*

Advancements in computational power are narrowing the gap between the performance of desktop applications and that of web-based applications. Indeed, many kinds of software typically used in the form of native applications are also available as web applications (e.g. word processors, photo editors). The same is true for interactive music applications that are widely used, like digital audio workstations (DAWs). Audiotool, for example, is a nearly full-fledged digital audio workstation with virtual instruments (AudioTool 2010). Noteflight, on the other hand, is a browser-based music notation application with an integrated online community for browsing and sharing notated music (Noteflight 2008).

The migration of existing musical applications is not limited to software; some audio hardware can also be emulated within a web browser. For example, the concept of a loop-based step sequencer (often called a "tone matrix") in the form of a grid-shaped controller (e.g. Tenori-on, Monome, Ableton Push) is widely used in multiple interactive music applications on the web (AudioTool 2008; Tan and McDonald 2016; Brower 2014). Tanguy is another example that emulates an analog synthesizer purely on the web (Teaford 2016). Some applications can be extended with actual hardware, like a MIDI keyboard. Pedalboard.js allows guitar players to connect their guitar inputs to their computers and use a set of virtual guitar effects running in a web browser (Amcalar 2013). The major motivation behind reproducing existing applications in a web browser is often to make music applications more readily accessible to a broader audience.

Another trend in web-based music applications involves supporting simple musical interaction with a minimal user interface design. This is similar to how interactive mobile music applications have proliferated with the appearance of smartphones. Oftentimes, a web-based music application emphasizes the process of musical interaction itself, rather than concentrating on a particular musical context, such as performance, composition or listening. The cross-context property may account for the frequently ambiguous purpose of web-based media. In addition, design patterns in web-based interactive music applications are subject to fewer restrictions compared to desktop software, which users expect will adhere to certain design norms (menu bars, buttons, panels, layout). Interestingly, web browsers, which serve as "shells" for websites, have a fairly standard set of widgets, menus and an address bar in a common layout across web applications and websites. The browser window provides a clean slate that discourages developers from nesting another traditional-looking application within it, encouraging them instead to utilize this space in creative ways.

As the Web Audio API enables controlling audio on a web page, developers and artists have begun making simple but compelling web-based applications to demonstrate the capabilities of Web Audio, highlighting the value of interactive music applications on the web as demonstrative, educational tools. The Chrome Music Lab showcases a "collection of experiments that let anyone, at any age, explore how music works," ranging from a simple step sequencer to an emulation of a physical string (Google 2016). The applications exhibited in the Music Lab are great tools not only for music educators to show interactive demos anywhere on the fly, but also for students to try out on their own without having to own any audio equipment, as long as they have access to a device that can run a web browser. For example, STRINGS helps users to understand the relationship between the length of a string and its pitch, a concept integral to such instruments as the guitar, the harp and the piano. Similarly, the website of tone.js, a popular Web Audio API wrapper, showcases a range of interactive music applications built with the library (Mann 2015). SOLARBEAT, for example, is an interactive data sonifier and visualizer that produces generative music based on the orbital frequencies of planets in the solar system (Twyman 2010). The application clearly demonstrates openness and fluidity in the purpose and design of web-based interactive media; the application can be seen as a piece of generative music, a data visualization or an interactive sound installation. Lastly, http://webaudio. github.io/demo-list also provides links to a collection of Web Audio-based applications (Lowis 2013).

The basic architecture of web-based media, which is designed to create content in a web browser, interactive or otherwise, requires creating a web page that hosts content (HTML), interactivity (JavaScript) and its appearance (CSS). While this structure works well for building interactive music applications, it can be unwieldy when the goal is purely to generate music and sound, especially for the existing computer music community. The

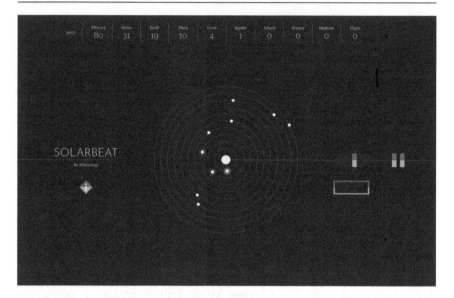

Figure 10.3 *SOLARBEAT*: An Interactive Visualization and Sonification of Planets in the Solar System.

majority of existing communities have been using programming or patching environments in which they can quickly build an audio processing program (or sound patch) on the fly and dynamically modify the program to make sound. However, an algorithm written in JavaScript can be updated only by modifying the underlying code and refreshing the page—thereby discarding the current program state. This may immediately feel limiting to computer music experts. Naturally, researchers and audio programmers would feel a need for a "musical playground" in which users can directly write programs in a dynamic, expressive music-making language, as if patching cables on an analog synthesizer. Olos is one attempt to create such an environment (Sigal 2016). It provides two contrasting representations of existing Web Audio API elements: graphical patching and textual coding. A set of nodes can be virtually connected by drag-and-drop gestures between inlets and outlets, and one can modify the configuration of each node in JavaScript code. This approach offers a gateway for the community that has expertise in a patching audio language, like Pure Data, to explore web-based audio programming without having to learn a new language or build a website. In addition, it provides a stand-alone environment in which audio code can directly produce a sonic outcome with the flexibility of using the Web Audio API, and without having to be hosted somewhere as a web page.

10.3.2. *Collaborative Music Making on the Web*

Much of our computer-mediated communication now resides on the internet (email, instant messaging, social media, collaborative software, online code repositories), and the internet provides an effective platform for remote communication and collaboration. In addition, the inherent networking capability of web-based media naturally encourages developers to have multiple users connected and to facilitate collaborative music making. Technologically supported musical interaction in groups dates back to the long tradition of network music, which is beyond the scope of this chapter but is thoroughly discussed in (Barbosa 2003). The specifics of designing systems for collaborative music making are well-documented in (Fencott and Bryan-Kinns 2013).

Indeed, people have imagined an equivalent of Google Docs for music, and music software migrated to the web platform has differentiated itself from native applications by supporting real-time collaboration. SoundTrap, for example, appears initially to be a plain DAW-type piece of software, but it supports collaborative composition from within the DAW (Lind and MacPherson 2017). One can seek and invite collaborators from the online community and share a multitrack DAW project with them. Support for videoconferencing is integrated into the system, and users can listen to the project audio together in real time. Similarly, flat.io is a music notation platform that supports real-time collaboration (Flat 2014).

There are a number of technologies that can support connectivity in collaboration—databases, WebSocket, WebRTC and cloud services, for example. Selecting which technology to use in supporting collaboration largely depends on two factors: the scale of data being exchanged and the extent to which minimal latency is essential to a given collaborative context. For example, WebRTC is designed for high-performance communication of audiovisual data, and is best suited to transmitting real-time audio from one site to another. However, if the data is transmitted on a symbolic level, like MIDI notes (noteOn/Off messages), WebSocket can adequately support such real-time communication between nodes. As an example, Dinahmoe has demonstrated compelling interactive music collaboration scenarios through the development of Plink (Dinahmoe 2010) and JAM with Chrome (Dinahmoe 2012), exchanging symbolic data via WebSockets, with each client machine performing sound synthesis independently. Supporting collaboration in music making can occur offline with a greater ecosystem of curricula, interactive applications and organizational efforts. In the following case study, we review a text-based music programming environment on the web with a strong motivation of making programming education accessible.

10.3.2.1. Case Study: EarSketch—Teaching Programming in a Web-Based DAW

EarSketch is a web-based programming environment that helps students learn programming languages like Python and JavaScript while making music (Mahadevan et al. 2015). The motivation of the system is to engage underrepresented groups in programming education through the collaborative creation of musical remixes, with a vision of "using hip-hop to teach computer science." A commonly envisioned workflow in a DAW involves importing (or recording), placing and adjusting audio samples in multiple tracks (*y*-axis) unfolded over time (*x*-axis). In EarSketch, this interaction is replaced with the ability to place audio clips into a DAW timeline using computer code, which reduces the manual effort required to organize sound clips. This computational approach is especially effective when there are certain patterns that can be conveniently expressed as algorithms (abstraction, repetition, condition). The underlying audio pipeline of the playback engine is built with the Web Audio API.

Interestingly, the initial version of EarSketch was a plugin added on top of DAW software. However, the native application plugin was isolated from the other modules of the EarSketch system: online tutorials, an audio library and an online community for sharing music. With the advancement of the Web Audio API, the creators of the system decided to migrate the whole system to the web, including building a DAW from scratch as a web application. The motivation behind migrating EarSketch to a web platform is well justified by the benefits of web-based interactive media. Target users—in this case, students in classrooms—can access the software immediately without first going through an installation process. In addition, given that such systems can be used on a large scale, updating the entire system does not involve any distribution beyond updating the website on which the software is hosted.

EarSketch is an example of a successful, web-based interactive music application. The system has been used in workshops, after-school programs and academic courses. It reflects two significant trends of early interactive music on the web: building a meta-programming environment that supports live composition, and migrating existing music software (DAWs) to the web platform. The combination of these two aspects has yielded a reliable, integrated system in which non-expert programmers (and non-expert musicians) can learn both musical and computational concepts.

10.3.3. The Browser as a Site for Sound Installation

As artists and composers in the twentieth century turned to making sound installations and site-specific projects, the internet too was adopted as a

Figure 10.4 *EarSketch*: Teaching Programming Through Remixing Music Tracks in DAW.

site for installation. These projects ask an important question: what does it mean to make a sound installation that is site specific to the internet? In exploring the web browser as a platform for sound installations, sound artists have integrated some of the web's core characteristics into their work: its persistence, connectivity, interactivity, graphics capabilities and data.

In its persistence, the web offers an installation site that can endure without requiring space in a gallery. This situation affords web-based works unique opportunities. For example, crowd-sourced compositions in which each visitor contributes a sound or an interaction can flourish in this paradigm, as the work can gradually accumulate contributions over many months or years. Online sound installations also have the ability to evolve over long periods of time. This is the case for *Bits & Pieces* (Traub 1999), which ran for over a decade, from 1999 to 2011. Each morning, the website created a daily sound collage from random audio recordings pulled from around the web. Rather than searching for specific types of sounds, the site searched for general terms, such as "sound file" or "music aiff," in order to grab a truly random sample of the sound files that exist on the internet. In this way, the piece acted as a sonic snapshot of the web and evolved over time as the sound content on the internet changed.

As a space for interactive media, the web has led sound artists to develop new genres of interactive music specifically for the web. In general, interactions in web-based sound installations fall into three categories: controlling audiovisual content with a mouse, touchpad or keyboard; contributing content in the form of sound recordings or data; and creating a momentary structure by exploring a series of pages through hyperlinks. Some projects have even used the browser itself as a control interface. Two such projects are *Jazz.Computer* (Mann and Rothberg 2015) and *Scrolling Through Sound* (Ziya 2015), which both use the browser's scrollbar as a handle to navigate through a musical composition. In this way, they repurpose the "endless scrolling" interaction paradigm of social media, cleverly turning it into a musical gesture.

The introduction of more powerful graphics capabilities within the browser in recent years has also contributed to the web's attractiveness for audiovisual installations. HTML5 canvas elements, the SVG format and WebGL combine to offer real-time synthesis of two- and three-dimensional graphics, allowing sound artists to create real-time visualizations of audio, or visuals to accompany an audio installation. Such is the case with Juno Brandel's *Patatap*, a site which the viewer activates by typing or touching (Brandel 2012). Each interaction triggers both a sound sample and an accompanying visual event. Part instrument, part art program, the result approaches a simulation of synesthesia.

Lastly, web-based sound installations have access to the vast amounts of media and data available on the web. Audio and video files on remote servers can be freely accessed and embedded in a sound installation,

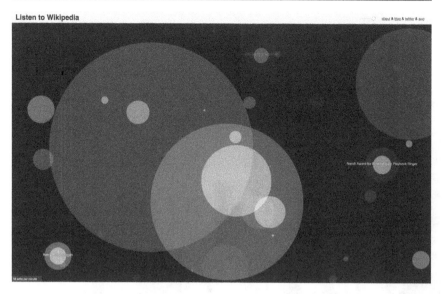

Figure 10.5 *Listen to Wikipedia*: Real-Time Sonification of Wikipedia Contributions.

giving an internet sound artist a virtually infinite library of media to sample and collage. Furthermore, many installations use web APIs to query real-time data from other sites to be used within the sound installation. *Listen to Wikipedia* is one such artwork, using as its score the constant stream of real-time contributions to the crowd-sourced encyclopedia Wikipedia (LaPorte and Hashemi 2013).

In these internet sound installations, we see artists take advantage of the characteristics of the internet and the web browser in order to make new types of sound art. Here, the internet is used not only as a distribution platform, but also as a source of inspiration. The web browser is not simply a host, but also a muse. At times, this requires us to broaden our view of musical interfaces, musical form or musical content. When we see hyperlinks and scroll bars used as musical interfaces, or when we explore a network of websites as a single musical composition, we are certainly making music through unorthodox means. All the same, this has been a consistent pattern when presenting sound installations on a platform that was not originally designed for them.

10.3.3.1. Case Study: ItSpace

Peter Traub's *ItSpace* presents an interactive site in which the viewer unwittingly creates a sound collage by browsing a social media websites (Traub 2011). *ItSpace* is an alternate version of MySpace, which was a

itspace
where objects play

by Peter M Traub
www.fictive.org

About | Objects | Participate | FAQ | License | Gallery ItSpace | ItSpace on NPR

About

ItSpace creates a network of pages within the social networking site MySpace. Instead of featuring people, the pages feature everyday household objects. Each page has a photo of the object, a description, and most importantly, a 1-minute piece of music composed of recordings of the object being struck and resonated in various ways. All the pages, or objects, are 'friends' with each other, so that visitors who discover one object may jump to the others by clicking on the 'friends' pictures at the bottom of each page.

You are invited to create new ItSpace pages with pieces made from your own household objects and link those in as 'friends' of the original set of objects. You are also invited and encouraged to remix and combine existing objects into new musical compositions.

Please note that the core nine ItSpace objects will only be 'friends' with other ItSpace sounding objects. They will not respond to friend requests from people or organizations. To learn how to make an ItSpace object, read on (and please read the FAQ too).

ItSpace is a 2007 commission of New Radio and Performing Arts, Inc., for the Networked_Music_Review. It was made possible with funding from the New York State Music Fund, established by the New York State Attorney General at Rockefeller Philanthropy Advisors.

Participate

While most people will just listen to the pieces (which is expected) everyone is invited to participate in building the ItSpace network. Participating in ItSpace will take a bit of effort, but together we can build an extended network of sounding objects inside of MySpace. There are two ways to participate: making new objects or remixing existing objects.

Making New Objects

1. Choose an inanimate object in your home. **This means no animals and no people.** Preferably the object should be something you wouldn't normally associate with music or sound.

2. Photograph your object.

Cool New Objects   [click on them]

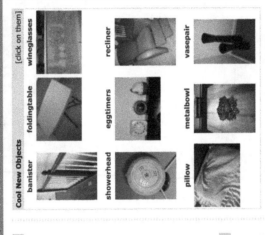

banister   foldingtable   wineglasses

showerhead   eggtimers   recliner

pillow   metalbowl   vasepair

FAQ

Will ItSpace objects be friends with MySpace people or bands?

What does your license mean?

Why don't you tell me what tools to use to make a new piece?

Figure 10.6 ItSpace: A Social Network of Sounds.

popular social network in the early 2000s known especially for facilitating music sharing. MySpace users could upload songs that they had written to their profiles, which could then be streamed by visitors. *ItSpace*, on the other hand, presents a social network of objects, rather than of people. Members of *ItSpace* are not people and their songs; instead, they are objects and their sounds, which have been contributed by participants. The installation comes to life when a viewer browses through the different profiles. Browsing becomes an act of composition; as the viewer opens browser tabs and browses to different profiles, he or she unwittingly composes a sound collage from the sounds within *ItSpace*.

ItSpace exemplifies many different aspects of online sound installations. It is inspired by the internet, repurposing a social network as an open-form composition. Hyperlinks, perhaps the most defining component of the internet, become a means of creating an indeterminate composition. Users interact with the piece in two ways: they help create the piece by contributing sounds, and they help to perform the piece by navigating from page to page and listening. Lastly, the work's success was made possible by its persistence on the web and its ability to accumulate new sound profiles over a long period of time. Installations like *ItSpace* help light the way to show us how we can explore the affordances of the web, even those that are not obvious. In doing so, we can make installations that take full advantage of the internet as an artistic medium.

10.3.4. The Browser as a Platform for Music Performances

The web browser has also been adopted as a platform on which a musical instrument can be built for the live performance of electronic music and media collages. The exact nature of this practice varies because Web Audio is capable of such a variety of styles. Web Audio has proven to be a fruitful platform for creative coders to experiment with in creating new instruments, new live coding languages and new types of media art.

Web Audio performers commonly choose to design and program their own instruments. For example, *Loop Drop* is a Web Audio synthesizer and sequencer instrument (McKegg 2016). The instrument is a simulation of an analog step sequencer with a graphical interface to create beats, change parameters, save presets and connect to external MIDI interfaces. While this is an example of a standard synthesizer instrument, others have designed audiovisual instruments, such as *Live Writing*, which uses WebGL to add visuals that respond to music on top of a poem, written live on stage (Lee, Essl and Martinez 2016).

Live coding has emerged as one of the popular interfaces for web audio performance. Live coding is a multifaceted practice involving writing code live to generate sound and/or visuals (Collins et al. 2003). Live

Figure 10.7 *Live Writing: Gloomy Streets*, Audiovisual Performance Interface for Live Poetry.

coding is an attractive interface for Web Audio performance because it gives the performer direct control over audio synthesis without first needing to develop a point-and-click interface, and it can let the performer control sound, visuals and browser actions from the same interface. Live coding languages developed for Web Audio include Roberts' *Gibber* (see a case study on *Gibber* below) and Stetz's language Lissajous (Stetz 2015).

Live coding has also been used in media collages that use the browser more holistically. By treating web media as flexible materials to loop, chop, rearrange and layer, a performer is able to compose new audiovisual art, drawn directly from the fabric of modern digital communications. In Taylor's *The Last Cloud* (2016), the performer uses live coding to collage multiple browser windows containing web media such as audio and video players, images, GIFs and text (Taylor 2016). These web media are layered, looped and processed in an audiovisual mashup. *The Last Cloud* uses browsing—loading and viewing web content—as an expressive compositional activity. The work situates Web Audio not as a stand-alone synthesis engine, but as a sound engine that is entangled with its broader context on the internet.

Performers have occasionally used the web as a source of found sounds and other found media. Sang Won Lee's performance *Live Coding YouTube* treats YouTube videos as sound sources to be sampled and remixed (Lee, Essl and Martinez 2017). In the performance, Lee uses live coding to load up to 16 YouTube videos and loop them rhythmically. He uses YouTube as a seemingly infinite supply of source material. By interacting directly with YouTube, a web media platform, Lee reveals its potential as an instrument for live sound collage, rather than as a distribution platform exclusively.

10.3.4.1. Case Study: Gibber

One of the first substantial web audio projects and toolkits, *Gibber* is an environment for live coding web audio in the browser (Roberts and Kuchera-Morin 2012). It offers an exceptionally terse syntax and many shortcuts that allow users to write complex audio expressions quickly. Over time, *Gibber* has expanded to include graphical interfaces, visual generators and connections to other programs. *Gibber* author Charlie Roberts has performed regularly by live coding in *Gibber*. Stylistically, Roberts's performances focus on probabilistic drum patterns, generative melodies, and the manipulation of harmonic patterns.

Gibber takes advantage of being situated on the web in several ways. In the system, users can live code together through real-time collaboration within the live coding text editor. It uses web graphics engines to visualize rhythms and sound events as the user codes them. *Gibber* lets users search for and retrieve sound samples on FreeSound, an online database of sound samples, and load them directly into their instruments. Lastly, users can also publish their musical code on *Gibber*'s website for others to remix.

10.3.5.   *Music Apps in Audiences' Palms at Live Concerts*

The proliferation of the smartphone brings with it a new situation in music performances: virtually every audience member has a computer and speaker in their pocket, all of which are connected to the internet. Out of this condition, a new performance genre has emerged in which an audience's smartphones function as a collective speaker system. This performance genre, sometimes referred to as "distributed music," has emerged at the nexus of several events: the advent of the smartphone, the emergence of Web Audio as a powerful synthesizer and the rise of JavaScript server development toolkits. Together, these events have led to a new performance paradigm in which a performer can access an array of intelligent, networked Web Audio synthesizers running in an audience's mobile browsers. Audience participation performances with mobile devices generally fall into one of the following four distribution schemes:

- *1 to N:* One performer on stage sends audio control messages to N audience devices running Web Audio engines.
- *N to 1:* Audience members use a web interface on their devices to send control messages to a single, central sound source. (No Web Audio involved.)
- *N to N:* Audience members use a web interface on their devices to create sound and send control messages to other audience members' sound engines, according to network logic.
- *Standalone*: The audience loads a static, stand-alone Web Audio instrument (or group of instruments) to play; no networking is involved, and the audience may be directed by a score or conductor.

Among these, schemes that make use of audience speakers offer perhaps the most opportunities for new sound experiences (Taylor 2017).

Although Web Audio and mobile computing currently enable distributed music, the genre predates the smartphone by several decades. Filipino composer José Maceda offers the first framework for large-scale participation of audience devices in his masterwork *Ugnayan*, a composition transmitted on 30 radio stations to the radios of audience members (Macada 1974). All residents of the city of Manila were encouraged to bring their radios out into the streets, turn their volume up and tune to different radio stations, creating a massive, participatory sound collage. When cell phones became common in the late 1990s, Golan Levin and a group of collaborators imagined a similar work using cell phone ringtones. The result was *Dialtones: A Telesymphony*, first performed at Ars Technica in 2001 (Levin 2001). From the stage, performers used a graphical interface to dial the phones of audience members, creating a musical composition from phones ringing across the audience.

In these and more recent distributed music compositions, composers use the audience as a form of audio processor, in some cases mimicking common studio effects. In *Ugnayan*, Maceda's audience simulated tape collage methods by accessing different radio streams at different times. In a more recent work, *Drops*, the composers create a delay effect by having melodies pass from phone to phone (Robaszkiewicz and Schnell 2015). In *Concert for Smartphones*, Bundin creates additive synthesis across the audience by having each audience member's phone play a single overtone of a sound; collectively, the audience creates a tone that no single device generates on its own (Bundin 2016). These efforts display the potential of distributed music to transform common electronic music techniques into interactive, spatial and crowdsourced activities.

10.3.5.1. *Case Study:* Fields

One of the successful distributed music compositions via audience's smartphones to date is *Fields* (Shaw, Piquemal and Bowers 2015). In the work, performers use the audience as a passive, but intelligent, speaker system. The composition recreates the sound of natural landscapes by turning each audience device into an object in the landscape. In a series of overlapping scenes, each audience device makes the sounds of a cricket, wind gust or raindrop. The composition is therefore designed to take advantage of the environment in which a mass of weak speakers exist dispersed throughout the audience. The piece additionally uses large house speakers for added power on low-frequency tones.

Fields is performed by two performers: one uses a web interface to send sound events to the audience, while the other uses a Pure Data interface to create sound on speakers on stage. Sound events are sent to the audience via a server built on Node.js, using Piquemal's own Rhizome server

Figure 10.8   *Fields*, Distributed Web Audio Agents as a Medium for Sound Diffusion.

tool kit for distributed performances (Piquemal 2014). In *Fields'* configuration, Rhizome serves a NexusUI (Taylor et al. 2014) interface to the
performer, and serves an audio engine running WebPD to each audience
member (Piquemal 2017). Interactions with the NexusUI interface are sent
as OSC messages through the server via WebSockets, and are received
by the WebPD patches on audience phones, where they manipulate audio
parameters. Before each performance, the performers set up multiple dedicated wi-fi routers, to ensure stable wireless connections for streaming
audio commands to the audience.

10.4.   Conclusion: The Web as a Platform for Interactive Media

In this chapter, we have examined both the basic capabilities of Web Audio
(section 10.2), as well as engaged with a range of projects that leverage Web
Audio as a platform (section 10.3). This gives us a broad view of what Web
Audio is: a unique platform for creating audio-based artistic projects.
Web Audio is a part of web technology and shares many of its advantages.
Access is as easy as navigating to a web page. This greatly simplifies the
process of distribution—and by extension, of participation. It provides a tre-
mendously accessible canvas. It is also a highly networked medium, allowing for interconnectivity, collaboration and participation to come into play.
    There are a few areas in which we have provided a start. With the material
in section 10.2, you should be ready to build Web Audio applications, though

many specific capabilities have not been covered. We recommend that you dive in, and if a capability is not covered here, explore the Web Audio API.

The potential for Web Audio applications is vast. In section 10.3 we structured important current Web Audio applications into five broad areas: musical interaction (10.3.1), collaboration (10.3.2), sound installation (10.3.3), performance (10.3.4) and audience participation (10.3.5). This is informed by the current main territories that have been explored by the Web Audio community. However, there is no doubt that the potential for new application areas is limited only by our imaginations. We recommend engaging with these chapters for much more material on how to dig deeper in all these application areas.

Notes

1. The low-level audio pipeline is rendered in a separate thread for the purposes of performance and the programmers do have access, although limited, to the audio rendering thread; for example, AudioWorklet lets programmers write sample-level synthesis algorithm (Choi 2017, 2018).
2. For some browsers, creating a new AudioContext before a user interacts with the web page will suspend the context by default. See autoplay policy for web audio for more detail (Beaufort 2017).

References

Amcalar, A., 2013. *Pedalboard.js* [online]. Viewed Sept 5, 2018 https://github.com/dashersw/pedalboard.js

AudioTool, 2008. *The Tonematrix, a Pentatonic Step Sequencer by Audiotool*. Viewed Sept 5, 2018 http://tonematrix.audiotool.com/

AudioTool, 2010. *AudioTool—Free Music Software—Make Music Online in Your Browser*. Viewed Sept 5, 2018 www.audiotool.com/

Barbosa, Á., 2003. Displaced Soundscapes: A Survey of Network Systems for Music and Sonic Art Creation. *Leonardo Music Journal*, 13, pp. 53–59.

Beaufort, F., 2017. *Autoplay Policy Changes* [online]. Viewed Sept 5, 2018 https://developers.google.com/web/updates/2017/09/autoplay-policy-changes#webaudio

Brandel, J., 2012. *Patatap* [online]. Viewed Sept 5, 2018 www.patatap.com/

Brower, G.M., 2014. *Rhythm* [online]. Viewed Sept 5, 2018 https://musiclab.chrome experiments.com/Rhythm

Bundin, A., 2016. Concert for Smartphones. In: *Web Audio Conference (WAC)*. Atlanta, Georgia, [online]. Viewed Sept 5, 2018 https://smartech.gatech.edu/bitstream/handle/1853/54639/WAC2016-15.pdf

Choi, H., 2017. *Enter AudioWorklet* [online]. Viewed Sept 5, 2018 https://developers.google.com/web/updates/2017/12/audio-worklet

Choi, H., 2018. AudioWorklet: The Future of Web Audio. In: *International Computer Music Conference*. Daegu, South Korea.

Choi, H. and Berger, J., 2013. WAAX: Web Audio API Extension. In: *International Conference on New Interfaces for Musical Expression (NIME)*. Daejon, South Korea, pp. 499–502.

Collins, N., McLean, A., Rohrhuber, J. and Ward, A., 2003. Live Coding in Laptop Performance. *Organised Sound*, 8, no. 3, pp. 321–330.

Dinahmoe, 2010. *Plink*. Viewed Sept 5, 2018 http://dinahmoelabs.com/plink

Dinahmoe, 2012. *Jam with Chrome*. Viewed Sept 5, 2018 www.dinahmoe.com/work/jam

Duckworth, W., 2005. *Virtual Music: How the Web Got Wired for Sound*. New York: Routledge.

Fencott, R. and Bryan-Kinns, N., 2013. Computer Musicking: HCI, CSCW and Collaborative Digital Musical Interaction. In: *Music and Human-Computer Interaction*. London: Springer, pp. 189–205.

Flat, 2014. *Online Collaborative Music Notation Software—Flat*. https://flat.io/

Google, 2016. *Chrome Experiments—Music Lab*. Viewed Sept 5, 2018 https://musiclab.chromeexperiments.com/

Haverbeke, M., 2014. *Eloquent JavaScript*, 2nd edition. San Francisco: No Starch Press.

Hedstrom, M., 1997. Digital Preservation: A Time Bomb for Digital Libraries. *Computers and the Humanities*, 31, no. 3, pp. 189–202.

Hindle, A., 2014. Cloudorch: A Portable Soundcard in the Cloud. In: *International Conference on New Interfaces for Musical Expression (NIME)*. London, pp. 277–280.

Hindle, A., 2015. Orchestrating Your Cloud Orchestra. In: *International Conference on New Interfaces for Musical Expression (NIME)*. Baton Rouge, Louisiana, pp. 121–125.

LaPorte, S. and Hashemi, M., 2013. *Listen to Wikipedia* [online]. Viewed Sept 5, 2018 http://listen.hatnote.com/

Lazzarini, V., Costello, E., Yi, S. and ffitch, J., 2015. Extending Csound to the Web. In: *Web Audio Conference (WAC)*. Paris, France.

Lee, S.W., Bang, J. and Essl, G., 2017. Live Coding YouTube: Organizing Streaming Media for an Audiovisual Performance. In: *International Conference on New Interfaces for Musical Expression (NIME)*. Copenhagen, Denmark, pp. 261–266.

Lee, S.W., Essl, G. and Martinez, M., 2016. Live Writing: Writing as a Real-Time Audiovisual Performance. In: *International Conference on New Interfaces for Musical Expression (NIME)*. Brisbane, Australia.

Letz, S., Orlarey, Y. and Fober, D., 2017. Compiling Faust Audio DSP Code to WebAssembly. In: *Web Audio Conference (WAC)*. London.

Levin, G., 2001. *Dialtones—a Telesymphony* [performance].

Lewis, P., 2015. *Guitar Tuner* [online]. Viewed Sept 5, 2018 https://guitar-tuner.appspot.com/

Lind, F. and MacPherson, A., 2017. Soundtrap: A Collaborative Music Studio with Web Audio. In: *Web Audio Conference (WAC)*. London.

Lowis, C., 2013. *Web Audio/MIDI Demo List* [online]. Viewed Sept 5, 2018 http://webaudio.github.io/demo-list

Macada, J., 1974. *Ugnayan*. [performance]. Manila.

Mahadevan, A., Freeman, J., Magerko, B. and Martinez, J.C., 2015. Earsketch: Teaching Computational Music Remixing in an Online Web Audio Based Learning Environment. In: *Web Audio Conference (WAC)*. Paris, France.

Mann, Y., 2015. Interactive Music with tone.js. In: *Web Audio Conference (WAC)*. Paris, France.

Mann, Y. and Rothberg, S., 2015. *Jazz.Computer* [online]. https://jazz.computer

McKegg, M., 2016. Building Desktop Apps Using Web Audio. In: *Web Audio Conference (WAC)*. Paris, France.

Noteflight, 2008. *Online Music Notation Software*. Viewed Sept 5, 2018 *www*.noteflight.com/

Piquemal, S., 2014. *Rhizome* [online]. Viewed Sept 5, 2018 https://github.com/sebpiq/rhizome

Piquemal, S., 2017. *Webpd* [online]. Viewed Sept 5, 2018 https://github.com/sebpiq/WebPd

Rawlinson, H., Segal, N. and Fiala, J., 2015. Meyda: An Audio Feature Extraction Library for the Web Audio API. In: *Web Audio Conference (WAC)*. Paris, France.

Robaszkiewicz, S. and Schnell, N., 2015. Soundworks—A Playground for Artists and Developers to Create Collaborative Mobile Web Performances. In: *Web Audio Conference (WAC)*. Paris, France.

Roberts, C. and Kuchera-Morin, J., 2012. Gibber: Live Coding Audio in the Browser. In: *International Conference on New Interfaces for Musical Expression (NIME)*. Ann Arbor, MI.

Roberts, C., Wakefield, G. and Wright, M., 2013. The Web Browser as Synthesizer and Interface. In: *International Conference on New Interfaces for Musical Expression (NIME)*. Daejon, South Korea, pp. 313–318.

Shaw, T., Piquemal, S. and Bowers, J., 2015. Fields: An Exploration into the Use of Mobile Devices as a Medium for Sound Diffusion. In: *International Conference on New Interfaces for Musical Expression (NIME)*. Baton Rouge, Louisiana.

Sigal, J., 2016. Olos: Visual Music Programming. In: *Web Audio Conference (WAC)*. Atlanta, GA.

Stetz, K., 2015. Lissajous: Performing Music with JavaScript. In: *Web Audio Conference (WAC)*. Paris, France.

Tan, M. and McDonald, K., 2016. *The Infinite Drum Machine—Ai Experiments* [online]. https://experiments.withgoogle.com/ai/drum-machine

Taylor, B., 2016. *The Last Cloud* [composition].

Taylor, B., 2017. A History of the Audience as a Speaker Array. In: *International Conference on New Interfaces for Musical Expression (NIME)*. Copenhagen, Denmark, pp. 481–486.

Taylor, B., Allison, J.T., Conlin, W., Oh, Y. and Holmes, D., 2014. Simplified Expressive Mobile Development with NexusUI, NexusUp, and NexusDrop. In: *International Conference on New Interfaces for Musical Expression (NIME)*. London, pp. 257–262.

Teaford, L., 2016. *Tanguy* [online]. Viewed Sept 5, 2018 http://tanguysynth.com/

Traub, P.M., 1999. *Bits and Pieces: A Sonic Installation for the World Wide Web*. Master's Thesis, Dartmouth College.

Traub, P.M., 2011. ItSpace. *ACM SIGGRAPH Art Gallery*, Vancouver, British Columbia, Canada: ACM, pp. 376–377.

Twyman, L., 2010. Solarbeat [online]. Viewed Sept 5, 2018 http://whitevinyldesign.com/solarbeat/

Wang, G., 2008. A History of Programming and Music. In: *Cambridge Companion to Electronic Music*. Cambridge, UK: Cambridge University Press.

Wilson, C., 2013. *A Tale of Two Clocks—Scheduling Web Audio with Precision* [online]. Viewed Sept 5, 2018 www.html5rocks.com/en/tutorials/audio/scheduling/

Ziya, E., 2015. Scrolling Through Sound. In: *Web Audio Conference (WAC)*. Paris, France.

Auditory Cue Design

David McGookin

11.1. Introduction

Just as not all graphical user interfaces are composed of text, auditory user interfaces do not need to be composed solely of speech. In this chapter we introduce auditory cues as one potential way of encoding information in an auditory display (Shinn-Cunningham et al. 1997). Similar to graphical icons, such cues play the same roles in auditory displays: being able to quickly communicate short, key information to a user in an effective manner. As with visual icons, there are multiple forms and ways to construct auditory cues, and choices a designer must make. Our goal is to not only present these different forms, but also allow you to reason over them, understanding their benefits and drawbacks to determine which might be best in a given situation. It is also important to consider what we don't know. Not all auditory cue designs have extensive research to validate their use. By understanding auditory cues in this way, you will gain good practical knowledge of how to use them and hopefully will be inspired to contribute to that body of knowledge.

11.2. Semiotic Issues in Auditory Cue Design

To understand and reason over the various cue types we will discuss, it is important to consider their general similarities and underlying structure. In all cases, auditory cues are examples of signs. A full discussion of semiotics is beyond the scope of this chapter. However, a basic consideration allows us to more easily reason over the differences between different cue designs, and better select an appropriate cue type for a given purpose.

Visual signs are common in everyday life and are perhaps most often encountered when driving. For example, Figure 11.1 shows a road sign

Figure 11.1   An Example of Three Road Signs from the UK (left), Finland (middle) and the United Stated of America (right): Although all indicate pedestrians crossing (signified), they have a different visual appearance (signifier).

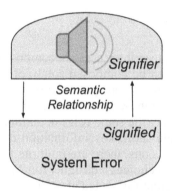

Figure 11.2   An Illustration of an Auditory Sign: Note that the sign is composed of two parts. The signifier, the sound itself (or image in the case of a visual icon), has a variable semantic relationship with the signified (what the sound actually represents). It is the characteristics of this relationship that define the benefits and drawbacks of each auditory cue type.

indicating a pedestrian crossing from three different countries (the U.K., Finland and the United States of America). While each of these signs look different, they all mean the same thing and should be interpreted in the same way: a car driver should slow down, take care and give way to pedestrians crossing the road. We can consider that the road sign has two component parts: the image depicted on the sign and how that sign is interpreted. Semantically, the image (or physical manifestation) of the sign is called the *signifier*, while what that image represents is called the *signified* (see Figure 11.2).

It is the relationship between these two, how it is created and expected to be understood by users, that impacts our consideration of auditory cue types. Charles Sanders Peirce considered that there are three broad categories of relationship between the signifier and the signified: iconic, symbolic and indexical (Chandler 2002). Iconic and symbolic relationships are perhaps easiest to understand. Iconic representations illustrate a resemblance or imitation of the signified. For example, the Finnish pedestrian crossing sign in Figure 11.1 visually resembles a pedestrian crossing the road. Even without knowing what the signifier (road sign) should signify, it is likely that an individual could make an informed guess as to what the sign represented. Symbolic relationships, on the other hand, are essentially arbitrary, with no assumption an individual will understand the relationship between the signifier and signified without being told. Keeping with our road sign example, Figure 11.3 shows a UK road sign. The sign indicates that the national speed limit (depending on the type of vehicle being driven) applies on this section of road. Without being told, it is unlikely someone would identify the signified, given their arbitrary relationship. Iconic and Symbolic relationships can be considered as the end points on a continuum representing a more to less obvious relationship between the signifier and the signified. However, indexical relationships are much less clear. We can consider that indexical signs have a more causal relationship between the signifier and signified. For example, smoke is a signifier of fire (signified): the fire *causes* the smoke. Such a relationship is not employed so much in auditory cue design. However, we can consider the bottling plant process of the ARKola simulation (Gaver, Smith and O'Shea 1991) (see Section 3.2.2) or spearcons (Walker et al. 2013; Walker et al. 2006) (See Section 3.3) as examples where the auditory cue is caused (or intended to be caused) by the information signified.

Before moving on we should note some important considerations. Firstly, as noted by Hawkes, signs may have mappings to more than one

Figure 11.3 The UK Road Sign Representing That the National Speed Limit Applies: The relationship between the signifier and its meaning is essentially arbitrary and must be learned.

category, with the category that dominates being defined mostly by use (Hawkes 1972). Which category dominates can also change over time, such as a symbolic sign that becomes embedded within a culture and begins to gain new meaning to become more iconic. The image of a floppy disk in a computer interface once had an iconic relationship with saving onto a floppy disk. However, over time, the relationship has become more metaphorical, with the floppy disk image acting as a sign for general saving. It is important therefore to consider that mappings are also culturally situated, and although there is very little work considering cross-cultural study of auditory cue design (Hussain et al. 2016 being a notable exception), cues designed within one cultural context may not work well in another.

11.3. Auditory Cue Design Techniques

In the following section we outline the main types of auditory cues that have been proposed. There is a variable amount of work on the study of each, and we present them in order from most to least studied.

11.3.1.   Earcons

Earcons (Blattner, Sumikawa and Greenberg 1989) were first proposed by Blattner, Sumikawa and Greenberg at the same time Gaver proposed "auditory icons" (Gaver 1989) (see section 11.3.2). Earcons can be described as "abstract, synthetic tones that can be used in structured combinations to create auditory messages" (Brewster 1994). Given our previous discussion on semiotics, earcons lie towards the symbolic end of the mapping continuum, essentially having an arbitrary relationship between the sound and what it signifies. This provides some level of privacy or security to the sound and its representation (if the meaning of the cue is to be kept private), but it also means that the relationship between audio and its meaning must be learned. Blattner, Sumikawa and Greenberg proposed that earcons be constructed from musical principles, and employ musical motives, "brief successions of pitches arranged to produce a rhythmic and tonal pattern sufficiently distinct to function as an individual recognizable entity," as their core form (Blattner, Sumikawa and Greenberg 1989). Such motives were already common in musical composition, originating from Wagner's theory of the music drama (Gryzanowski 1877). Sergei Prokofiev, for example, used different musical motives to represent each of the characters in *Peter and the Wolf* (Wikipedia Foundation 2018c). Similarly, motives have been used in advertising and to represent companies. The "Intel-Inside" motive used to represent Intel microprocessors for example (Intel Corporation 2014).

Such examples represent what Blattner et al. termed "one-element ear-cons" (Blattner, Sumikawa and Greenberg 1989). These are the most basic form and represent a single mapping between one motive and the infor-mation it represents. For example, the sounds used to represent a text mes-sage, new email and a calendar reminder on a user's smartphone, could be considered as one-element earcons. Each sound has a largely (at least initially) symbolic relation with its meaning. More importantly, there is no relationship between the sounds. One-element earcons are atomic and cannot be broken down into smaller parts. By knowing the meaning of the text message sound, it is not possible to guess or interpret the mean-ing of a new email or calendar sound. Each sound has a unique meaning that must be learned independently. While useful for simple sounds, such as audio notifications, one-element earcons do not scale well to seman-tically richer messages. The three other types of earcons Blattner et al. proposed—transformational earcons, hierarchical earcons and compound earcons—deal with this issue (Blattner, Sumikawa and Greenberg 1989).

11.3.1.1. Transformational Earcons

Transformational earcons use a set of rules, or "grammar," to construct motives by systematically mapping data parameters (signified) onto sound attributes (signifiers). The use of systematic mappings means that users must learn only a small set of mapping rules to understand a relatively large set of individual earcons. Transformational earcons are most useful when parameterized data (or data that can be parameterized) are to be communicated. For example, message notification sounds in a smartphone could be parameterized into the source "app" (text message, Facebook or email), the type of person that sent the message (family, close friends, others) and the length of the message sent (short, medium and long). With transformational earcons, each of the parameters of the message is mapped to a different auditory attribute. For example, notification source might be mapped to musical timbre (instrument), the person type mapped to the musical rhythm, and the size of message mapped to the musical register the rhythm is played in. Figure 11.4 illustrates how this relatively small set of rules can generate a set of 27 individual earcons.

11.3.1.2. Hierarchical Earcons

Hierarchical earcons are closely related to transformational earcons. How-ever, they have a hierarchical relationship with each other. Each earcon is a node in the tree and inherits the properties of the earcon above it. This makes them useful to support context in tree structures, such as menu-based systems (Brewster, Raty and Kortekangas 1996). For example,

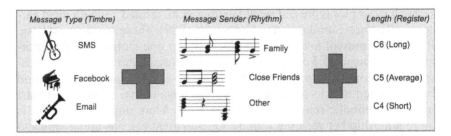

Figure 11.4 An Example of How a Set of Transformational Earcons Representing Messages in a Mobile Phone Can Be Created: By combining the type of message with its sender and length, a set of 27 earcons can be easily generated.

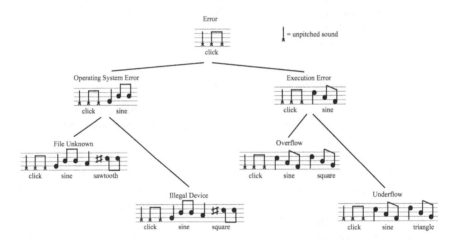

Figure 11.5 An Example of Hierarchical Earcons Representing Common Errors: Each earcon inherits and modifies the earcons in the node above.

Source: Adapted from Brewster et al. (Brewster, Wright and Edwards 1994)

Figure 11.5 illustrates a hierarchical representation for a family of earcons representing errors in a computer system. Each level of the hierarchy inherits and modifies the motives on the path from the root node to represent a wide range of possible Errors.

One practical issue with hierarchical earcons is that as earcons are extended at lower levels of the hierarchy, they can take a long time to play. Brewster, in his work on evaluating earcons (Brewster, Wright and Edwards 1994), argued that it was important for sound to keep up with interactions in a user interface, and it should not lag behind the interactions a user was performing. Therefore, he played only the last motive of

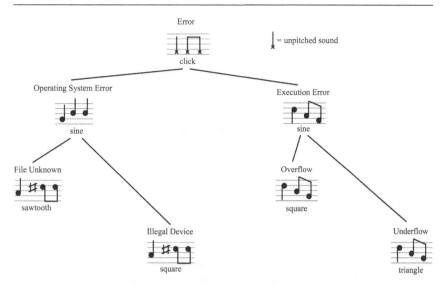

Figure 11.6   An Example of How the Hierarchical Earcons in Figure 5 Can Be Shortened to Reduce the Playing Time By Using Only the Modified Motive.

Source: Adapted from Brewster et al. (Brewster, Wright and Edwards 1994)

the leaf node earcon, making hierarchical earcons essentially the same as transformational earcons (see Figure 11.6).

11.3.1.3.   Compound Earcons

We can consider that a set of earcons derived from the same "grammar" as a family, sharing a common set of characteristics. The number of individual earcons in each family is in practice limited by the number of sound parameters that can be used (e.g. timbre, register, rhythm), and the degree to which each can be modified to be uniquely identifiable by a user. For example, a user might be able to identify if a motive is played with a piano or a trumpet, but might not be able to identify whether a motive is played with a trumpet or a French horn (Brewster, Wright and Edwards 1995a). To increase the range of messages that can be communicated, earcons from different families can be combined to generate semantically richer messages. For example, one-element earcons representing "save," "close" and "open" operations could be combined with one-element earcons representing "file," "folder" and "application" objects to generate Compound earcons representing "save file," "open application," and so on (see Figure 11.7). As with hierarchical earcons, compound earcons can become long, and care needs to be taken to avoid them becoming too long.

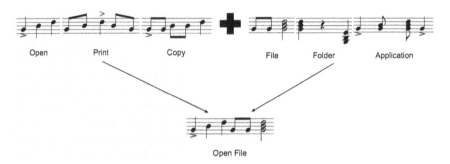

Figure 11.7 An Example of How Earcons From Two Families Can Be Combined by Being Played Consecutively to Form Compound Earcons.

Source: Adapted from Brewster (Brewster, Wright and Edwards 1994).

Brewster found that for compound earcons composed of two one-element earcon parts, presentation time could be halved by playing each simultaneously in different ears without reducing accuracy of identification (Brewster, Wright and Edwards 1995b).

11.3.1.4. Earcon Design Guidelines

Significant work was initially carried out by Brewster to determine how to design effective earcons, and where earcons may be beneficial in a user interface (Brewster, Wright and Edwards 1994, 1995b; Brewster, Raty and Kortekangas 1996; Brewster, Wright and Edwards 1995a). His studies covered both hierarchical and compound earcons, as well as how training impacts on performance. His studies showed that earcons should be created with a musical approach, using musical timbre and complex rhythmic pitch structures. He further identified that register (the base pitch used to play the earcon) was least well identified by users and accounted for most errors. Providing the maximum possible variation between different auditory parameters helped reduce the likelihood that earcons would be misidentified. In practice, this means that an earcon family is constrained to between 27 and 36 earcons using three (or four, for timbre) values for each of the three main auditory attributes (timbre, melody and register). Further work by Leplâtre found that the use of more subtle musical elements, such as attack, sustain and harmonic progression, could increase the number of earcons significantly further (Leplâtre and Brewster 2000). However, here, earcons supported the hierarchical position in a menu structure and did not need to be directly recalled. McGookin and Brewster further expanded on Brewster's parallel presentation work (McGookin and Brewster 2004b). Using a family of transformational earcons, they investigated how concurrent (overlapping) presentation of the earcons impacted identification. While a number of acoustic modifications (such as staggering the onset times of earcons, and presenting them in a spatialized audio environment) improved identification (McGookin and Brewster 2004a, 2004b), concurrently presenting earcons from the same family significantly reduced identification. Table 11.1 provides an overview of

Table 11.1 A Table of Earcon Design Guidelines.

Auditory Parameter	Guideline for Effective Earcons
Timbre	Musical timbres are preferred, and more effective, than simple sinusoidal tones. Timbres that are subjectively easy to tell apart should be used. Timbres from the same instrument family can be confused. However, if playing earcons concurrently with the same timbre, using different timbres from the same musical family can improve identification (McGookin and Brewster 2004b).
Rhythm and Pitch	Rhythms should use different numbers of notes and use different pitch structures to aid discrimination. The pitch structure used in the earcon can also be used to support more metaphorical mappings with data (Hankinson and Edwards 2000). Earcons must also be short enough to keep up with interactions in the user interface. Brewster notes earcons with up to six notes (as short as 0.03 seconds long) can be effective (Brewster, Wright and Edwards 1995a). Tempo is also effective at differentiating the earcons.
Register	Do not use register if absolute judgements of pitch are to be made. If register is used to represent a data parameter on its own, then only a few should be incorporated, and large differences (of several octaves) should be used. McGookin and Brewster found that earcons with register encoded information could be identified with around 70% accuracy (McGookin and Brewster 2004b). Removing the need to recall register increased this to around 90% accuracy.
Intensity	The intensity, or loudness, of a sound is not effective as an earcon parameter. More generally, sounds that are too loud will be annoying, whilst sounds that are too quiet will be ineffective. Overall sound level should be under the control of the user.
Spatial Location	Spatializing the presentation of earcons from different families is effective and can be used to reduce the time taken to present compound earcons. However, although concurrent spatial presentation of earcons from the same family can help improve performance (McGookin and Brewster 2004a), that improvement may not be enough to support effective earcon identification.

Source: Adapted from (Brewster, Wright and Edwards 1995a)

key earcon design recommendations adapted and revised from Brewster (Brewster, Wright and Edwards 1995a).

While good work on how to design earcons has been carried out, less detailed consideration has been given to how data parameters should be mapped to auditory attributes. Walker and Kramer distinguish between categorical sound dimensions (such as timbre) and linear sound dimensions (such as pitch) (Walker and Kramer 1996). They propose that categorical dimensions of data (e.g. the application type from our previous messaging example) should be mapped to a similar categorical auditory attribute, while continuous dimensions of data (such as the message length) should be mapped to a linear sound attribute. By doing so we can incorporate metaphorical aspects into the data—sound mapping, making the earcon more metaphorical, than purely symbolic (Hankinson and Edwards 2000). For example, McGookin and Brewster used the tempo of an earcon melody to represent the intensity of a theme park ride, and the register in which that melody was played to represent the cost of the ride (increasing register indicating higher cost) (McGookin and Brewster 2004b). Although they did not explicitly study the effect of this, in terms of learning the earcons, it is a promising approach.

11.3.1.5. Learning Earcons

As earcons have an essentially arbitrary relationship between the sound and its meaning, it is important to consider how users will learn this mapping. In Brewster, Wright and Edwards' work investigating earcon design, they carried out a dedicated training phase, with participants hearing, and having to learn, the earcons and their meaning before taking part in the study (Brewster, Wright and Edwards 1993). In later work Brewster compared different approaches to training participants (Brewster 1998). These ranged from one-on-one tutorial sessions on the earcon family, to a "worst case" scenario, where participants did not hear the earcons and read only a basic description of each. Their results found that not being able to hear the earcons caused a significant reduction in performance. They further identified that individuals will create their own meanings for the sound if none is provided. For example, participants interpreted that the use of rising and falling rhythms might map to items on the left and right of the menu hierarchy used in the study (Brewster 1998). Brewster also highlighted the role of presenting the structure of the earcons as a learning aid. This approach was later used by McGookin and Brewster, who used a ten-minute free listening session with an outline of the earcon family structure, before asking participants to correctly identify three random earcons as a testing phase (McGookin and Brewster 2004b). Although the studies were different, earcon identification performance was similar to those of Brewster, Wright and Edwards (Brewster, Wright and Edwards 1994). Hoggan and Brewster investigated training for cross-modal icons: icons that can be presented as either earcons or tactons (tactile icons; Brewster and Brown 2004; Hoggan and Brewster 2010). They used a novel game approach to training, with levels of initial recognition similar to those in Brewster (Brewster, Wright and Edwards 1994) and McGookin and Brewster (McGookin and Brewster 2004b). However, identification rose to 100% after 40 minutes. While a well-designed earcon set can be recalled with high accuracy, many use cases may not support the training needed to achieve this.

11.3.2. Auditory Icons

In everyday life we use sound in a variety of ways. When closing a car door we often use the sound it makes to determine if the door is closed fully. In my office I know the heating system is working because of the noise it makes, and the occasional noise of a colleague's door opening and closing gives me awareness of others in the building. Close your eyes for a moment, what do you hear? What does it tell you about the world? People often make use of these "everyday" sounds to support awareness of the world around them. Sounds can blend into the background, so we don't notice them, and only become apparent when they change.

Around the same time as Blattner was developing earcons, Gaver was considering how these everyday sounds could be employed in computer interfaces (Gaver 1989). He proposed "auditory icons" as "everyday sounds mapped to computer events by analogy with everyday sound-producing events" (Gaver 1997). Gaver considered that auditory icons were the auditory equivalent of visual icons, and like those visual icons the relationship between the auditory signifier and its signified was largely focused towards the iconic end of the signifier/signified relationship. For example, the use of breaking glass could act as a signifier of an error occurring in a computer process (Gaver 1989). The advantage of auditory icons, in comparison to earcons, is that they are easily learned, with an "intuitive" mapping between signifier and signified, overcoming the need to learn "arbitrary" mappings as with earcons.

11.3.2.1. Sonic Finder

Gaver demonstrated use of auditory icons in the SonicFinder (Gaver 1989). This was an auditory augmentation of the graphical user interface (GUI) for an Apple Macintosh computer. Operations in the user interface (opening windows, dragging files, emptying the trash can, etc.) were mapped to representative sounds, providing an auditory background to complement vision. While iconic mappings to represent events, such as a camera shutter sound being used to represent a screenshot being taken, or crumpling paper being used to represent the trash can being emptied,[1] are common examples of auditory icons, Gaver took an ecological approach in deciding the sounds to use (Gaver 1993a). This focused on "everyday listening," where we do not so much hear the musical properties of sounds (as in musical listening exploited by earcons; Blattner, Sumikawa and Greenberg 1989), but rather use the sounds to infer physical properties of the objects that we interact with. For example, the noise made by tapping a bottle will reveal if it is made by glass or metal, and whether it is full or empty. SonicFinder used these principles and created an ecological soundscape (Gaver 1989) that augmented how on-screen graphical objects would interact if they existed in the real world. Different on-screen objects (icons, folders, disks, applications, etc.) were mapped to a physical material.[2] How these objects were manipulated by the user mapped to sounds that reflected analogous interactions with those objects in the real world. For example, if the user dragged an icon (represented by metal) to the trash can, clicking on the icon would generate a tapping on metal sound. The size of the file might then be represented through the volume, or size, of the metal tap sound. Dragging this file would then cause the sound of metal being scrapped along a surface. Dropping the file in the trash would generate a breaking sound to act as a confirmation it was dropped in the trash can. The sounds reflected those generated by physical manipulation of the materials used. The analogy of mapping sounds to everyday sound-producing events was reflected in other places, such as the sound of water pouring from one container into another acting as an analogy of copying files, with the rate of water flow and the size of the container that water was poured into, respectively, representing the transfer speed and amount of data to be transferred.

11.3.2.2.   ARKola

Gaver, Smith and O'Shea expanded on these ecological characteristics to investigate the role of auditory icons in a collaborative task (Gaver, Smith and O'Shea 1991). The ARKola bottling plant simulated machines in a cola bottling plant. Raw materials arrive (bottles, sugar, etc.) and go through various processes (melting, bottling, packing, labeling, etc.) to produce bottles of cola. All machines must be kept working, running at the correct rate, and supplied with material to ensure bottles of cola are produced. However, the plant was split in two and required two users, in different parts of the factory, to monitor and control the processes. An audio and video link were provided to allow both controllers to communicate. Gaver, Smith and O'Shea compared how successfully pairs of participants could keep the plant running with and without auditory icons (Gaver, Smith and O'Shea 1991).

Gaver, Smith and O'Shea developed a set of auditory icons to represent actions by the machines (e.g. the capping machine putting a cap on a bottle) (Gaver, Smith and O'Shea 1991). These sounds were played repeatedly, and rhythmically. They argued that the overall sound was like a car engine, with the cadence of each machine adding to an auditory whole that would provide an awareness of the health of the plant. Additional sounds indicating machine breakdowns, as well as sounds to indicate wasted materials when a machine was running too fast or slow, were added. For example, the sound of spilled cola if the bottling machine was supplying bottles too slowly, or the sound of broken glass if it was running too fast. Sounds were designed to avoid perceptually masking each other if they overlapped. Gaver, Smith and O'Shea found that the auditory feedback supported more effective background monitoring of the process and allowed effective awareness and shared reference points between participants. Auditory icons supported individuals to become quickly aware of an error, and provided an indication of what and where the error was (in comparison to conventional error sounds). They did note however, that the urgency of the alarm sounds did not always reflect the urgency of the underlying issue. For example, the sound of breaking glass often caused participants to try to stop the sound (by turning off the machine) before understanding the problem and correcting it.

11.3.2.3. Creating Auditory Icons

In later auditory icon examples (Ankolekar, Sandholm and Yu 2013), they are largely restricted to single event sounds, where a sound is iconically or metaphorically mapped to a computer event, losing the ecological relationship argued by Gaver et al. For example, the "whooshing" sound played when an email is sent on an Apple iPhone is an auditory icon. Most of the work described in section 11.4 and 11.5 reflects this more limited use of auditory icons. In part, this is due to the limited ability to manipulate sounds across everyday dimensions. While musical sounds, like earcons (Blattner, Sumikawa and Greenberg 1989), can be easily created using music editing software, computational generation of everyday sound dimensions to create more parameterized auditory icons (Gaver 1993b) is much more challenging; it was particularly so in the late 1980s, when both earcons and auditory icons were introduced. Attempts to support this kind of audio manipulation have been made. Gaver proposed synthesis algorithms to outline how the kinds of auditory icons used in ARKola (Gaver, Smith and O'Shea 1991) and SonicFinder (Gaver 1989) systems could be created (Gaver 1993b). Nicol, Brewster and Gray developed a prototype system to support manipulation of sounds (Nicol, Brewster and Gray 2005, 2004) according to material properties using timbre spaces (Štěpánek, Otčenášek and Melka 1999).

The lack of readily available tools means that most work on auditory icons still relies on sampled sound files (either recorded from real-world objects or synthesized using Foley techniques (Donaldson 2014)). A notable exception is Williamson, Murray-Smith and Hughes who developed the Shoogle system (Williamson, Murray-Smith and Hughes 2007). This used inertial sensors to monitor the orientation and movement of a mobile device. This movement was then linked to a physics-based simulation that represented some state of the device. For example, each SMS message was represented by a ball, with the material and size of the ball mapped to the relationship and length of the SMS message respectively. As the user shook the device, the balls would bounce off each other, providing a lightweight auditory overview of the messages in the device. Similarly, they metaphorical mapped remaining battery life to the sound of liquid in a container, such that a full battery would sound like "a full bucket of water sloshing around," to a few drips if the battery was almost expired. Williamson, Murray-Smith and Hughes used granular synthesis, where small samples are taken from a recording of a particular sound (e.g. a metal impact sound), and combined to manipulate the acoustical parameters of those sound samples (O'Modhrain and Essl 2004).

11.3.2.4. Designing Auditory Icons

In comparison to earcons, there is much less guidance on how to select sounds to use as auditory icons. Gaver provides a hierarchical description and map of the kinds of sounds objects in different states (gas, liquid, solid) would make when they are manipulated (Gaver 1993a). For example, liquids might make dripping or splashing sounds. These would be influenced by the viscosity of the liquid, and so forth. This provides a good starting point to consider parameterizing sounds. More concretely, Mynatt proposed a set of design principles to create an effective set of auditory icons (Mynatt 1994b). She employed these in her work to develop a GUI that was usable by blind individuals (Mynatt 1997). She identified four factors that influenced the usability of auditory icons: identifiability, conceptual mapping, physical parameters and user preferences (see Table 11.2).

Mynatt further proposed a methodology on how to apply these considerations in Auditory Icon design (Mynatt 1994b). Most notably, she argues that user testing should be done to determine if users can identify the sounds selected (i.e. what the sound is of, rather than its mapping in a user interface) using free form answers. That auditory icons should be tested in isolation, outside of their use in the auditory display, to determine how

Table 11.2 A Outline of Key Considerations That Should Be Considered to Design Effective Auditory Icons.

Factor	*Description*
Identifiability	Auditory icons fundamentally exploit user's existing understanding of the auditory world to link sound referents to meaning. Therefore, users must be able to recognize what the sound means in the real world before that understanding can be exploited. If the sound is not common, then individuals might not recognize it. If the sound is not unique (in comparison to other sounds) it might be confused. For example, Cohen used the sound of a key turning in a lock to represent a user logging in to a system (Cohen 1994). Although there is a clear metaphor between the key turning in a lock and logging in, the sound could not be identified as a key turning in a lock, making the mapping irrelevant.
Conceptual Mapping	Considering the mapping between sound and what it represents is obviously important. Mynatt provides little guidance on this but notes the importance of cultural understanding. Is it likely users will have the domain knowledge needed to understand the link between signified and signifier? Cohen notes how he used sounds from the *Star Trek* TV series to represent file-sharing events in an auditory display (Cohen 1993). He noted how participants' ability to grasp the mappings of these depended on their familiarity with the TV series. As we discuss later in musicons, Mynatt notes how individuals often build complex stories around the sounds they hear (Mynatt 1994b).
Physical Parameters	As discussed more fully under earcon design, the parameters of the sound, including length, intensity, quality, bandwidth and frequency all impact on the identifiability of the sound. If sounds share similar bandwidths and are played at the same time, they are likely to partly mask each other and impair identification. Gaver, Smith and O'Shea noted that they were careful to avoid these issues in the design of their auditory icons (Gaver, Smith and O'Shea 1991).
User Preferences	How users emotionally respond to sounds also has a strong impact. Sounds that are harsh, irritating or annoying will not be used by users. They may just switch sound off.

Source: Adapted from (Mynatt 1994a)

easily each maps to what it represents. Then usability testing with auditory icons in the final use context should be carried out. Mynatt's approach was evaluated through her Mercator work (Mynatt 1997; Edwards and Mynatt 1994), and while appears to work well, it is time consuming to implement.

11.3.3. Spearcons

Researchers have spent considerable time comparing auditory icons and ear-cons to determine which is "best," with varying results (Bussemakers and Haan 2000; Lemmens, Bussemakers and de Haan 2001). A consideration of earcons as more symbolic representations, and auditory icons as more semantically iconic representations, however, highlights that the advantages of one are often the drawbacks of the other. Earcons are easy to create, but that means they must be learned. Auditory icons can remove the need to learn the mappings, but only if an appropriate sound (with an appropriate mapping to what it represents) can be found. While both can be combined (such as using auditory icons as part of a compound earcon), few research-ers have considered this (Dingler, Lindsay and Walker 2008). One auditory cue that is both easily learned, and easy to find mappings for, is synthesized speech. However, speech can be slow to listen to. Walker, Nance and Lind-say proposed spearcons as a way to overcome these issues (Walker, Nance and Lindsay 2006). Spearcons are synthesized text-to-speech recordings, but are presented at a high rate, such that they may only be heard as a unique sound without the spoken words being discernible. In a sense, they repre-sent the indexical mapping discussed in section 11.2, where the original text "causes" the auditory spearcon sound. Both are intrinsically linked.

Spearcons were original conceived to support navigation for visually impaired individuals. Although existing work on earcons can support structural awareness of a user's location in a menu system (e.g. Leplâtre and Brewster 2000), and auditory icons can support awareness of items in the menu (e.g. the currently selected menu item), both are "brittle" (Walker, Nance and Lindsay 2006). They do not support users well when the menu is changed (either adding or removing items, or changing the structure). Changes need to be manually made by a designer, requiring users to learn the new structure or items in the menu. The impact of this depends on the type of cue and how the mapping has been derived. As with the work of Leplâtre (Leplâtre and Brewster 2000), Walker (Walker et al. 2013) considered that spearcons would function as a prefix to a spo-ken menu item. Thus, the menu item "save file" would be heard as the spearcon of "save file," followed by the spoken text-to-speech "save file." Although Spearcons cannot be heard as the words they are derived from, similar words will share acoustic similarities (e.g. "save" and "save as" will share an acoustically similar first part), and will vary on length based on the words. In this way, understanding of the spearcon could be learned over time, allowing users to become faster and preempt the spoken text, speeding up interaction over time.

Although spearcons have mostly been studied in menu-based structures, where the goal of users is to locate a specific item in a list, studies that have applied them outside this context have also found benefits. Hussain et al. evaluated how spearcons could be used to provide audio walking directions for a pedestrian (Hussain et al. 2016). A notable point of Hussain et al.'s work, is that they compared Chinese spoken instructions to both English and Chinese derived spearcons. Both spearcon sets provided similar performance, supporting significantly faster completion than spoken instructions. All participants in their study were native Chinese speakers, and their findings imply that spearcons can work cross-culturally. Such findings highlight the need to better consider how all forms of auditory cues work cross-culturally.

11.3.4. Spindex

In considering the benefits of spearcons with menus, and how words with similar prefixes would sound similar when converted to spearcons, Jeon and Walker developed spindex cues (Jeon and Walker 2011). These exploited this property to speed navigation through alphabetically ordered lists. Instead of prefixing with a spearcon, spindex prefixed each entry with its first letter. This provided structural level support of the menu (or list). In this way, as users quickly scroll through a list, they can identify both when the spindex changes (indicating they are browsing entries with a new letter), and over time (as users learn the spindex sound generated for a given letter) have place on where in the list they are. Jeon and Walker found that spindex significantly reduced the time taken to find an item in an auditory list. As spindex requires the list to be sorted to be beneficial, and this would cause the same spindex to be played for each menu item, Jeon and Walker adjusted the cues based on subsequent presentation (i.e. the first instance of "A" would be played, but subsequent menu items would have the "A" modified). They identified that attenuating the amplitude of the second to the last instance of a spindex cue (either by a constant amount, or gradually) had no negative impact on performance, but was preferred by participants.

11.3.5. Musicons

Another approach to overcome the learnability issues of earcons, yet retain the easy learnability of auditory icons, are musicons (McGee-Lennon et al. 2011). Musicons represent short snippets (between 0.2 seconds and 1–2 seconds long) of music tracks (such as pop songs, film soundtracks or classical compositions). McGee-Lennon et al. were inspired by drop needle tests, common in musicology classes to test the ability of students to name a track from a short sample (McGee-Lennon et al. 2011). They were focused on the role musicons could play in home reminder systems, such as those that might be used by elderly or cognitively impaired persons to support independent living. Such reminders may be presented over a home voice assistant device (e.g. Amazon Echo (Wikipedia Foundation 2018a) or Google Home (Wikipedia Foundation 2018b)), and may contain information that is sensitive, and which a person may wish to keep private. Therefore, like earcons (see section 11.3.1), musicons should incorporate some level of privacy. McGee-Lennon et al. compared musicons to spoken reminders for everyday household tasks. While participants could identify speech with 100% accuracy, musicons had an identification rate around 95%, with musicons of 0.5 seconds having both high accuracy and being most preferred by participants. Participants still had recognition accuracy of over 80% when retested after one week.

McGee-Lennon et al. intended that mappings between musicons and what they represented should be symbolic, similarly to earcons. However, they found participants used both the lyrics of the song, as well as its cultural use, to help remember what the musicon represented. For example, in using the theme from the TV show *Friends*, participants associated actors entering and leaving the scene via a door to help remember that the musicon mapped to a "lock the door" reminder. The main ability for musicons to act as good cues is therefore likely to be both in individuals being able to firstly identify the song used, as well as being able to exploit the content and context of the music to create meaningful and memorable associations between the musicon signifier and what it signifies. This reflects the design process for auditory icons discussed earlier (Mynatt 1994b). The ability to do this is clearly impacted by both cultural convention and prior individual knowledge. Therefore, musicon privacy is likely impacted by the common knowledge of the music used.

McLachlan, McGee-Lennon and Brewster sought to investigate these issues by asking users to select musicons from tracks in their own music collections (McLachlan, McGee-Lennon and Brewster 2012). For each track participants selected a five-second section representing their favorite part of the track, and a section that represented its "essence." From these sections musicons were generated. However, in comparison to control tracks, which were drawn from popular culture, the user selected tracks were less well identified (although these were still identified with high accuracy) but were preferred.[3] McLachlan, McGee-Lennon and Brewster further studied the characteristics of the musical section selected by participants. They identified vocal content, as well as content from the chorus and any riffs, as being most common, and derived guidelines on how to select candidate musical sections from a track to create musicons.

While work has shown musicons as a potential auditory cue type, so far there has been little work to study how to best create them, particularly in comparison to more established cue types, such as earcons, auditory icons and spearcons. While the initial guidelines of McLachlan, McGee-Lennon and Brewster provide a good start to select parts of music tracks to form musicons (McLachlan, McGee-Lennon and Brewster 2012), there are currently no guidelines on how those musical snippets should be best mapped to data. Investigating this is an area ripe for further study.

11.3.6. Other Cues

Beyond the clearly classified auditory mappings that have been introduced, several instances of other cue types have been proposed by researchers. None of these have yet been well studied, but they do provide further ways sound might be used to effectively communicate data.

Ferati et al. proposed "audemes," described as "a type of non-speech sounds whose meaning is derived from the combination of sounds" (Ferati, Mannheimer and Bolchini 2009). They provide the example of combining the sound of jangling keys and a car engine starting to represent the meaning of driving. Ferati et al. argued that audemes differ from auditory icons in that auditory icons focus on causal relationships (e.g. a tapping sound means an object has been selected), while audemes are argued to support the listener to consider the wider socio-technical context in which the sound exists. That is, the mapping of data to sound is both metaphorical, as with auditory icons, but also evocative (e.g. a lion roar may convey both danger, but also power or stature). In this way, audemes are best considered as providing a feeling, or sense, of information. Ferati, Mannheimer and Bolchini, for example, describe how audemes could be used to support blind users in browsing a bookshelf, with audemes representing the genre of a selected book (Ferati et al. 2012).

Other work has considered the convenience of emotions through non-speech audio. Such work often parallels the notion of emoticons in text messaging, where small images of facial expressions can be incorporated into sent text. Froehlich and Hammer compared speech and non-speech, human-like sounds (e.g. a giggle) to convey emoticons in a visually impaired email reading task (Froehlich and Hammer 2004). They identified that users significantly preferred the use of human-like sounds to express these. Wersényi conducted further studies using auditory emoticons that are non-speech human sounds (Wersenyi 2009). A survey with 50 users found most were positive towards them. Jeon has proposed "lyricons" (Jeon 2013), non-speech cues (similar to earcons) that are phonetic representations of spoken commands. For example, the command "Function On" would be represented by three notes that follow the same pitch structure as the words if spoken. Initial studies (Sun and Jeon 2015) have been positive, but no further evaluations have been carried out.

11.4. Comparing Cue Types

Given the wide variety of ways data can be encoded in auditory cues, an obvious question is: "Which one is best?" There is relatively little work that focuses on direct comparison of cue types. An overview of the main cue types is given in Table 11.3. Most work focuses on the differences between auditory icons and earcons. From our previous discussion, auditory icons, with iconic and metaphorical mappings, are more "intuitive," and don't require explicit training. However, it can be difficult to identify real-world sounds that provide such a mapping for many concepts. It is also practically difficult to manipulate auditory icons along real-world sound-producing dimensions (as outlined in Gaver 1989). Well-designed earcons can effectively represent any concept (as the mapping is largely arbitrary). However, this means users must be trained in some way to identify the earcons. In specialized applications (e.g. an airplane cockpit) where training is provided, this may be possible. In consumer technology, this is much less likely. In practical use both earcons and auditory icons are very simple versions, and often lack the more complex relationships previously described. While auditory icons may be seen as a better option, this may not always be the case. Sikora, Roberts and Murray (Sikora and Roberts 1997) and Roberts and Sikora (Roberts and Sikora 1997) compared environmental and musical sounds in terms of agreement of function, appropriateness and pleasantness. Musical sounds were rated as more pleasant and appropriate than environmental sounds. However, both were rated as less appropriate and pleasant than speech.

More recently, spearcons have been found to offer advantages over both earcons and auditory icons. Walker et al. compared spearcons to a set of both auditory icons and earcons in a selection task with a set of four drop-down menus (similar to the menus in a menu bar on a desktop computer application) (Walker et al. 2013). All three were presented as a prefix to the menu items being spoken with text-to-speech software and were also compared to text-to-speech without a prefix. They found spearcons were both significantly faster than earcons and auditory Icons, and were more accurate. No significant effect was found between spearcons and menus items being read as speech. However, Walker et al. identified that the text-to-speech part of the cue was only listened to on average in 11% of trials, meaning that participants were largely judging the menu item based on the spearcon (Walker et al. 2013). Dingler, Lindsay and Walker have compared learnability between earcons, auditory icons, spearcons and speech (Dingler, Lindsay and Walker 2008). They developed a set of 20 possible audio cues that represented environmental objects (trees, fire hydrants, etc.) that might be encountered in a visually impaired navigation context.

Table 11.3 An Overview of the Main Auditory Cue Types and Their Key Advantages and Disadvantages.

Cue Type	Description	Advantages	Disadvantages
Earcons	Short structured music-like sounds that can be manipulated using musical rules to encode information.	• Relatively rich messages can be generated from a set of simple rules. • As the encoding between sound and what it signifies is arbitrary, any signified can be represented. • Earcons have been extensively evaluated, and there are good guidelines for their creation and use.	• As the encoding between sound and signified is arbitrary it must be learned, requiring training. • The need to have sounds be perceptually different can limit the size of a set of possible earcons.
Auditory Icons	Everyday sounds used to represent computer events by analogy of everyday sound-producing events.	• The relationship between sound and what it represents is based on common knowledge, so training is not required. • Auditory icons can employ more dynamic elements, such as physics simulations. • Guidelines are available to support effective auditory icon creation.	• It can be hard to identify sounds that map to all needed events. • Manipulating sound across real-world parameters is nontrivial. • Auditory icons assume domain knowledge of the everyday use of sounds, it is not clear if this will work cross-culturally.
Spearcons	Speech significantly speeded up to sound like an auditory texture.	• Can easily create spearcons for required signifieds • Sounds can be trivially generated. • Can be used to overcome the "brittle" nature of other cues in structured menus.	• Evaluation of spearcons has largely been focused on menu structures, it is not clear how spearcons will perform in many of the situations auditory icons and earcons work.
Spindex	A modification to spearcons to emphasize the first letter and provide quicker navigation through alphabetized menus.	• Can improve the access speed in alphabetized menu structures.	• Is limited to supporting alphabetized menus and is not a general-purpose technique (as with auditory icons, earcons and potentially spearcons). • A spoken (or visual) menu item is still required. Spindex only provides a faster way to navigate through.
Musicons	Small snippets of music used to represent information associated with the music or lyrics.	• Supports individuals to create their own link between the musicon and referend. • Can potentially provide a degree of privacy in the meaning of the musicon.	• There is limited study and guidelines on selecting effective musicons.

Participants were first trained by listening to each sound and being told what it represented, then were tested on each of the sounds by being presented with a sound and selecting what it represented. If the participant did not get all of the cues correct, the training and testing were continued until all 20 cues were correctly identified. Spearcons were found to be as easy to learn as speech, requiring on average one training cycle. Auditory icons required on average three training cycles, while earcons required on average eight.

While spearcons would therefore seem to be the best option, their evaluation has largely been focused on list and menu item browsing. It is also unknown how spearcons would perform when encoding multiple parameter data (which might be represented as a relatively long textual phrase), or where the concept or feature might not be able to be communicated in one to two words. Unfortunately, beyond auditory icons, earcons and spearcons, we have few good guidelines on how to design other cue types, and therefore no studies that seek to compare their relative benefits and drawbacks.

11.5. Applications of Cues

In considering the range of auditory cues that can be used, it is also worth considering the areas where they have been deployed and found useful. It should be noted that these areas are not exhaustive, but represent areas where auditory cues have been shown to benefit, or are emerging into.

Auditory cues have been found useful in cases where the user cannot devote their visual attention to a GUI. ARKola (Gaver, Smith and O'Shea 1991), as previously discussed, represents such an example. It wasn't possible for either user to monitor the whole bottling plant visually, with audio helping to provide awareness. Such ideas also extend to entirely non-visual scenarios. Mynatt et al. developed Audio Aura, an early ubiquitous computing system that employed active badges and other location sensing technology to track employees in a building (Mynatt et al. 1997, 1998). They developed a set of parameterized auditory cues (earcons, auditory icons and speech) that allowed individuals to gain high-level information about their colleagues. For example, passing by a colleague's office might present an overview of how long it has been since their colleague was last there, and by inference when they would be likely to return. These ideas were extended by Sawhney and Schmandt who developed an auditory display for personal information (Sawhney and Schmandt 2000). They employed the idea of using different cue types to represent increasing prominence of data. Their Nomadic Radio system would firstly introduce information (e.g. a new text message) as an ambient sound that would evolve through multiple cue types, each becoming more detailed, until the full text message was read. In this way, information would emerge from the background, and could be attended to or dismissed by users, avoiding the need for all information to be read out as speech.

More recently, McGee-Lennon and Brewster have applied auditory cues to support home reminders (like remembering to take keys before leaving one's house, or remembering to take out the trash), with the objective of supporting the elderly to lead independent lives for longer (McGee-Lennon and Brewster 2011). Their work considered both the effectiveness of cues, but also the social context of their use. For example, many reminders may be private or embarrassing if individuals have a visitor. Having cues that are discrete, or have meanings that are unknown to visitors, may be beneficial. This was one of the motivations in creating musicons, that they would be somewhat private (McGee-Lennon et al. 2011).

A final example is in supporting individuals who are visually impaired. While modern smartphones can "read out" information as a user places his or her finger on the touchscreen, the Mercator project (Edwards, Mynatt and Stockton 1995) employed a variety of auditory cues to help blind users navigate and understand graphical user interfaces. In many ways an extension of Gaver's SonicFinder (Gaver 1989), Mercator incorporated filters and other audio processing techniques to change the sounds to reflect their on-screen state, for example, low-pass filtering an auditory icon to indicate it was "greyed-out" and not selectable. Extensive work has also been carried out to support blind individuals in accessing and manipulating mathematical equations, with auditory cues used to provide contextual information on brackets, sub-expressions and other obvious visual mathematical notation that becomes cumbersome when accessed non-visually (Stevens, Edwards and Harling 1997). Murphy, Bates and Fitzpatrick have begun to compare earcons and spearcons to determine which is more effective in communicating mathematical notation (Murphy, Bates and Fitzpatrick 2010). Beyond application to GUIs, other work has considered how auditory cues can be used within sonification systems. Here, data directly drives a sonified output. The most common (and basic) approach being to map musical pitch to a continuous data value, making it possible to present that data (e.g. a line graph) in sound (Mansur 1985). Such sonifications can be complex and multidimensional, necessitating the need to mark and refer to particular data points easily (Kildal and Brewster 2007). Beacons—audible markers—can support users to mark and refer to these points later (Kramer et al. 1999). While beacons do not specify the kinds of sound cues that can be used, all of the types discussed could fulfill this role. Beacons can also be used outside of data visualizations. The locative cues used by Audio Aura can be considered a kind of beacon, while spearcons were originally developed to form part of Wilson et al.'s SWAN navigation system (Wilson et al. 2007) for visually impaired users. This was designed to both support navigation and increase awareness of the environment around a visually impaired person. Spearcon beacons were identified as being able to effectively support use of the SWAN system (Dingler, Lindsay and Walker 2008).

While sound is clearly beneficial when there is no visual interface, it also brings benefits when the amount of screen space is reduced. In such cases on-screen widgets can become smaller, making them difficult to hit. Brewster applied simple earcons to button events in a touchscreen keypad (Brewster 2002). He found that the application of earcons allowed the size of the buttons to be significantly reduced, without negatively impacting performance.

A final area that is of increasing importance is in the application of auditory cues to support driverless and driver-assisted cars (Siwiak and James 2009). Particularly as cars move from internal combustion engines, which have natural auditory cues, towards electric cars with fewer such auditory cues, there is a need to consider how to best present information to the driver. Beattie, Baillie and Halvey compared different auditory cues for replacing vehicle sounds (Beattie, Baillie and Halvey 2015). They found that earcons were comparable in both user workload and usability to existing sounds that cars make. They further found that auditory icons and speech were less usable, with auditory icons also being distracting to the driving task. Gabel et al. applied spindex to the selection of a music track on a mobile phone during a simulated driving task (Gable et al. 2013). They found that spindex cues reduced the time the driver had to take their eyes off the road and resulted in better driving than when no audio cues were provided. While using a mobile phone while driving is not recommended, their results highlight how the addition of sound to various in-car controls could help reduce the time drivers need to remove attention from the road. Such cues have also been studied to provide awareness of potential collisions, awareness of hazards for driving and other alerts for drivers to alter their behavior. An overview of cues in this area is presented by Nees and Walker (Nees and Walker 2011).

11.6. Conclusions

In this chapter we have tried to give a broad overview of the main audio cue types that can be used in an auditory display, how they relate to each other and how to best design them. However, it is also worth discussing some of the limitations of this work and things that we do not know. A common question to ask is "Which cue type is best?" While there are comparative studies (Dingler's comparative work on spearcons is probably among the most broad; Dingler, Lindsay and Walker 2008), such comparisons are only as good as the design choices made in creating the cues. Sometimes, guidelines of best practice are not followed. Unlike the premade toolkits for GUIs, which allow developers to use high-quality interface widgets without the need to design their visual look and feel, there are no similar solutions for auditory cue designers. All cues must be hand crafted. This requires an understanding of both their effective design and their aesthetic design to make sounds pleasant. Solutions such as Nicol, Gray and Brewster (Nicol, Brewster and Gray 2005), as well as attempts to create sound-enhanced widget toolkits, have been developed (Crease, Gray and Brewster 2000), but sound generation programs like Apple's GarageBand (Apple Inc. 2018) are still the most common way to generate auditory cues.

Another relevant point is the cultural impact of cue design. As we discussed, metaphorical and iconic mappings are often derived from existing real-world knowledge or conventions. However, most work has been undertaken on US and European participants. There has been little work on cross-cultural studies to consider how effective techniques are. Are individual earcons or Musicons effective across different cultures? With the notable exception of Hussain et al. (Hussain et al. 2016), who investigated Chinese language spearcons, we do not know. Such study is an another area ripe for further investigation.

Notes

1. This was originally part of Gaver's SonicFinder (Gaver 1989) but is still used in the current Mac OS operating system.
2. Not every auditory icon had an intuitive meaning, in this case the mapping is largely arbitrary and needs to be learned.
3. McLachlan, McGee-Lennon and Brewster (2012), however, did not provide a mapping from the musicons to some data. In essence, there was no signified. Therefore, identification was based on recall of the musical track, rather than what it represented.

References

Ankolekar, A., Sandholm, T. and Yu, L., 2013. Play It by Ear: A Case for Serendipitous Discovery of Places with Musicons. In: *Proceedings of CHI 2013*, pp. 2959–2968. Viewed Jul 1, 2018 http://dl.acm.org/citation.cfm?id=2481411

Apple Inc., 2018. Garageband. Viewed Jul 1, 2018 www.apple.com/lae/mac/garageband

Beattie, D., Baillie, L. and Halvey, M., 2015. A Comparison of Artificial Driving Sounds for Automated Vehicles. In: *Proceedings of the 2015 ACM International Joint Conference on Pervasive and Ubiquitous Computing*, Springer Verlag, pp. 451–462.

Blattner, M.M., Sumikawa, D.A. and Greenberg, R.M., 1989. Earcons and Icons: Their Structure and Common Design Principles. *Human Computer Interaction*, 4, no. 1, pp. 11–44.

Brewster, S., 2002. Overcoming the Lack of Screen Space on Mobile Computers. *Personal Ubiquitous Computing*, 6, no. 3, pp. 188–205. Viewed Jul 1, 2018 http://dx.doi.org/10.1007/s007790200019

Brewster, S. and Brown, L.M., 2004. Tactons: Structured Tactile Messages for Non-Visual Information Display. In: *Proceedings of the Fifth Conference on Australasian User Interface—Volume 28. AUIC '04.* Darlinghurst, Australia: Australian Computer Society, Inc., pp. 15–23. Viewed Jul 1, 2018 http://dl.acm.org/citation.cfm?id=976310.976313

Brewster, S.A., 1994. *Providing a Structured Method for Integrating Non-Speech Audio into Human-Computer Interfaces.* PhD thesis, Department of Computer Science, University of York. Viewed Jul 1, 2018 http://citeseerx.ist.psu.edu/viewdoc/download?doi=10.1.1.157.33&rep=rep1&type=pdf

Brewster, S.A., 1998. Using Nonspeech Sounds to Provide Navigation Cues. *ACM Transactions on Computer-Human Interaction*, 5, no. 3, pp. 224–259. Viewed Jul 1, 2018 http://doi.acm.org/10.1145/292834.292839

Brewster, S.A., Raty, V.P. and Kortekangas, A., 1996. Earcons as a Method of Providing Navigational Cues in a Menu Hierarchy. In: *Proceedings of HCI 96.* London: Springer, pp. 167–183.

Brewster, S.A., Wright, P.C. and Edwards, A.D.N., 1993. An Evaluation of Earcons for Use in Auditory Human-Computer Interfaces. In: *Proceedings of the INTERACT '93 and CHI '93 Conference on Human Factors in Computing Systems. CHI '93.* New York: ACM, pp. 222–227. Viewed Jul 1, 2018 http://doi.acm.org/10.1145/169059.169179

Brewster, S.A., Wright, P.C. and Edwards, A.D.N., 1994. A Detailed Investigation into the Effectiveness of Earcons. In: Kramer, G. (Ed.), *Auditory Display: Sonification, Audification, and Auditory Interfaces.* Reading, MA: Addison-Wesley, pp. 471–498.

Brewster, S.A., Wright, P.C. and Edwards, A.D.N., 1995a. Experimentally Derived Guidelines for the Creation of Earcons. In: *Proceedings of BCS-HCI 95.* Huddersfield, UK: Springer, pp. 155–159.

Brewster, S.A., Wright, P.C. and Edwards, A.D.N., 1995b. Parallel Earcons: Reducing the Length of Audio Messages. *International Journal of Human-Computer Studies*, 43, no. 2, pp. 153–175.

Bussemakers, M.P. and Haan, A.D., 2000. When It Sounds Like a Duck and It Looks Like a Dog . . . Auditory Icons vs. Earcons in Multimedia Environments. In: *ICAD 2000.* Atlanta, GA: ICAD.

Chandler, D., 2002. *Semiotics: The Basics.* New York: Routledge.

Cohen, J., 1993. "Kirk Here": Using Genre Sounds to Monitor Background Activity. In: *INTERACT '93 and CHI '93 Conference Companion on Human Factors in Computing Systems.* CHI '93. New York: ACM, pp. 63–64. Viewed Jul 1, 2018 http://doi.acm.org/10.1145/259964.260073

Cohen, J., 1994. Monitoring Background Activities. In: *Santa Fe Institute Studies in the Sciences of Complexity Proceedings.* Reading, MA: Addison-Wesley.

Crease, M., Gray, P. and Brewster, S., 2000. A Toolkit of Mechanism and Context Independent Widgets. In: *International Workshop on Design, Specification, and Verification of Interactive Systems*, pp. 121–133.

Dingler, T., Lindsay, J. and Walker, B.N., 2008. Learnability of Sound Cues for Environmental Features: Auditory Icons, Earcons, Spearcons, and Speech. In: *Proceedings of the 14th International Conference on Auditory Display*, June 24–27. Paris, France. Viewed Sept 5, 2018 http://sonify.psych.gatech.edu/publications/pdfs/2008ICAD-DinglerLindsayWalker.pdf

Donaldson, L.F., 2014. The Work of an Invisible Body: The Contribution of Foley Artists to On-Screen Effort. *Alphaville: Journal of Film and Screen Media*, 7. Viewed Jul 1, 2018 https://cora.ucc.ie/handle/10468/5846

Edwards, K.W. and Mynatt, E.D., 1994. An Architecture for Transforming Graphical Interfaces. In: *Proceedings of UIST 94*. Marina del Ray, CA: ACM Press, pp. 39–47.

Edwards, K., Mynatt, E.D. and Stockton, K., 1995. Access to Graphical Interfaces for Blind Users. *Interactions*, 2, no.1, pp. 56–67.

Ferati, M. et al., 2012. Audemes at Work: Investigating Features of Non-Speech Sounds to Maximize Content Recognition. *International Journal of Human-Computer Studies*, 70, no. 12, pp. 936–966.

Ferati, M., Mannheimer, S. and Bolchini, D., 2009. Acoustic Interaction Design Through Audemes: Experiences with the Blind. In: *Proceedings of the 27th ACM International Conference on Design of Communication*, ACM Press, pp. 23–28.

Froehlich, P. and Hammer, F., 2004. Expressive Text-to-Speech: A User-Centred Approach to Sound Design in Voice-Enabled Mobile Applications. In: *Second Symposium on Sound Design*.

Gable, T.M. et al., 2013. Advanced Auditory Cues on Mobile Phones Help Keep Drivers' Eyes on the Road. In: *Proceedings of the 5th International Conference on Automotive User Interfaces and Interactive Vehicular Applications*, ACM, pp. 66–73.

Gaver, W.W., 1997. Auditory Interfaces. In: Helander, M.G., Landauer, T.K. and Prabhu, P.V. (Eds.), *Handbook of Human-Computer Interaction*. Amsterdam: Elsevier, pp. 1003–1041.

Gaver, W.W., 1993a. How Do We Hear in the World? Explorations in Ecological Acoustics. *Ecological Psychology*, 5, no. 4, pp. 285–313.

Gaver, W.W., 1993b. Synthesizing Auditory Icons. In: *Proceedings of INTERCHI'93*. Amsterdam, The Netherlands: ACM Press, pp. 228–235.

Gaver, W.W., 1989. The SonicFinder: An Interface That Uses Auditory Icons. *Human Computer Interaction*, 4, no. 1, pp. 67–94.

Gaver, W.W., Smith, R.B. and O'Shea, T., 1991. Effective Sounds in Complex Systems: The ARKOLA Simulation. In: *Proceedings CHI'91*. CHI '91. New York: ACM, pp. 85–90. Viewed Jul 1, 2018 http://doi.acm.org/10.1145/108844.108857

Gryzanowski, E., 1877. Richard Wagner's Theories of Music. *The North American Review*, 124, no. 254, pp. 53–81. Viewed Jul 1, 2018 www.jstor.org/stable/25109997

Hankinson, J.C.K. and Edwards, A.D.N., 2000. Musical Phrase-Structured Audio Communication. In: *Proceedings of ICAD 2000*. Atlanta, GA: ICAD.

Hawkes, T., 1972. *Metaphor*. London: Methuen.

Hoggan, E. and Brewster, S.A., 2010. Crosstrainer: Testing the Use of Multimodal Interfaces in Situ. In: *Proceedings of the SIGCHI Conference on Human Factors in Computing Systems*. CHI '10. New York: ACM, pp. 333–342. Viewed Jul 1, 2018 http://doi.acm.org/10.1145/1753326.1753378

Hussain, I. et al., 2016. Chinese-Based Spearcons: Improving Pedestrian Navigation Performance in Eyes-Free Environment. *International Journal of Human-Computer Interaction*, 32, no. 6, pp. 460–469.

Intel Corporation, 2014. Intel Bong Still Going Strong. Viewed Jul 1, 2018 https://newsroom.intel.com/editorials/intel-bong-chime-jingle-sound-mark-history/

Jeon, M., 2013. Lyricons (Lyrics+ Earcons): Designing a New Auditory Cue Combining Speech and Sounds. In: *International Conference on Human-Computer Interaction*. HCI International 2013—Posters' Extended Abstracts. Berlin: Springer, pp. 342–346.

Jeon, M. and Walker, B.N., 2011. Spindex (Speech Index) Improves Auditory Menu Acceptance and Navigation Performance. *ACM Transactions on Accessible Computing*, 3, no. 3, pp. 1–26. Viewed Jul 1, 2018 http://portal.acm.org/citation.cfm?doid=1952383.1952385

Kildal, J. and Brewster, S., 2007. EMA-Tactons: Vibrotactile External Memory Aids in an Auditory Display. In: Baranauskas, C. et al. (Eds.), *Human-Computer Interaction—INTERACT2007*. Lecture Notes in Computer Science. Berlin and Heidelberg: Springer, pp. 71–84. Viewed Jul 1, 2018 http://dx.doi.org/10.1007/978-3-540-74800-7_6

Kramer, G., 1994. An Introduction to Auditory Display. In: Kramer, G. (Ed.), *Auditory Display: Sonification, Audification and Auditory Interfaces*. Reading, MA: Addison-Wesley, pp. 1–77.

Kramer, G. et al., 1999. *Sonification Report: Status of the Field and Research Agenda*, ICAD.

Lemmens, P.M.C., Bussemakers, M.P. and de Haan, A., 2001. *Effects of Auditory Icons and Earcons on Visual Categorization: The Bigger Picture. ICAD 2001*. Espoo, Finland: ICAD, pp. 117–125.

Leplâtre, G. and Brewster, S.A., 2000. Designing Non-Speech Sounds to Support Navigation in Mobile Phone Menus. In: *Proceedings of ICAD 2000*. Atlanta, GA: ICAD, pp. 190–199.

Mansur, D.L., 1985. Graphs in Sound: A Numerical Data Analysis Method for the Blind. *Journal of Medical Systems*, 9, no. 3, pp. 163–174.

McGee-Lennon, M. et al., 2011. Name That Tune: Musicons as Reminders in the Home. In: *Proceedings of the SIGCHI Conference on Human Factors in Computing Systems*. CHI '11. New York: ACM, pp. 2803–2806. Viewed Jul 1, 2018 http://doi.acm.org/10.1145/1978942.1979357

McGee-Lennon, M.R. and Brewster, S., 2011. Reminders That Make Sense: Designing Multimodal Notifications for the Home. In: *Pervasive Computing Technologies for Healthcare (PervasiveHealth), 2011 5th International Conference on*, IEEE, pp. 495–501.

McGookin, D.K. and Brewster, S.A., 2004a. Space the Final Frontearcon: The Identification of Concurrently Presented Earcons in a Synthetic Spatialised Auditory Environment. In: *Proceedings of ICAD 2004*. Sydney, Australia: ICAD.

McGookin, D.K. and Brewster, S.A., 2004b. Understanding Concurrent Earcons: Applying Auditory Scene Analysis Principles to Concurrent Earcon Recognition. *ACM Transactions on Applied Perception*, 1, no. 2, pp. 130–155.

McLachlan, R., McGee-Lennon, M. and Brewster, S., 2012. The Sound of Musicons: Investigating the Design of Musically Derived Audio Cues. In: *Proceedings of the 18th International Conference on Auditory Display*, June 18–21. Atlanta, GA, pp. 148–155.

Murphy, E., Bates, E. and Fitzpatrick, D., 2010. Designing Auditory Cues to Enhance Spoken Mathematics for Visually Impaired Users. In: *Proceedings of the 12th*

International ACM SIGACCESS Conference on Computers and Accessibility. ASSETS '10. New York: ACM, pp. 75–82. Viewed Jul 1, 2018 http://doi.acm. org/10.1145/1878803.1878819

Mynatt, E.D., 1994a. Designing with Auditory Icons: How Well Do We Hear Auditory Cues? In: *Proceedings of CHI 94*. Boston, MA: ACM Press, pp. 269–270.

Mynatt, E.D., 1994b. Designing with Auditory Icons. In: *Proceedings of ICAD 94*. Santa Fe, NM, pp. 109–119.

Mynatt, E.D., 1997. Transforming Graphical Interfaces into Auditory Interfaces for Blind Users. *Human—Computer Interaction*, 12, no. 1–2, pp. 7–45.

Mynatt, E.D. et al., 1997. Audio Aura: Light-Weight Audio Augmented Reality. In: *Proceedings of UIST 97*. Banff, Canada: ACM, pp. 211–212.

Mynatt, E.D. et al., 1998. Designing Audio Aura. In: *Proceedings of CHI 98*. CHI '98. New York: ACM Press/Addison-Wesley Publishing Co., pp. 566–573. Viewed Jul 1, 2018 http://dx.doi.org/10.1145/274644.274720

Nees, M.A. and Walker, B.N., 2011. Auditory Displays for In-Vehicle Technologies. *Reviews of Human Factors and Ergonomics*, 7, no. 1, pp. 58–99.

Nicol, C., Brewster, S.A. and Gray, P.D., 2004. Designing Sound: Towards a System for Designing Audio Interfaces using Timbre Spaces. In: *Proceedings of ICAD 04-Tenth Meeting of the International Conference on Auditory Display*, Jul 6–9. Sydney, Australia.

Nicol, C., Brewster, S.A. and Gray, P.D., 2005. A System for Manipulating Audio Interfaces Using Timbre Spaces. *Computer-Aided Design of User Interfaces*, IV, pp. 361–374.

O'Modhrain, S. and Essl, G., 2004. PebbleBox and CrumbleBag: Tactile Interfaces for Granular Synthesis. In: *Proceedings of the 2004 Conference on New Interfaces for Musical Expression*, NIME, pp. 74–79.

Roberts, L.A. and Sikora, C.A., 1997. Optimizing Feedback Signals for Multimedia Devices: Earcons vs Auditory Icons vs Speech. In: *IEA97*. Tampere, Finland, pp. 224–226.

Sawhney, N. and Schmandt, C., 2000. Nomadic Radio: Speech and Audio Interaction for Contextual Messaging in Nomadic Environments. *ACM Transactions on CHI*, 7, no. 3, pp. 353–383.

Shinn-Cunningham, B. et al., 1997. Auditory Displays. In: Gilkey, R.H. and Anderson, T.R. (Eds.), *Binaural and Spatial Hearing in Real and Virtual Environments*. Mahwah: Lawrence Erlbaum, pp. 611–633.

Sikora, C.A. and Roberts, L.A., 1997. Defining a Family of Feedback Signals for Multimedia Communication Devices. In: *INTERACT '97*. Sydney, Australia: Chapman & Hall, pp. 374–380.

Siwiak, D. and James, F., 2009. Designing Interior Audio Cues for Hybrid and Electric Vehicles. In: *Audio Engineering Society Conference: 36th International Conference: Automotive Audio*, AES.

Štěpánek, J., Otčenášek, Z. and Melka, A., 1999. Comparison of Five Perceptual Timbre Spaces of Violin Tones of Different Pitches. *The Journal of the Acoustical Society of America*, 105, no. 2, p. 1330.

Stevens, R.D., Edwards, A.D.N. and Harling, P.A., 1997. Access to Mathematics for Visually Disabled Students Through Multimodal Interaction. *Human-Computer Interaction*, 12, no. 1, pp. 47–92.

Sun, Y. and Jeon, M., 2015. Lyricon Lyrics + Earcons Improves Identification of Auditory Cues. In: *Proceedings, Part II, of the 4th International Conference on Design,*

User Experience, and Usability: Users and Interactions—Volume 9187. New York: Springer-Verlag New York, Inc., pp. 382–389. Viewed Jul 1, 2018 http://dx.doi.org/10.1007/978-3-319-20898-5_37

Walker, B.N. et al., 2013. Spearcons (Speech-Based Earcons) Improve Navigation Performance in Advanced Auditory Menus. *Human Factors: The Journal of the Human Factors and Ergonomics Society*, 55, no. 1, pp. 157–182. Viewed Jul 1, 2018 http://journals.sagepub.com/doi/10.1177/0018720812450587

Walker, B.N. and Kramer, G., 1996. Mappings and Metaphors in Auditory Displays: An Experimental Assessment. In: *ICAD 96*. Palo Alto, CA, pp. 71–80.

Walker, B.N., Nance, A. and Lindsay, J., 2006. Spearcons: Speech-Based Earcons Improve Navigation Performance in Auditory Menus. In: *Proceedings of ICAD 2006*, ICAD, pp. 63–68.

Wersenyi, G., 2009. Evaluation of Auditory Representations for Selected Applications of a Graphical User Interface. In: *Auditory Display, 6th International Symposium, CMMR/ICAD 2009*. Copenhagen, Denmark, May 18–22, 2009.

Wikipedia Foundation, 2018a. Amazon Echo. Viewed Jul 1, 2018 https://en.wikipedia.org/wiki/Amazon_Echo

Wikipedia Foundation, 2018b. Google Home. Viewed Jul 1, 2018 https://en.wikipedia.org/wiki/Google_Home

Wikipedia Foundation, 2018c. Peter and the Wolf. *Wikipedia*. Viewed Jul 1, 2018 https://en.wikipedia.org/wiki/Peter_and_the_Wolf

Williamson, J., Murray-Smith, R. and Hughes, S., 2007. Shoogle: Excitatory Multimodal Interaction on Mobile Devices. In: *Proceedings of the SIGCHI Conference on Human Factors in Computing Systems*. CHI '07. New York: ACM, pp. 121–124. http://doi.acm.org/10.1145/1240624.1240642.

Wilson, J. et al., 2007. Swan: System for Wearable Audio Navigation. In: *11th IEEE International Symposium on Wearable Computers, 2007*, IEEE, pp. 91–98.

12

Soundscape Generation Systems

Miles Thorogood

12.1. Soundscape Generation Models

When creating soundscapes for video games, animation or virtual reality, a sound designer performs a set of established tasks to reach the desired goal. Typically, the desired goal can be written down as a description of the intended environment that is to be represented. A commonplace of such a description is the close reading of a script that indicates the sound sources and mood of the scene. From this description, the sound designer knows the learned skills to be engaged.

Soundscape, researcher and soundwalk artist McCartney (2002) broadly outlines the necessary tasks for creating a soundscape composition as follows:

1. *Specify an environmental context*. The specification is a description of the environment to be represented in the sound design.
2. *Sound file retrieval*. Sound files matching the specification, or parts thereof, will be retrieved from a database using semantic criteria, such as keyword search.
3. *Listen for salient regions in recordings*. A sound file is segmented for parts that hold an aesthetic interest. These segments are extracted for processing and mixing.
4. *Mix and sequence*. Regions are sequenced on a timeline, and panning, attenuation and sound effects applied to the sound design output.

Each task in the list requires time to learn, but once a level of mastery is achieved, then some actions become repetitive. For example, searching through an extensive database and extracting regions of audio recordings, and sequencing alternative sound designs is time consuming. We will explore how these processes are automated with the computer.

12.2. Soundscape Composition and Simulation

Designing a soundscape for a fixed-length animation or an open-world video game requires different artistic and technical approaches. These approaches lend themselves to the broad categories of composition or simulation.

The composition approach reflects the process of digital audio workstations (DAWs; e.g Audacity, Reaper, Pro Tools), which have an arbitrary number of audio tracks with audio clips sequenced along a timeline. Audio clips are processed using adjustments to amplitude, fade in/out and effects applied. Different levels of abstracting the soundscape are achieved by processing sounds with the goal of evoking the listener's associations, memories and imagination related to the soundscape. Figure 12.1 shows a simple representation of a layered mix with a long background audio clip, and shorter event sounds on other tracks at irregular intervals. This approach calls for creativity in selecting and mixing sounds and is the primary means of production in sound design for linear media such as film and animation.

Soundscape simulation is a different approach to sequencing and mixing than composition. The simulation approach uses a physical model to place a virtual listener in a simulated three-dimensional environment where sounds occur at different locations and are possibly moving. Figure 12.2 shows a simple representation of the physical model approach, with a three-dimensional space containing a listener, event sounds in fixed and moving positions and an omnidirectional background sound. Industry standard software, such as FMOD and Wwise, support soundscape simulation, and is the typical approach to sound design in video games and virtual reality applications.

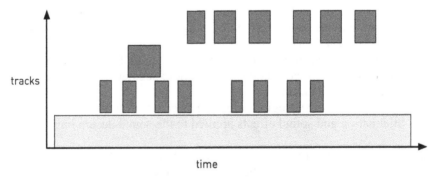

Figure 12.1 The Layered Model Approach to Sequencing and Mixing: A long background audio clip in grey acts as an ambient background for the shorter event sounds in dark grey.

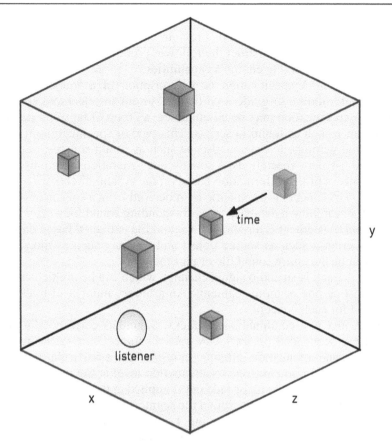

Figure 12.2 The Physical Model Approach to Sequencing and Mixing: A listener is positioned in the three-dimensional space with an omnidirectional background sound (grey), and shorter event sounds (dark grey) in fixed and moving positions.

12.3. Generative Systems

Generative soundscape systems automate one or more of the tasks (i.e. search, segmentation and mixing) typical of sound design. Such systems automate these tasks by executing algorithms from machine learning, music information retrieval and expert systems. These might take the form of, for example, a library of sound files indexed by user-contributed tags, automatic classification and segmentation of sound files, or automated mixing.

One challenge for designing a generative soundscape system is formalizing the creative tasks into a set of useful definitions. By outlining the creative process, we are in a position to model the tasks that a computer program can represent. From our previous definition of the soundscape composition tasks, we construct the following set of definitions necessary for determining a mixing engine's capabilities.

Specification. A specification is a description of a soundscape that includes information to guide a particular system to select and combine sounds. A specification may be an utterance, a group of tags or a descriptive sentence such as found in scripts. This type of specification includes information germane to the soundscape such as sound sources, duration, mood and listener expectations. A soundscape generation system uses one or more pieces of this information based on the system directive.

Concept. A concept is a semantic term derived from a specification that is used for searching a database for corresponding sound files.

Segment. A segment is a region of a sound file extracted based on some semantic criteria such as source object and saliency such as mood. The region can be the entire sound file or part thereof.

Track. A track is an audio signal channel, which can be multichannel or mono, hosting one or more segments. An arbitrary number of tracks can be created for each concept.

Mix. A mix is a combination of tracks. Segments corresponding to a concept are sequenced on tracks. A track is subject to processing with effects, pan and attenuation, affecting a segment at a particular time. For example, applying an envelope to attenuation level at the beginning of a segment results in a fade in, or fade out if applied at the end of a segment.

The specification, concepts, sound file segments and track configuration affect the design of the mixing engine. Looking at this set of definitions, we can associate these with the three principal tasks: search (specification concept); segmentation (segment); and mixing (track mix).

A soundscape generation system aims to automate one or more of these tasks. The balance between automated and human involved tasks directs the system toward different use cases. While some systems run in real time, others render audio offline or non-real time. Similarly, it is possible for systems to have varying degrees of uncertainty in the output.

Before discussing their specific implementations, we introduce a categorization of soundscape generation systems based on distinguishing features.

12.3.1. *Non-Autonomous, Semiautonomous or Autonomous*

These terms identify the degree to which a generative music system is responsible for making decisions and executing tasks. Non-autonomous systems are those requiring human operators to make decisions and execute actions at all levels of the composition. An example of a non-autonomous system is a DAW. Many tools available with DAWs do much to simplify production tasks. However, the system only executes instructions as directed by the sound designer.

A semiautonomous system controls some decisions, but the human operator is needed for others. For example, a semiautonomous system may make decisions on applying effects and mixing sounds. However, sounds may need to be curated, and the composition evaluated by human ears. This approach has the advantage that mundane tasks such as segmenting sound files can be under the control of computational assistive tools, but actions humans are better at, such as making qualitative judgments, can still be administered by the sound designer.

Autonomous soundscape generation systems are computer controlled at every level of the composition. These systems use a multidisciplinary approach to address a broad range of tasks such as soundscape context formulation, analyzing sound files and executing composition decisions.

12.3.2. Real Time, or Offline

Systems executing instructions in real time generate audio for time-critical applications such as live performance or game environments. Popular tools such as SuperCollider (Wilson, Cottle and Collins, 2011) and Max/MSP (Manzo, 2011) are used in many of these generative systems as they afford real-time access to manipulating sound. However, they offer little in the way of non-real-time alternatives.

Systems employing off-line approaches can generate soundscapes faster or slower than real time. Systems with faster than real-time capabilities can output a higher number of examples for tasks such as machine evaluation or situations with parallel requests.

12.3.3. Stochastic, or Deterministic

A stochastic system is a system that produces results having a random probability distribution or pattern that may be analyzed statistically but not predicted precisely. These systems will have a possibly different output each time.

A deterministic system produces an output that may be predicted precisely given a set of input conditions. Traditional sound-design tools, such as DAWs, are typically deterministic. Similarly, code-based tools for designing soundscapes are typically deterministic. However, non-deterministic behavior can be imitated by randomly selecting sound assets, as is typical practice in video game sound.

Some advanced autonomous systems are capable of learning. Learning systems evolve their initial knowledge base, resulting in new decisions, not in the original set of instructions. For example, Stockholm and Pasquier (2009) describe a mood-based file retrieval system that improves its accuracy based on feedback from users in an interactive music environment. Applying learning capabilities to soundscape generation systems would endow the system with the potential of creating new compositions and stylistic decisions not thought of by the system designer.

We will return to these categories for mapping the different types of systems that are presented in the following survey.

12.4. Soundscape Generation Systems

This section presents a survey of soundscape generation systems. Different systems aim to automate different tasks usually undertaken by sound designers for soundscape composition and simulation. The following systems are separated into two groups, either soundscape composition or simulation, surveyed regarding the typology specified in section 12.2, and listed in the typology grid at the end of this section.

12.4.1. Systems Generating Soundscape Composition

The following systems constitute soundscape composition systems, as the range of output goals favor creativity and listener interpretation. The typical approach for these systems is to create a database of audio files labeled according to some composition protocol that is used by an automatic selection service.

For example, Birchfield, Mattar and Sundaram (2005) report on a semi-autonomous system that uses a selection of user models that influence the sound selection and mixing parameters of the system. A database of three hundred sound files was curated and annotated by the researchers according to spatial placement, location and soundscape contexts. The system automatically selected sound files and mixed these based on the user model, taking into account a user's contextual expectations of sounds. A sound designer then made decisions on applying effects.

Cano et al. (2004) describe a system of file selection, audio track sequencing and mixing for semiautomatic ambiance generation to address the information retrieval problems and homogeneity of standard sound effects libraries. The authors augmented a commercial database of sound files by indexing simple tags as high-level concepts using WordNet. The system then selects sounds from text-based utterances using a keyword spotting technique that links concepts of the keywords and returns a randomized set of sound files with the same concept from the database.

Sound files with a long duration are considered as ambient tracks, whereas short-duration sound files are sound events.

Cano et al. outline a soundscape model of an extended ambiance with short event sounds occurring at intervals corresponding to moments of less energy in the ambiance. Their system sequences sound files by selecting a long ambient loop track and inserting shorter event tracks based on a probabilistic criterion of the inverse of frame-based energy from the ambient loop track. The effect of this sequencing method is a background with event sounds occurring at some moments of less energy in the ambiance. Additionally, their system applies automatic panning, attenuation and equalization to sequences. The pan position of the ambient track is centered in the mix while the event sounds are panned left or right to greater degrees depending on how close they are in the sequence. Attenuation is applied to event sounds so that they are 3dB above the mean level of the ambient track. Lastly, equalization is applied to frequency bands of event sounds that overlap considerably with the frequency components of the ambient track. The system described by Cano et al. generates soundscapes that have a high degree of spectral separation.

Using a similar model of sequencing sounds, Rossignol et al. (2014) describe the soundscape generation system SimScene for creating material in psychology studies on soundscape perception. Rossignol et al. manually label sets of sounds for textures and events using an urban sound taxonomy, organizing sounds based on the domain, category and sound class. They define the SimSound soundscape model as background texture with multiple types of event sounds occurring at distinct intervals. A user selects a sound event and texture class through a graphical interface. SimSound generates a texture by concatenating texture files for that class, and sequences all sounds of the selected event class according to parameters set by the user, for instance, duration, interval time, the intensity of sound events and a variability factor for those properties.

Using a different approach, Eigenfeldt and Pasquier (2011) designed a system that combined autonomous software agents and audio signal analysis to generate a continuously evolving soundscape composition. A database of sound files was curated and selected by the researchers. Software agents generated a layered soundscape composition in real time using sound file selection criteria negotiated by semantic tags and spectral attributes. The system described by Lopes, Liapis and Yannakakis (2016) for generating game levels in the horror genre uses a database of audio files labeled with tension values. The system selects and layers sounds based on feedback it receives from the game engine. As with the other systems above, a set of audio files is curated and labeled by the researchers, resulting in an adequate yet small number of options for the system to select from.

Using an autonomous approach to creating the database of audio files, Thorogood, Pasquier and Eigenfeldt (2012) describe a system the curates and selects sound files based on text analysis of social media posts and sound file descriptions. After that, sound file recommendations are pushed to a human performer for mixing. Thorogood and Pasquier (2013a) later describe the Audio Metaphor system for automating the tasks of human performers. Sound file recommendations returned after text analysis are segmented and classified by the system using a machine learning algorithm and psychoacoustic model. The system then selects and applies effects for a layered soundscape based on a set of rules informed by production notes from the soundscape composition *Island* by Canadian composer Truax (2009).

Another method of selecting, annotating and mixing sound files is to capitalize on user's propensity for generating data in online media sharing platforms (Roma et al. 2012). Roma Herrera and Serra (2009) used the sound file dataset and users' behavior from the file-sharing platform Freesound (Akkermans et al. 2011) for generating continuous mixes of sound files. Through the web interface, users linked sounds together from the database into simple sound design patches. These patches created a directed graph where sounds were vertices and edges, the play order path. A linear cross-fade applied between sounds completed the composition. Users then rated each other's patches that were processed using an interactive genetic algorithm to create new patches.

Tag-based selection grounded in formal sound design concepts provides a highly flexible and informed tool for soundscape composition systems that is rooted in practice.

12.4.2. Systems Simulating Soundscapes

The remainder of this section describes systems aimed at creating realistic soundscapes. Finney and Janer (2010) designed a semiautonomous system for generating soundscapes in virtual environments. Sounds used in their system were handpicked for audio quality and semantic attributes viewed as germane to the represented environment. Sounds were mixed using an interactive map interface, and combined into background or foreground layers. Further spatialization of sounds occurred conditionally based on user attributes in the virtual environment. Finney and Janer's system attempts to address the volatility of group dynamics and the questionable qualities of crowd-sourced tagging that is undesirable in an application requiring predictable outcomes.

Semantic tags do not account for the spectral properties of sound signals. Instead, it is preferable to apply methodologies from audio signal analysis and machine learning for modeling representations of the audio signal. Roma et al. (2010) used such a methodology for autonomously labeling sound files with high-level concepts from the sound taxonomy proposed by Gaver (1993). Using these autonomously labeled files, Janer et al. (2011) developed a semiautonomous system for augmented reality systems that generated soundscapes for virtual environments. Their strategy was to provide an authoring environment that lets users select sounds based on the taxonomy and semantic tags, and position sounds by hand using a map interface.

An issue when generating soundscapes in virtual environments is the limitation of representing physical phenomena, such as more extended temporal notions, times of the day and movement of sound sources. Two widely used game sound tools, by Firelight Technologies (2002) and Audiokinetic (2000), automate triggering curated sounds based on specific events. The tools use a physics engine for modulating effects, such as panning, brightness, Doppler shift and filtering.

Virtual environments provide opportunities to simulate auditory cues such as interaural time differences and sound reflections relative to the environment. Unrealistic soundscape can be avoided using cues such as those. Innami and Kasai (2011) propose a system for generating dynamic real-world soundscapes that take into account the complexities of the physical world including the weather, movement and geological features. Such an effort would provide significant opportunities for the immersive experience in virtual environments.

Works utilizing soundscape creation for virtual environments, such as from Tactical Sound Garden Toolkit by Shepard (2007) and the Urban Remix project by Freeman et al. (2011), move toward the development of collaborative recording, exploration and soundscape creation systems. The focus of their research toward augmenting soundscape in public spaces highlights the potential of soundscape generation systems in urban design. These systems provided participants with a mobile interface for recording and environmental tagging sounds on a map interface. Sounds were layered based on geographic locations and spatialized as a virtual soundscape based on user's proximity.

Another soundscape generation system is outlined by Casu, Koutsomichalis and Valle (2014), who describe a set of automated search and composition tools named SoDA for assisting sound designers with generating soundscapes. SoDA uses a database of sound files accompanied by a set of corresponding descriptive documents called RDF. The base RDF schema is a modification of the W3C Ontology for Media Resources, describing the content and media properties, and MPEG-7 audio standard for low-level audio features (e.g. centroid, loudness, spread). The user enters sound description fields by hand, and the machine automation enters analysis features. The search interface of SoDA uses a keyword spotting technique for retrieving semantically linked sound files. The authors do not make use of audio analysis features at the time of writing.

SoDA's soundscape model includes an omnidirectional background sound, and sequences of shorter event sound occur in a three-dimensional space relative to where the sounds occur in physical space. The sequencing engine reproduces a soundscape based on this model with a listener positioned in the space comprising distinct zones where sounds are produced. Stochastic algorithms sequence sound into patterns for a specific event type. A SoDA user specifies the attributes of a sound behavior model, including the zone positioning, texture or event sounds and the pattern sequencing algorithm. A semantic query triggers sounds and associated behavior models to be executed and added to the composition. The effect of a SoDA composition is a background sound texture with patterned event sounds panned in a three-dimensional space.

12.5. Comments Regarding the Survey

This section discusses the systems from the survey and identifies future directions in advancing the field of autonomous soundscape generation systems.

Researchers have adopted techniques from AI, machine learning and audio signal analysis for automating the generation of soundscapes. These techniques have been adopted for modeling soundscape audio signals and composition tasks. Some researchers have applied established methods from music information retrieval and theories from sound design, creating even greater autonomy for generating artificial soundscapes.

12.5.1. Autonomous Classification and Segmentation

Soundscape generation systems face many nontrivial challenges. The challenges become greater as the number of autonomous tasks increases and the complexities of modeling the nuances of human perception and environmental audio signals that are involved with sound design present themselves. Different systems demonstrate varying levels of autonomy. A non-autonomous system requires a human operator at every decision stage. A semiautonomous system is capable of generating solutions to decisions, while a fully autonomous system does not require a human operator in generating soundscapes.

One of the challenges of designing an autonomous soundscape generation system is representing the symbolic and the non-symbolic information of soundscape recordings. Symbolic information, such as semantic data in the form of annotations, does not account for the non-symbolic information, such as salient qualities in the audio signal. Annotations should give a clear indication of what object the audio recording represents, but that is not guaranteed when they are entered by sound designers.

The World Soundscape Tape Library (Truax 2015) is an outstanding example of annotated recordings. Items in the collection describe recordings in great detail. Similarly, the Freesound database (Akkermans et al. 2011) includes an annotation field, giving contributors an opportunity to add semantic information about the recording. However, annotations are regularly imprecise, and the data is noisy. Moreover, audio recordings can be long and contain more than one source sound at different times. Much of that information is missing as it is not practical to be entered by a human being. To overcome the varying quality of recordings and annotations, Freesound includes a popularity rating as a crowdsourcing methodology used to infer the quality of one audio recording over another. Recordings of higher quality and accuracy in annotations tend to have a higher popularity rating.

It is not always feasible to use crowdsourcing methodologies. Services such as Mechanical Turk (Buhrmester, Kwang and Gosling 2011) provide crowdsourcing solutions but at a high cost. Costly professionally curated commercial databases should ensure high-quality audio recordings. Commercial databases are collections of curated recordings regularly used by the film, animation and game industries. However, a useful survey of these databases reveals the semantic quality of metadata is neglected in favor of transferring curatorial responsibility to the sound designer. Sound designers are indeed masters in sourcing and segmenting audio recordings for production, and the question remains if a machine is capable of similar quality work.

Applying audio signal classification and segmentation methods from music information retrieval (MIR) (Downie 2004) could automate the sound designer's tasks, such as search and segmentation. The need for automation increases as the size and scope of audio recording databases grow. Thorogood and Pasquier (2013a) describe a methodology for segmenting sound files based on the classification of background and foreground sounds. While Thorogood and Pasquier' classification system is valuable for soundscape generation systems, it is not designed for recognizing particular sound sources in noisy data. For example, when requesting only a dog sound, the computer does not know a dog when segmenting a sound file with dogs and cats.

Therefore, a more general classification methodology is needed for making decisions about audio recordings. Research in auditory scene analysis (Wang and Brown 2006) points toward the potential for general sound object separation and classification algorithms that would disambiguate audio recordings. However, there are still challenges to realizing robust methodologies due to the large and complex search space. A naive solution is to weight sound file recommendations based on the number of terms in the metadata relative to the duration of the recording. However, there remains a problem of invalid metadata entries due to the human intelligence task, and such an algorithm remains imprecise.

Similarly, there is no guarantee on the accuracy of metadata referencing salient qualities of audio recordings. An essential feature of soundscapes is the mood it communicates (Botteldooren, Coensel and Muer 2006; Berglund, Nilsson and Axelsson 2007; Hall et al. (2011), which is reinforced by the notion of creating types of feelings in sound design for film or video games. For example, the particular quality of a soundscape in a haunted abandoned village is different from that of a soundscape associated with a delightful village on a sunny day.

To date, information such as the perceived mood of a sound is not accounted for in soundscape generation systems. (Barrington, Oda and Lanckriet 2009) discuss how mood is an essential factor of music recommender systems, which has been a focus of MIR research. Applying methodologies from MIR to the task of soundscape generation would factor salient qualities, such as mood, into search results. Thorogood and Pasquier (2013b) illustrate the feasibility of classifying the mood of soundscape composition. A further grounding of that research is articulated by Fan, Thorogood and Pasquier (2015). Their results show a promising direction in autonomous labeling of affect on audio recordings for soundscape generation systems.

12.5.2.   Sound File Search by Utterance

In consideration of soundscape generation systems, a typical search problem is when an utterance is forwarded as a query, either by a machine or human, and the system then makes a (possibly empty) set of sound file recommendations to fulfill the query. An utterance may contain many concepts that a single recommendation would not fulfill, and a query would need a broader set of recommendations to fulfill.

Regarding query-based systems with automatic search, these systems disambiguate an utterance using a keyword spotting technique. These systems handle keywords in different ways for searching. For example, the system described by Cano et al. (2004) maps keywords in an utterance to higher-level concepts and uses each concept to search for recommendations indexed from the same mapping technique. This technique has the advantage of returning files in the conceptual domain of the query but overlooks the specific sounds in a query. Another technique is described by Thorogood, Pasquier and Eigenfeldt (2012), who generate the set of all possible combinations of adjacent keywords and search for the smallest number of sound files to represent the query from a database indexed by word frequency and location. This practical approach acts as a type of dimension reduction by limiting the possible number of sound files for later processing and mixing by the system. Another approach is given by Casu, Koutsomichalis and Valle (2014), who define the SoDA system that retrieves sound files for each concept recognized in the query. A user can further enrich the search by adding concepts suggested by the system. The SoDA method refines a search toward a more specific set of sound file recommendations. Each of these search methods addresses the problem of returning sound files for processing and combining by a mixing engine.

12.6. Conclusion

Sound design involves many creative tasks. Some of these tasks are repetitive and can be automated. Automating sound design tasks involves formalizing the creative process to design computer programs that encode those processes. In this chapter, we have outlined the basic tasks involved in creating a sound design ambiance, such as a soundscape. After reviewing various systems that have approached the soundscape automation problem, we identified the different tasks that need to be solved. Systems take an approach to automate the layering of sounds to either reflect multitrack recording techniques or simulate the world in three dimensions. Drilling into these systems, we were able to define a set of the specific tasks that have been the focus of attention in the field of sound design and the challenges that this automation brings to light. Although it is challenging to fully take creative control from the sound designer, the development of new computationally assistive tools as discussed here will provide new methods of streamlining the repetitive tasks in sound design production and leave considerably more time for higher-level creative tasks. The desired outcome of this chapter, thus, is to engage the reader in reflecting on the creative process and orient future creation of new and exciting tools for sound design.

References

Akkermans, V., Font, F., Funollet, J., de Jong, B., Roma, G., Togias, S. and Serra, X., 2011. Freesound 2: An Improved Platform for Sharing Audio Clips. In: *International Society for Music Information Retrieval Conference.*

Audiokinetic, 2000. Wwise. Viewed Jan 12, 2015 www. audiokinetic.com

Barrington, L., Oda, R. and Lanckriet, G.R., 2009. Smarter Than Genius? Human Evaluation of Music Recommender Systems. *Proceedings of the International Symposium on Music Information Retrieval*, 9, pp. 357–362.

Berglund, B., Nilsson, M.E. and Axelsson, O., 2007. Soundscape Psychophysics in Place. In: Boone, R. (Ed.), *InterNoise 2001* [electronic resource]*: The 2001 International Congress and Exhibition on Noise Control Engineering*, Aug 28–30, 2001. The Hague, The Netherlands: Proceedings.

Birchfield, D., Mattar, N. and Sundaram, H., 2005. Design of a Generative Model for Soundscape Creation. *International Computer Music Conference*. Barcelona, Spain: ICMC.

Botteldooren, D., Coensel, B.D. and Muer, T.D., 2006. The Temporal Structure of Urban Soundscapes. *Journal of Sound and Vibration*, 292, no. 1–2, pp. 105–123.

Buhrmester, M., Kwang, T. and Gosling, S.D., 2011. Amazon's Mechanical Turk a New Source of Inexpensive, Yet High-Quality, Data? *Perspectives on Psychological Science*, 6, no, 1, pp. 3–5.

Cano, P., Fabig, L., Gouyon, F. and Loscos, A., 2004. Semi-Automatic Ambiance Generation. In: *Proceedings of 7th International Conference on Digital Audio Effects*, pp. 1–4.

Cano, P., Koppenberger, M., Herrera, P., Celma, O'. and Tarasov, V., 2004. Sound Effect Taxonomy Management in Production Environments. In: *25th AES International Conference. Metadata for Audio*, June 17–19. London.

Casu, M., Koutsomichalis, M. and Valle, A., 2014. Imaginary Soundscapes: The Soda Project. In: *Proceedings of the 9th Audio Mostly: A Conference on Interaction with Sound*. Aalborg, Denmark: ACM, article no. 5.

Downie, J.S., 2004. The Scientific Evaluation of Music Information Retrieval Systems: Foundations and Future. *Computer Music Journal*, 28, no. 2, pp. 12–23.

Eigenfeldt, A. and Pasquier, P., 2011. Negotiated Content: Generative Soundscape Composition by Autonomous Musical Agents in Coming Together: Freesound. In: *Proceedings of the Second International Conference on Computational Creativity*, pp. 27–32.

Fan, J., Thorogood, M. and Pasquier, P., 2015. Automatic Recognition of Eventfulness and Pleasantness of Soundscape. In: *Proceedings of the 10th Audio Mostly*, Oct 7–9. Thessaloniki, Greece.

Finney, N. and Janer, J., 2010. Soundscape Generation for Virtual Environments Using Community-Provided Audio Databases. In: *W3C Workshop: Augmented Reality on the Web*, June 15. Barcelona, Spain.

Firelight Technologies, 2002. FMOD. Viewed Jan 12, 2015 http:// www.fmod.org

Freeman, J., DiSalvo, C., Nitsche, M. and Garret, S., 2011. Soundscape Composition and Field Recording as a Platform for Collaborative Creativity. *Organized Sound*, 16, pp. 272–281.

Gaver, W.W., 1993. What in the World Do We Hear? an Ecological Approach to Auditory Event Perception. *Ecological Psychology*, 5, pp. 1–29.

Hall, D.A., Irwin, A., Edmondson-Jones, M., Phillips, S. and Poxon, J.E., 2011. An Exploratory Evaluation of Perceptual, Psychoacoustic and Acoustical Properties of Urban Soundscapes. *Applied Acoustics*, 74, no. 2, pp. 48–254.

Innami, S. and Kasai, H., 2011. On-Demand Soundscape Generation Using Spatial Audio Mixin. In: *International Conference on Consumer Electronics*, pp. 29–30.

Janer, J., Kersten, S., Schirosa, M. and Roma, G., 2011. An Online Platform for Interactive Soundscapes with User-Contributed Audio Content. In: *Audio Engineering Society Conference: 41st International Conference: Audio for Games,* Feb 2–4. Piccadilly, London.

Lopes, P.L., Liapis, A. and Yannakakis, G.N., 2016. Framing Tension for Game Generation. In: *Proceedings of the Seventh International Conference on Computational Creativity*, June, pp. 205–212.

Manzo, V., 2011. *Max/Msp/jitter for Music: A Practical Guide to Developing Interactive Music Systems for Education and More*. Oxford, UK: Oxford University Press.

McCartney, A., 2002. Soundscape Compositions and the Subversion of Electroacoustic Norms, in The Radio Art Companion. *New Adventures in Sound Art*, pp. 14–22. Viewed Jul 1, 2018 www.soundartarchive.net/articles/McCartney-2003-Soundscape%20 Composition%20and%20the%20Subversion%20of%20Electroacoustic%20Norms.pdf

Roma, G., Herrera, P. and Serra, X., 2009. Freesound Radio: Supporting Music Creation by Exploration of a Sound Database. In: *Computational Creativity Support Workshop CHI09.*

Roma, G., Herrera, P., Zanin, M., Toral, S.L., Font, F. and Serra, X., 2012. Small World Networks and Creativity in Audio Clip Sharing. *International Journal of Social Network Mining*, 1, no. 1, pp. 112–127.

Roma, G., Janer, J., Kersten, S., Schirosa, M., Herrera, P. and Serra, X., 2010. Ecological Acoustics Perspective for Content-Based Retrieval of Environmental Sounds. *EURASIP Journal of Audio Speech Music Process*, 7, pp. 1–11.

Rossignol, M., Lafay, G., Lagrange, M. and Misdariis, N., 2014. SimScene: A Web Based Acoustic Scenes Simulator. In: *1st Web Audio Conference (WAC)*, Jan 2015. Paris, France. Viewed Jul 1, 2018 https://hal.archives-ouvertes.fr/hal-01078098v2/document

Shepard, M., 2007. Tactical Sound Garden Toolkit. In: *ACM SIGGRAPH 2007 Art Gallery, SIGGRAPH '07.* New York: ACM, p. 219.

Stockholm, J. and Pasquier, P., 2009. Reinforcement Learning of Listener Response for Mood Classification of Audio. In: *Proceedings of the 2009 International Conference on Computational Science and Engineering 4*. IEEE Computer Society, pp. 849–853.

Thorogood, M. and Pasquier, P., 2013a. Computationally Generated Soundscapes with Audio Metaphor. In: *Proceedings of the 4th International Conference on Computational Creativity*, pp. 1–7.

Thorogood, M. and Pasquier, P., 2013b. Impress: A Machine Learning Approach to Soundscape Affect Classification for a Music Performance Environment. In: *Proceedings of the International Conference on New Interfaces for Musical Expression*. Daejeon, Republic of Korea, pp. 256–260.

Thorogood, M., Pasquier, P. and Eigenfeldt, A., 2012. Audio Metaphor: Audio Information Retrieval for Soundscape Composition. In: *Proceedings of the 6th Sound and Music Computing Conference*, pp. 372–378.

Truax, B., 2009. Island. In: *Soundscape Composition* DVD. DVD ROM (CSR-DVD 0901). Cambridge Street Publishing.

Truax, B., 2015. World Soundscape Project Tape Library. Viewed Jan 12, 2015 www.sfu. ca/sonic-studio/srs/index2. html

Wang, D. and Brown, G.J., 2006. *Computational Auditory Scene Analysis: Principles, Algorithms, and Applications*. New York: Wiley-IEEE Press.

Wilson, S., Cottle, D. and Collins, N., 2011. *The SuperCollider Book*. Cambridge, MA: The MIT Press.

13

Sound Effect Synthesis

David Moffat, Rod Selfridge and Joshua D. Reiss

13.1.   Introduction

Sound effects are commonly defined as non-musical, non-speech sounds used in some artificial context, such as theater, TV, film, video games or virtual reality. The purpose of a sound effect is typically to provide a diegetic context of some event or action, that is, a sound that exists within the narrative of the story line. A 1931 BBC white paper proposed that there were six types of sound effects (BBC 1931).

1. *Realistic, confirmatory effect*—The convincing sound of an object that can be seen to directly tie into the story (e.g. the sound of a gunshot when we see a gun being fired);
2. *Realistic, evocative effect*—A convincing sound within the landscape that cannot be directly seen (e.g. a bird in a forest tweeting off screen);
3. *Symbolic, evocative effect*—Sounds that don't actually exist within the narrative, designed to create an emotion within the listener (e.g. a swelling sound to build suspense);
4. *Conventionalized effect*—A sound that, though not entirely realistic, is perceived as realistic due to overuse and hyperrealism (e.g. the ricochet after a gunshot in a western film);
5. *Impressionistic effect*—Creating a general feeling or indication of an occurrence without an exact realistic example (e.g. a cartoon punch sound); and
6. *Music as an effect*—Producing a sound effect through some musical means (e.g. chimes to represent a transformation).

Sound effects can often be the linchpin of a sound scene, and different sounds and styles will vary drastically depending on the style and design of the medium, among other factors.

Sound synthesis is the technique of generating sound through artificial means, either in analog or digital or a combination of the two. Synthesis is typically performed for one of three reasons:

- To facilitate some interaction or control of a sound, whether for a performance or direct parameter driven control of a sound (see, e.g. Heinrichs, McPherson and Farnell 2014; Wilkinson et al. 2016).
- To help a sound designer searching for a suitable sound within a synthesis space rather than through a sound effect library (see, e.g. Hendry and Reiss 2010).
- To create something that does not exist, such as artificial sci-fi sounds, or to repair damaged sound files (see, e.g. Puronas 2014).

Public demand is increasing for instantaneous and realistic interactions with machines, particularly in a gaming context. Farnell (2007) defines procedural audio (PA) as "nonlinear, often synthetic sound, created in real time according to a set of programmatic rules and live input." As such, PA can be viewed as a subset of sound synthesis, where all sounds are produced in real time, with a particular focus on synthesis control and interaction. PA is fundamental to improving human perception of human-computer interactions from an audible perspective, but there are still many unanswered questions in this field (Fournel 2010). Bottcher and Serafin (2009) demonstrated subjectively that in an interactive game play environment, 71% of users found synthesis methods more entertaining than audio sampling. Users rated synthesized sound as higher quality, more realistic and preferable. From this, it is clear that user interaction is a vital aspect of sound synthesis.

Foley sound was created in the 1920s by Jack Foley. The premise is that a Foley artist or "performer" can produce a particular sound using any objects that may create the idea of the sound, rather than just use a recording of a real sound. The story goes that someone was looking for a bunch of chains to rattle to create a prison scene, and Jack just pulled out a bunch of keys and rattled them in front of a microphone recording the sound. When they listened back, they were happy with the results and the concept of Foley sound was born. The emphasis on creating a "larger than life" sound was one of the key founding aspects of Foley work. A sound does not need to be real, it just needs to be convincing. This has resulted in the idea of "hyperrealism," which is commonplace in much of Hollywood sound design (Puronas 2014). Hyperrealism is the idea that a sound must be bigger, more impressive and "more real" than the real-world sound, so as to create a level of excitement or tension (Mengual et al. 2016). This is particularly common in many TV and film explosion and gunshot sounds, where a real-world recording of a gunshot is considered too boring and mundane compared to the artificial gunshot, which is often some combination of much bigger sounds, such as a gunshot, an explosion, a car crash, a lion roar and a building collapse. Foley attempts to provide a similar idea with some performance sounds, where each action or idea is significantly over-performed and every action is made larger than the real-world case. Foley grew into an entire field of work, and professional "Foley artists" can still be found worldwide. Foley sound became prominent since it allowed a sound designer to perform, act or create the desired sound, and easily synchronize it with the action. The level of control that a Foley artist had over the sound was greater than ever before.

Much in the same way that Foley sound allowed for control, interaction and performance of a sound, sound synthesis can allow for control over digital sounds. Previously, the only way to digitize a sound was to record it. Now we can model a sound and control its parameters in real time. This creates a much more naturally occurring sound, as controls can be derived directly from the physical parameters, and thus the expectation of the listener is satisfied, when every small detail and interaction they have produces a predictably different sound. As such, in many ways sound synthesis, and especially procedural audio, can be considered digital Foley.

The key advantage of a synthesized sound effect over a recording is the ability to control and interact with the sound. This interaction creates a feeling of a realistic world Heinrichs and McPherson (2014). Immersion is a key goal in game design. A player feels more immersed in a game if they feel like they are actually situated in the game environment. Immersive sound can be created either through the use of three-dimensional sound, or by creating realistic interactions with sonic objects. Creating an immersive sound is an important aspect, as it will draw a user into the virtual environment and make them feel more like part of the game rather than like a spectator watching the game through a window. Realistic sonic feedback is a vital part of producing a believable and consistent immersive world.

13.2. Sound Effect Synthesis

There are many methods and techniques for synthesizing different sound effects, and each one has varying advantages and disadvantages. There are almost as many sound synthesis classification methods, but the most prominent was produced by Smith (1991). Sound synthesis can generally be grouped into the following categories: sample based, signal modeling, abstract and physical modeling synthesis.

13.2.1. Sample Based Synthesis

In Sample based synthesis, audio recordings are cut and spliced together to produce new or similar sounds. This is effective for pulse-train or granular sound textures, based on a given sound timbre.

The most common example of this is granular synthesis. Granular synthesis is the method of analyzing a sound file or set of sound files and extracting sonic "grains." A sound grain is generally a small element or component of a sound, typically between 10 and 200ms in length. Once a set of sound grains has been extracted, they can then be reconstructed and played back with components of the sound modified, such as by selecting a subset of grains for a different timbre, or changing the grain density or rate to alter the pitched qualities of the sound.

13.2.2. Signal Modelling Synthesis

Signal modelling synthesis is the method where sounds are created based on some analysis of real-world sounds, and this analysis is used to inform a resynthesize of the original waveform sound, not the underlying physical system. The premise of signal modeling is that through comparing and reproducing the actual sound components, we can extrapolate the control parameters and accurately model the synthesis system. The most common method of signal modeling synthesis is spectral modelling synthesis (SMS) Serra and Smith (1990). SMS assumes that sounds can be synthesized as a summation of sine waves and filtered noise. Spectral modeling is often performed by analyzing the original audio file, selecting a series of sine waves to be used for resynthesis and then creating some "residual" noise shape, which can be summed together to produce the original sound (Amatriain et al. 2002). SMS performs best on simple harmonic sounds. For less harmonic sounds, other analysis and resynthesis methods are preferable, such as nonnegative matrix factorization Turner (2010) or latent force modeling Wilkinson et al. (2017).

13.2.3. Abstract Synthesis

Abstract synthesis is the method where abstract methods and algorithms are used, typically to create entirely new sounds. A classic example of abstract synthesis is frequency modulation (FM) synthesis (Chowning 1973). FM synthesis is a method derived from telecommunications. Two sine waves are be multiplied together to create a much richer sound. These sounds can be controlled in real time, as computation is low, to create a set of sounds that do not exist in the natural world. A lot of traditional video game sounds and 1980s keyboard sounds were based on FM synthesis.

13.2.4. Physical Modeling Synthesis

In physical modelling synthesis, sounds are generated based on modeling of the physics of the system that created the sound. The more physics is incorporated into the system, the better the model is considered to be, however, the models often end up very computational and can take a long time to run. Despite the computational nature of these approaches, with Graphics Processing Units(GPU) and accelerated computing, physical models are beginning to be capable of running in real time. As such, physical models are often based on fundamental physical properties of a system and solving partial differential equations at each sample step (Bilbao 2009).

13.2.5. Synthesis Methods Conclusion

There is a range of different synthesis methods that can produce a range of different sounds, from abstract synthesis techniques that are lightweight and can be implemented on old hardware, to physical modeling techniques that require optimization and GPU and even then are only just able to operate in real time. Each one has its advantages and disadvantages. Misra and Cook (2009) perform a rigorous survey of synthesis methods and recommend different synthesis techniques for each type of sound to be produced. Abstract synthesis is great for producing artificial sounds, retro sounds from 1980s games and some musical sounds. Signal modeling can produce excellent voiced sounds and environmental sounds. Physical models are great for impact or force driven sounds, such as the pluck of a string. Sound textures and environmental sounds are often best produced by sample-based models. A summary of recommendations as to a method of synthesis that would work for each type of sound class can be found in Table 13.1.

13.3. Evaluation

The aims of sound synthesis are to produce realistic and controllable systems for artificially replicating real-world sounds. Evaluation is vital, as it helps us understand both how well our synthesis method performs and how we can improve our system. Without a rigorous evaluation method, we cannot

Table 13.1 Recommendation of Synthesis Method for Each Sound Type.

Sound Type	Synthesis Method
Sci-Fi/Technology Sounds	Abstract Synthesis
Environmental Sounds	Sample Based Model/Signal Models
Impact Sounds	Physical Models/Signal Models
Voiced Sounds	Signal Models
Sound Textures/Soundscapes	Sample Based Models

understand if our synthesis method performs as required or where it fails. Evaluation of a sound synthesis system can take many different forms. Jaffe (1995) presented ten different methods for evaluating synthesis techniques. There are many examples of these evaluation methods being employed in literature, including evaluation of controls and control parameters (Rocchesso et al. 2003; Merer et al. 2013; Selfridge et al. 2017b), human perception of different timbre (Merer et al. 2011; Aramaki et al. 2012), sound identification (Ballas 1993; McDermott and Simoncelli 2011), sonic classification (Gabrielli et al. 2011; Hoffman and Cook 2006; Moffat et al. 2017) and sonic realism (Moffat and Reiss 2018; Selfridge et al. 2018a, 2017c). Evaluation methods can be broken down into one of two approaches.

13.3.1. Evaluation of Sonic Qualities

One of the most important aspects of evaluating a synthesis method is evaluating the sonic quality of the sound produced. Does the produced sound actually sound as intended? If you cannot create the sound you prefer, then no quantity of sound interaction will make a synthesis model effective. Generally, this evaluation needs to be performed with human participants, where recorded samples of a given sound can be compared to samples rendered from a synthesis method, and the two compared by users in a multistimulus perceptual evaluation experiment (Moffat and Reiss 2018; Bech and Zacharov 2007). This evaluation comparison method will assess synthesized sounds and compare them against recordings, in the same contextual environment. This method of evaluation can be applied to a range of different sounds (Mengual et al. 2016; Selfridge et al. 2017a, 2017b, 2017c, 2017d, 2018a).

It is important that similar sounds are compared, and that participants are asked suitable questions. Generally, participants are asked to evaluate how real or how believable a given sound is. This is important because although participants may have a strong idea of what a sound is, this does not mean that their impression of a real sound is correct. It has often been the case that a participant will rate a synthetic sound as "more realistic" than a real recording of a sound, especially with less common sounds. This is due to the hyperrealism effect discussed earlier. As people are generally expecting explosions and gunshots to be "larger than life," when they hear a real recording versus a synthesized sound, the recording just seems flat and boring compared to a synthesized sound (Mengual et al. 2016).

However, despite the importance of perceptual listening assessments, there is rarely effective perceptual evaluation of synthesis methods. Schwarz (2011) noted in a review of 94 published papers on sound texture synthesis that only seven contained any perceptual evaluation of the synthesis method.

13.3.2. Evaluation of Control and Interaction

Evaluating the control and interaction of a synthesis engine is a vital aspect of understanding in which environment the sound can be used. Much in the same way that a Foley is the performance of acoustic sounds, synthesis is the performance of digital sounds, and the control interaction is key. However, in most cases, the physical interaction that creates the sound will not be suitable for directly driving the individual synthesis parameters, and as such, some mapping layer for parameters and physical properties of a game will be required (Heinrichs, McPherson and Farnell 2014; Heinrichs and McPherson 2014). There are numerous methods for evaluating these sonic interactions, and in many cases, the control evaluation has to be custom designed according to the synthesis methods and parametric controls (Heinrichs, McPherson and Farnell 2014; Heinrichs and McPherson 2014; Turchet et al. 2016; Selfridge et al. 2017b). User listening tests, where participants are able to interact with the synthesis engine through some mapping layer, can be performed to evaluate a series of criteria. Key aspects of synthesis control systems to evaluate are

- Intuitiveness—How intuitive and interpretable are the controls? Can a user easily find the exact sound they want?
- Perceptibility—How much can someone perceive the impact each control makes, at all times, so as to understand what each control does?
- Consistency—Do the controls allow for consistent reproduction of sound, or is there some control hysteresis?
- Reactiveness/Latency—Do the controls immediately change the sound output, or is there a delay on control parameters that impact the ease of usability? Typically 20ms of latency is acceptable in most cases, so long as the latency is consistent Jack, Stockman and McPherson (2016).

13.4. Example Design Process

A number of synthesis techniques have been identified and here we illustrate how to apply these principles to design our own sound effect. We looked at designing a sword sound effect, initially answering a number of questions:

- What synthesis technique shall we use to implement the effect?
- Are we going to design from scratch or based on samples?
- Do we want real-time operation?

- Are we going to use specialist hardware?
- What software will we use to implement the effect?
- How do we want to control the effect?

For this example, we wanted our sound effect to be able to be used as part of a procedural audio implementation and to be able to capture elements of natural behavior. This meant some sort of physical model was preferred. Such physical models generally involve synthesis from scratch since they are based on the physics that produces the sound rather than analysis or manipulation from a sound sample. From the definition of procedural audio, real-time operation is key to enabling the effect to adapt to changing conditions.

The use of specialist hardware, graphics processing units (GPUs) or field programmable gate arrays (FPGAs), are mostly used for musical instruments rather than sound effects Bilbao (2009). Due to the complex nature of the computations, these are necessary for real-time operation. It was not our intention to require specialist hardware in order for the model to operate in real time, which indicated that the model should avoid highly complex computations. However, simplifications that result in far weaker audio quality or realism, as deployed for dynamic level of detail Durr et al. (2015), should also be avoided.

The choice of software to implement the effect was based on a number of factors, including programming experience, licence required or open source, complexity of the model and efficiency of the language. The open-source programming language Puredata has proven to be excellent at developing the sound effects via a graphical syntax (Farnell 2010), though more recent approaches have used the Web Audio API and JSAP plugin standard Jillings et al. (2016b) for browser-based sound synthesis Bahadoran et al. (2018).

When developing a sound effect it is of value to look at other state of the art synthesis techniques used to create similar sound effects. A number of sword models have been developed and are listed in Table 13.2. A signal-based approach to a variety of environmental sound effects, including sword whoosh, waves and wind sounds, was undertaken in Marelli et al. (2010). Analysis and synthesis occur in the frequency domain using a sub-band method to produce narrow-band colored noise.

Four different sword models were evaluated in Bottcher and Serafin (2009). Here, the application was for interactive gaming, and the evaluation was focused on perception and preference rather than accuracy of sound. The user was able to interact with the sound effect through the use of a Wii controller. One model was a band-filtered noise signal with the center frequency proportional to the acceleration of the controller. A physically inspired model replicated the dominant frequency modes extracted

Table 13.2 Table Highlighting Different Synthesis Methods for Swing Sounds.

Reference	Synthesis Method	Parameters	Comments
Marelli et al. (2010)	Frequency domain, signal-based model	Amplitude control over analysis and synthesis filters	Operates in real time
Bottcher and Serafin (2009)	Granular	Accelerometer speed	Mapped to playback speed
	Sample based	Accelerometer speed	Triggered by threshold speeds
	Noise shaping	Accelerometer speed	Mapped to band-pass center frequency
	Physically inspired	Accelerometer speed	Mapped to the amplitude of frequency modes
Dobashi et al. (2003)	Computational fluid dynamics	Length, diameter and swing speed	Real time but requires initial off-line computations

from a recording of a bamboo stick swung through the air. The amplitude of the modes was mapped to the real-time acceleration data.

The other synthesis methods in Bottcher and Serafin (2009) both mapped acceleration data from the Wii Controller to different parameters; one used the data to threshold between two audio samples, and the other used a granular synthesis method mapping acceleration to the playback speed of grains. Tests revealed that the granular synthesis was the preferred method for expression and perception. One possible reason that the physical model was less popular could be that the band-filtered noise frequency was not correlated with the speed of the swing of the Wii controller. However, this correlation between frequency pitch and speed was present in the granular model.

A physical model of sword sounds was explored in Dobashi et al. (2003). Here, off-line sound textures were generated based on the physical dimensions of the sword. The sound textures were then played back with speed proportional to the movement. The sound textures were generated using off-line computational fluid dynamics software (CFD), solving the fundamental fluid dynamics equations. In this model Dobashi et al. (2003), the sword was split into a number of compact sound sources, spaced along the length of the sword. As the sword was swept through the air, each source moved at a different speed; therefore, the sound texture for each source was adjusted accordingly. The sounds from each source were combined and presented to the listener.

The desire for our sword example was to create a sound effect in which parameter changes modified the output sound in real time, requiring no off-line processing, while simultaneously producing plausible sounds within a procedural audio context. These requirements meant that a physically inspired synthesis model was the most suitable.

13.4.1. Aeroacoustics

The sound of a sword swinging through the air comes from a class of sounds called aeroacoustics. Aeroacoustics is the name given to the field of study that determines the sounds produced by air, for example a boiling kettle or helicopter rotor. Aeroacoustics has a number of fundamental tones, which individually and collectively can replicate a number of common sound effects. A basic taxonomy of these is shown in Figure 13.1. We can see from this diagram that the main fundamental tone that a swinging sword produces is the Aeolian tone.

It was shown in Curle (1955) and confirmed in Gerrard (1955) that aeroacoustic sounds, in low flow speed situations, could be modeled by the summation of compact sources, namely monopoles, dipoles and quadrupoles. An acoustic monopole, under ideal conditions, can be described as a pulsating sphere, much smaller than the acoustic wavelength. This is shown in Figure 13.2a. A dipole, under ideal conditions, is equivalent

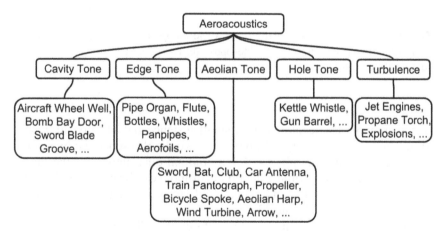

Figure 13.1 Simplified Taxonomy of Fundamental Aeroacoustic Sounds Including Examples of Each.

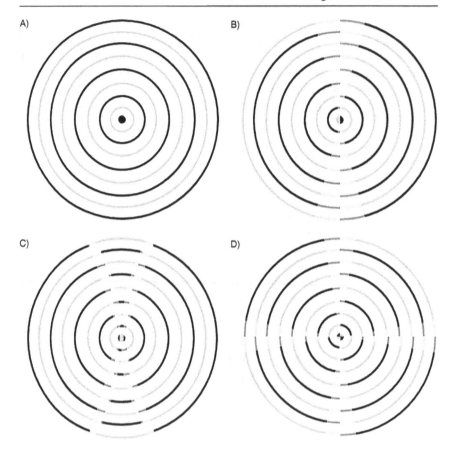

Figure 13.2   Ideal Radiation Pattern for a Monopole, Dipole, Lateral Quadrupole and Longitudinal Quadrupole, (a) Monopole, (b) Dipole, (c) Longitudinal Quadrupole, (d) Lateral Quadrupole.

to two monopoles separated by a small distance but of opposite phase, shown in Figure 13.2b. Quadrupoles are two dipoles separated by a small distance with opposite phases. A longitudinal quadrupole has the dipole axes in the same line while a lateral quadrupole can be considered as four monopoles at the corners of a rectangle, Crighton et al. (2012). These are shown in Figures 13.2c and 13.2d.

The sword model created by Dobashi et al. (2003) made use of compact sound sources to generate their aeroacoustic sounds. This required complex off-line CFD calculations to generate sound textures for each compact source and then the sound textures are concatenated together with playback speed mapped to the speed of the sword swing.

In our model, we also used compact sound sources but instead of off-line calculations we carried out research through equations, known as *semi-empirical equations*, where assumptions and generalizations have been made to simplify calculations or to yield results in accordance with observations. Although many of these equations may at first appear complicated, once all the relevant parameters are known, they produce exact results with errors only due to the approximations made during the equation derivation.

13.4.2. Aeolian Tone

To implement a physically inspired model, basic knowledge of the physics behind the fundamental tone that was being generated was required. Determining how much knowledge of physics the model possesses is often a balance between computationally complexity, the choice of software and hardware used for implementing the model and perceptual relevance. The first step was to gain basic understanding of the Aeolian tone and the parameters upon which it depends (Selfridge et al. 2016).

The Aeolian tone is generated when air flows around an object. The vast majority of research on this tone has been carried out on cylindrical objects, and we modeled our sword based on these results. When air passes around a cylinder, vortices are generated and then released, or shed, from the back of the cylinder. This is depicted in Figure 13.3 where we can see what is known as a vortex street behind the cylinder. It can also be seen that vortices are shed from alternate sides of the cylinder. This causes an oscillating lift force perpendicular to the flow. Normal to the flow there is a drag force at twice the frequency of the lift force. We are able to model each of the oscillating forces and their harmonics by dipole compact sound sources.

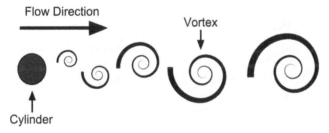

Figure 13.3 Example of Vortices Being Shed From Opposite Sides of a Cylinder Known as a Vortex Street.

13.4.2.1. Frequency Calculations

In 1878, Czech physicist Vincenc Strouhal carried out one of the first important studies into the frequency of a tone produced as air flows around a cylinder. He defined the formula given in Eqn. 1:

$$S_t = \frac{fd}{u} \tag{1}$$

where S_t is the Strouhal number, f is the tone frequency, d the cylinder diameter and u the airspeed. The fluctuating lift force caused by the vortex shedding is dominated by a fundamental frequency f_l with $S_t \approx 0.2$, the drag force is dominated by a fundamental frequency f_d, which is $2f_l$. From Eqn. 1, we can therefore calculate the tone frequencies due to the lift, drag and any harmonics.

13.4.2.2. Acoustic Intensity Calculations

The time-averaged Acoustic Intensity I_{1l} (W/m²) of the Aeolian tone lift dipole source was derived in Goldstein (1976). The full derivation is beyond the scope of this publication and the reader is referred to Goldstein (1976) if they wish to find out more. The Acoustic Intensity I_{1l} is proportional to:

$$I_{l1} \propto u^6 \sin^2\theta \cos^2\phi \tag{2}$$

where θ is the elevation angle and ϕ the azimuth angle.

13.4.2.3. Tone Bandwidth

The bandwidth around the fundamental frequency is affected by the airspeed and diameter, the higher the airspeed or diameter, the wider the tone bandwidth.

13.4.2.4. Wake Noise

When the airspeed or diameter increases the vortices produced by shedding diffuse rapidly, merging into a turbulent wake. The wake produces wide band noise modeled by quadrupole sources whose intensity varies with the flow speed to the power of 8, $I_w \propto u^8$ (Etkin, Korbacher and Keefe 1957). There is very little noise content below the lift dipole fundamental frequency f_l (Etkin, Korbacher and Keefe 1957). Above this frequency the roll off of the turbulent noise amplitude is $1/f^2$ (Powell 1959). The wake is modeled by a range of combinations between the various longitudinal and lateral quadrupoles.

13.4.3. Implementation of a Compact Sound Source

In section 13.4.2 values for the frequency of the lift dipole and drag dipole were identified along with the acoustic intensity values for the lift dipole and the wake noise. We also highlighted the fact that as the airspeed or diameter increases that the bandwidth around the tone frequencies increases. Considering this in relation to deciding which method to use to implement our physically inspired model, it is judged that subtractive synthesis would provide a suitable method.

13.4.3.1. Sampled Variables

As control parameters can be modified during operation of our synthesis method, we need to look at sampling the variables over discrete time [n]. We can record the control parameter at audio rate (44100Hz), but this requires performing every single calculation for every audio sample. This can result in a very computationally heavy process, which requires a lot of CPU to process. Alternatively we can update control parameters at a lower rate, such as every one thousand samples (441 Hz), however we then introduce the possibility of some parameter jumps causing audio glitches. This can potentially be resolved with smoothing control parameter values.

In our model, a number of variables were sampled at audio rate of 44100 Hz. This allows real-time performance but overuse can put a strain on the audio buffers causing drop outs. A balance between accuracy and perception has to be achieved if this occurs. A list of sampled variables includes:

- Airspeed $u[n]$
- Elevation $\theta[n]$
- Azimuth $\phi[n]$
- Distance $r[n]$

With these variables measured at discrete time [n] we were able to calculate the discrete parameters based on the semi-empirical equations described above. The list is certainly not exclusive and variables like length and diameter can be sampled to allow real-time morphing of the sword. The properties of air could also be varied to create realistic sounds depending on weather changes or even due to alien atmospheres.

13.4.3.2. Lift and Drag Dipoles

Subtractive synthesis is based on shaping the frequency response of a noise source by filtering. For our purposes, band-pass filtering was employed to obtain the required sound of the dipoles. Band-pass filters require a center frequency and the width of the peak. Using Eqn. 1, the center frequencies of band-pass filters representing the lift and drag dipoles were calculated.

In signal processing, the relationship between the peak frequency and bandwidth is called the Q value, ($Q = f_l/\Delta f$). The bandwidth around the tone is related to the Reynolds number. Data available in Norberg 1987 enabled a definition for the required Q values to be defined.

The output for the lift dipole is given as the output from a band-pass filter whose input was a white noise source, with center frequency f_l and predicted Q. Likewise, the drag dipole output and third harmonic were generated in the same manner.

13.4.3.3. Wake Quadrupole

The final aspect added into the compact sound source was the noise associated with the wake. The required noise profile required a roll off of $1/f^2$, which is known as a brown noise source. There is little wake contribution below the fundamental frequency (Etkin, Korbacher and Keefe 1957). Therefore, a high-pass filter was applied with the filter cut-off set at the lift dipole fundamental frequency, f_l. This produces the turbulent noise profile required.

13.4.3.4. Final output

The output was obtained by adding lift dipole, drag dipole, harmonic and wake, giving the final output, as shown in Figure 13.4.

13.4.4. Modeling a Sword

13.4.4.1. Physical Model of a Sword

The knowledge gained from understanding some of the aeroacoustic sound generating processes assists in design decisions on how to model the sword. If we wished to capture the physics of the entire sword we could place a number of compact sound sources all the way from the tip to the hilt. Dobashi et al. (2003) determines the output in exactly this manner with the output from each compact source obtained through off-line CFD calculations. To copy this in our model would have increased the complexity and the chance of audio drop outs. Instead we looked at Eqn. 2 and the other intensity equations while appreciating the characteristics of a swinging sword.

For the Aeolian tone the acoustic intensity of the dipoles is proportional to u^6 and the wake quadrupoles to u^8. This means that the majority of sound generated will be in the area that u is the greatest. In the case of a

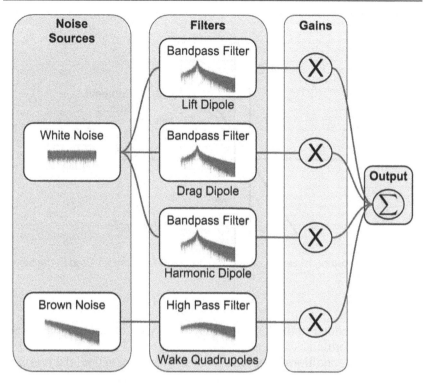

Figure 13.4   Flow Diagram Outlining the Signal Processing Chain Implementing the Aeolian Tone Compact Sound Source.

sword swinging this will be the tip. In our model we placed six compact sound sources at the tip of the sword, one at the hilt, one halfway between the hilt and one at the tip. This is illustrated in Figure 13.5.

13.4.4.2.   Modeling the Behavior of a Sword

Modeling the behavior of a sword swing is an important aspect of achieving a believable physical model sound effect. To model the sword behavior, a number of design decisions were made and implemented to give a limited range of motion to the user for the swing. It is feasible to attach the compact sound source models to a game object and calculate the sword dimensions from the graphics and airspeed from the animation.

The speed of the sword at the start and end of the swing were set to 0m/s, with the top speed (set by the user) at the halfway point. The track of the swing was set to be circular for ease of programming—in reality a swing may probably act more like a variety of arcs. The length of the

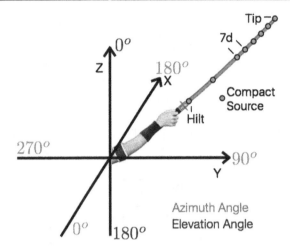

Figure 13.5 Position of Eight Compact Sources and Coordinates Used in the Sword Model.

forearm was added to the length of the sword with the elbow joint at the center of a coordinates system, shown in Figure 13.5.

With these conditions imposed on the sword swing action, the user was able to set a number of parameters prior to the swing. Parameters available to the user were:

- Position of the observer in three dimensions.
- Start position of the sword—azimuth and elevation.
- Thickness of the tip and hilt of the sword—diameter of each source extrapolated from these.
- Length of the sword.
- Top speed the tip will reach during the swing.

It is important when modeling the swing of an object that the Doppler effect is taken into consideration, as well as panning if a stereo sound effect is being produced. By adjusting the diameters of the compact sound sources we are able to replicate the geometry of a number of other objects. In order to replicate objects as accurately as possible, measurements of the dimensions of a number of objects were taken. We took the measurements of a baseball bat, 3-wood and 7-iron golf clubs, a broom handle, a wooden sword and a metal sword and programmed the physical model to give pre-set values that replicate these.

13.4.5. Evaluation of the Physical Model

13.4.5.1. Objective Evaluation

The objective evaluation focused on the compact sound source. There have been a number of studies giving results of the frequency of air passing at known airspeeds and diameters. These are either experimental from wind tunnel measurements or simulated from CFD calculations. The average absolute error for our model was 4.66%, while CFD has an average absolute error of 18.11%. The absolute error is calculated as the absolute difference between the recording or simulation and the model we are measuring.

13.4.5.2. Subjective Evaluation

To subjectively evaluate the sword model, listening tests were carried out asking participants to rate the sword sounds on how plausible they believed them to be. The Web Audio Evaluation Tool (Jillings et al. 2015, 2016a) was used to build and run listening tests in the browser. Each participant was presented with a page for each of the preset sound effects. The sounds generated by the physical model were compared to real recordings and alternative synthesis methods, including SMS (Amatriain et al. 2002) for all objects and samples from Bottcher and Serafin (2009) and Dobashi et al. (2003) for the metal sword.

To obtain recorded samples as close as possible to those we were attempting to replicate, recordings were captured by the authors of the actual objects we had measured and used to define our presets. These were recorded within the Listening Room, Electronic Engineering and Computer Science Department, Queen Mary University of London. They were recorded on a Neumann U87 microphone placed approximately 20 cm from the midpoint of the swing and at 90 degrees to the plane of the swing. The impulse response of the room was captured and applied to all other sounds in the listening test so that the natural reverb of the room would not influence the results (except samples from Bottcher and Serafin 2009; Dobashi et al. 2003).

The anchors were created from a real-time browser-based synthesis effect to allow a thorough comparison of how plausible the synthesis method is compared to the recorded sample. It was expected that a low-pass filtered sample, as used in the MUltiple Stimuli with Hidden Reference and Anchor (MUSHRA) standard, would still be considered plausible, whereas a low-quality anchor would encourage the full use of the scale and allow for better understanding as to the effectiveness of the synthesis method.

Rating the plausibility of sound from a physical model was the preferred judgement in Castagne and Cadoz (2003), who stated that a plausible sound was one that listeners thought "was produced in some physical manner." Box plots for all five objects are shown in Figure 13.6. The box plots are a visualization of the distribution of the result. It shows the median of the data, in the center black line, the upper and lower bounds of the box showing the first and third quartiles. This means the boxed area shows 50% of the data within the box, and the "whiskers," or the lines from the box show the last 25% of the data, on each edge. Identified outliers are marked with an o. It can be seen from Figure 13.6, that our physical model outperforms the alternative synthesis methods on all of the objects except the metal sword. The metal sword performed poorly for plausibility in this test, with the model using added cavity tones performing slightly better.

13.4.6. Discussion

We can see from Figure 13.6 that results from the listening test indicate that overall our model performs well compared to other synthesis models. It has exceptional performance for the broom handle, baseball bat, golf club and wooden sword objects, where participants found sounds generated by our model were as plausible as real recordings. The exception to this was the metal sword physical model sound effect. It is important once we have designed the sound effect that we evaluate it through objective and subjective testing, to try and understand the difference between results and recorded sounds. Having an understanding of this or a hypothesis of the reasons behind the differences shows us areas where the model can be improved in the future.

The broom handle, baseball bat and golf club objects were all cylindrical with thickness-to-width ratios of 1:1. For the wooden sword this ratio decreases to approximately 0.37:1 and for a metal sword to approximately 0.14:1. The Aeolian tone model is designed around vortex shedding from cylindrical objects and it is reasonable to assume additional discrepancies may exist when there is a deviation from the thickness-to-width ratio of a cylinder.

The metal sword clearly has the poorest performance of all the modeled objects. Objects thicker than the sword have greater wake noise, which may influence the plausibility of the sounds. SMS performs identification of the specific harmonic sinusoidal components and reproduces them with pure tones. Thinner objects produce sounds closer to pure tones and hence are better synthesized using SMS than those of thicker objects.

A possible reason for the poor rating of the metal sword object compared to the other objects is that the number of participants who have swung a real sword and heard the sound may well be less than those who have perhaps swung a golf club and some of the other objects. Memory plays an important role in perception Gaver and Norman (1988). If participants

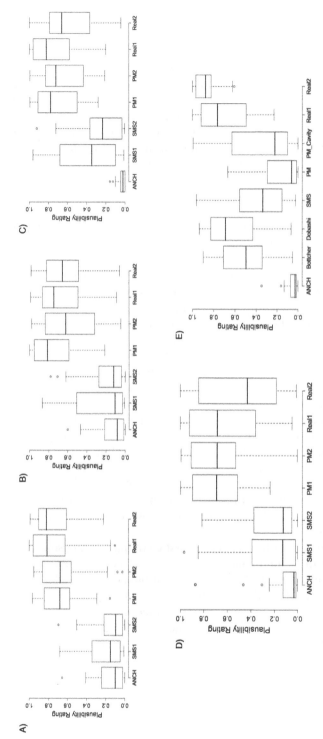

Figure 13.6 Box Plots Showing Plausibility Results for the Preset Objects. ANCH, Anchor; SMS, Spectral Modelling Synthesis; PM, Physical Model; PMCavity, physical model including cavity tone; Real, recorded sample. (a) Broom Handle, (b) Baseball Bat, (c) Golf Club, (d) Wooden Sword, (e) Metal Sword.

have heard a Foley sound effect for a sword more often than an actual sword sound, this may influence their perception of the physical model.

In contrast, it can be argued that participants will have more likely heard the actual sounds of a golf club at a live sporting event or within sporting broadcasts and hence their memory of these sounds would be closer to the physical model. Since all participants were from the UK, the baseball bat would most likely not be as familiar to them as other objects and hence they might not have as strong a memory of the sound made by this object. This would diminish the difference between a memory of a Foley sound and that of an actual sound.

We developed a method for modeling the behavior of the sword that includes mapping the compact sound sources directly to graphical objects in a game engine. Care has to be taken in relation to this style of mapping as often the movement of the graphics extends what is physically possible. An example of this is when a person swinging a sword generates speeds that are much higher than physically possible. Remembering that acoustic intensity of the dipoles is proportional to u^6, the signal level goes extremely high and will clip at the output.

The model presented here offers a unique and novel approach to synthesizing aeroacoustic sound effects. Instead of starting by modeling the sound of the entire sword, we investigate and model compact sound sources, which can then be manipulated to model the sword in addition to a range of objects that would not otherwise have been possible. As well as being used to model clubs and bats, the compact sound sources can be extended to model other items. In Selfridge et al. (2017a) compact sound sources are used to generate the vortex sounds that are an integral component of the sound of a propeller spinning. The coupling of vortex shedding and mechanical vibrations was illustrated in a physical model of an Aeolian harp (Selfridge et al. 2017d), where the output sound is mainly caused by the vibration of strings. In this model, the vortex shedding causes the strings to vibrate with the compact sound source providing control data for this process.

The development of further aeroacoustic compact sound sources offers an increase in the sounds and objects that can be modeled. The cavity tone model (Selfridge et al. 2017e) was implemented to add a grooved profile to the sword used in our listening tests. A physically informed synthesis model of an edge tone was presented in Selfridge et al. (2018b) through an approach that illustrates the use of machine learning techniques to provide information in circumstances where the physics are not fully understood. The use of compact sound sources can be extended to a number of other sound effects, which are listed in Chanaud (2010) and given in Table 13.3.

Table 13.3 Source and Effect Types.

Source Type	Sound Effect
Monopole	pistons, exhaust flows, propane torch (combustion), weapon discharges (explosions), drums (membranes), automobile tire sounds, bubbles, splashes, waterfalls, electrical sparks, kettle whistles, corrugated pipe tone
Dipole	airfoil sounds, propellers, bicycle spokes, exhaust flows, Aeolian tone, cavity tone, ring tone, edge tone, vortex whistle, bottles, police whistle, Levavasseur whistle, screech tone (supersonic jets)
Quadrupole	subsonic jets, wakes, supersonic jets

13.5. Conclusions

In this chapter we have given a comprehensive overview of sound effect synthesis and control methods. We have contextualized and shown the importance of sound effects, showing how they are subcategorized, as well as the importance of Foley artists. An in-depth review of the wide range of sound effect synthesis techniques has been given, highlighting the strengths and weaknesses of different synthesis methods.

It was seen that control of a synthesized sound effect is paramount in generating the desired sound and exploring the nuances of the range of sounds a model can produce. Evaluation techniques of synthesis models have been described along with reasons why evaluation is essential when deciding which sound effect we should use and how research will develop in the future.

We have looked at the definition of procedural audio, drawing attention to why this is an area being developed for games and virtual reality environments. A sample design process has been given for a physically inspired sound synthesis model, which can be integrated as a procedural sound effect. Understanding of physics can be incorporated into the model and how this can influence design techniques. This is one such approach that can effectively model a given sound. This approach can extend to a range of physically derived synthesis models, though this approach is not guaranteed to work in every single case, with modifications, a similar approach can be used for developing synthesis methods.

It is never expected that in the future for every single sound within a game to be produced by a synthesis approach. The subtle aspects of sound design will result in a fusion of different types of sounds and tools, and synthesis is one such approach that can be integrated into the sound designed.

It is clear that sound effects add vital cues for listeners and viewers across a range of media content. The development of more realistic, interactive, immersive sound effects is an exciting and growing area of research, with many research questions still to be answered.

References

Amatriain, X. et al., 2002. Spectral Processing. In: Zolzer, U. (Ed.), *DAFx: Digital Audio Effects*, chapter 10. Chichester, UK: John Wiley and Sons, Ltd, pp. 373–438.

Aramaki, M. et al., 2012. Perceptual Control of Environmental Sound Synthesis. In: *Speech, Sound and Music Processing: Embracing Research in India*. Springer, pp. 172–186.

Bahadoran, P. et al., 2018. Fxive: A Web Platform for Procedural Sound Synthesis. In: *Proceedings of the 144th Audio Engineering Society Convention*. Milan, Italy.

Ballas, J.A., 1993. Common Factors in the Identification of an Assortment of Brief Everyday Sounds. *Journal of Experimental Psychology: Human Perception and Performance*, 19, no. 2, pp. 250–267.

BBC. *The BBC Year Book 1931*, 1931. The Use of Sound Effects, pp. 194–197. British Broadcasting Corporation.

Bech, S. and Zacharov, N., 2007. *Perceptual Audio Evaluation-Theory, Method and Application*. Chichester: John Wiley & Sons.

Bilbao, S., 2009. *Numerical Sound Synthesis: Finite Difference Schemes and Simulations in Musical Acoustics*. Chichester: Wiley Online Library.

Bottcher, N. and Serafin, S., 2009. Design and Evaluation of Physically Inspired Models of Sound Effects in Computer Games. In: *35th Audio Engineering Society Conference: Audio for Games*. London.

Castagne, N. and Claude Cadoz, C., 2003. 10 Criteria for Evaluating Physical Modelling Schemes for Music Creation. In: *6th Digital Audio Effects Conference*, London: Digital Audio Effects (DAFx).

Chanaud, R.C., 2010. *Tools for Analyzing Sound Sources*. Essex, CT: CCR Associates, LLC.

Chowning, J.M., 1973. The Synthesis of Complex Audio Spectra by Means of Frequency Modulation. *Journal of the Audio Engineering Society*, 21, no. 7, pp. 526–534.

Crighton, D.G., Dowling, A.P., Williams, J.E.F., Heckl, M.A. and Leppington, F.A., 2012. *Modern Methods in Analytical Acoustics: Lecture Notes*. New York: Springer Science & Business Media.

Curle, N., 1955. The Influence of Solid Boundaries Upon Aerodynamic Sound. In: *Proceedings of the Royal Society of London A: Mathematical and Physical Sciences* 231, no. 1187, pp. 505–514.

Dobashi, Y. et al., 2003. Real-Time Rendering of Aerodynamic Sound Using Sound Textures Based on Computational Fluid Dynamics. *ACM Transactions on Graphics (TOG)*, 22, pp. 732–740. ACM.

Durr, G. et al., 2015. Implementation and Evaluation of Dynamic Level of Audio Detail. In: *56th Audio Engineering Society Conference: Audio for Games*, Audio Engineering Society.

Etkin, B., Korbacher, G.K. and Keefe, R.T., 1957. Acoustic Radiation from a Stationary Cylinder in a Fluid Stream (Aeolian Tones). *The Journal of the Acoustical Society of America*, 29.

Farnell, A., 2007. An Introduction to Procedural Audio and Its Application in Computer Games. In: *Audio Mostly Conference*. Ilmenau, Germany: ACM, pp. 1–31.

Farnell, A., 2010. *Designing sound*. Cambridge, MA: The MIT Press.

Fournel, N., 2010. Procedural Audio for Video Games: Are We There Yet? In: *Game Developers Conference 2010*. San Francisco, CA: Sony Computer Entertainment Europe, Game Developers Conference.

Gabrielli, L. et al., 2011. A Subjective Validation Method for Musical Instrument Emulation. In: *131st Audio Engineering Society Convention*. New York.

Gaver, W.W. and Norman, D.A., 1988. *Everyday Listening and Auditory Icons*. PhD thesis, University of California, San Diego, Department of Cognitive Science and Psychology.

Gerrard, J.H., 1955. Measurements of the Sound from Circular Cylinders in an Air Stream. *Proceedings of the Physical Society*, pp. 453.

Goldstein, M.E., 1976. *Aeroacoustics*. New York: McGraw-Hill International Book Co.

Heinrichs, C. and McPherson, A., 2014. Mapping and Interaction Strategies for Performing Environmental Sound. In: *1st Workshop on Sonic Interactions for Virtual Environments at IEEE VR 2014*.

Heinrichs, C., McPherson, A. and Farnell, A., 2014. Human Performance of Computational Sound Models for Immersive Environments. *The New Soundtrack*, 4, no. 2, pp. 139–155.

Hendry, S. and Reiss, J.D., 2010. Physical Modeling and Synthesis of Motor Noise for Replication of a Sound Effects Library. In: *Audio Engineering Society Convention 129*. Los Angeles, CA.

Hoffman, M.D. and Cook, P.R., 2006. Feature-Based Synthesis: A Tool for Evaluating, Designing, and Interacting with Music ir Systems. In: *International Symposium on Music Information Retrieval (ISMIR)*, International Society on Music Information Retrieval (ISMIR), pp. 361–362.

Jack, R.H., Stockman, T. and McPherson, A., 2016. Effect of Latency on Performer Interaction and Subjective Quality Assessment of a Digital Musical Instrument. In: *Proceedings of the Audio Mostly*. Norrkoping, Sweden: ACM, pp. 116–123.

Jaffe, D.A., 1995. Ten Criteria for Evaluating Synthesis Techniques. *Computer Music Journal*, 19, no, 1, pp. 76–87.

Jillings, N. et al., 2015. Web Audio Evaluation Tool: A Browser-Based Listening Test Environment. In: *Proceedings of the 12th Sound and Music Computing Conference*, Sound and Music Computing (SMC).

Jillings, N. et al., 2016a. Web Audio Evaluation Tool: A Framework for Subjective Assessment of Audio. In: *Proceedings of the 2nd Web Audio Conference*, Apr. Atlanta.

Jillings, N. et al., 2016b. JSAP: A Plugin Standard for the Web Audio API with Intelligent Functionality. In: *Audio Engineering Society Convention 141*, Audio Engineering Society.

Marelli, D. et al., 2010. Time-Frequency Synthesis of Noisy Sounds with Narrow Spectral Components. *IEEE Transactions on Audio, Speech and Language Processing*, 18, 2010.

McDermott, J.H. and Simoncelli, E.P., 2011. Sound Texture Perception Via Statistics of the Auditory Periphery: Evidence from Sound Synthesis. *Neuron*, 71, no. 5, pp. 926–940.

Mengual, L. et al., 2016. Modal Synthesis of Weapon Sounds. In: *Proceedings Audio Engineering Society Conference: 61st International Conference: Audio for Games*, Feb. London: Audio Engineering Society.

Merer, A. et al., 2011. Abstract Sounds and Their Applications in Audio and Perception Research. *Exploring Music Contents*, pp. 176–187.

Merer, A. et al., 2013. Perceptual Characterization of Motion Evoked by Sounds for Synthesis Control Purposes. *ACM Transactions on Applied Perception (TAP)*, 10, no, 1, pp. 1–24.

Misra, A. and Cook, P.R., 2009. *Toward Synthesized Environments: A Survey of Analysis and Synthesis Methods for Sound Designers and Composers*. Ann Arbor, MI: MPublishing, University of Michigan Library.

Moffat, D. and Reiss, J.D., 2018. Perceptual Evaluation of Synthesized Sound Effects. *ACM Transactions on Applied Perception (TAP)*, 15, no. 2.

Moffat, D. et al., 2017. Unsupervised Taxonomy of Sound Effects. In: *Proceedings 20th International Conference on Digital Audio Effects (DAFx-17)*, Sept. Edinburgh, UK, DAFx-17.

Norberg, C., 1987. *Effects of Reynolds Number and a Low-Intensity Freestream Turbulence on the Flow Around a Circular Cylinder*. Goteborg, Sweden: Chalmers University, Technological Publications.

Powell, A., 1959. Similarity and Turbulent Jet Noise. *The Journal of the Acoustical Society of America*, 31.

Puronas, V., 2014. Sonic Hyperrealism: Illusions of a Non-Existent Aural Reality. *The New Soundtrack*, 4 no. 2, pp. 181–194.

Rocchesso, D. et al., 2003. Sounding Objects. *IEEE MultiMedia*, 10, no. 2, pp. 42–52.

Schwarz, D., 2011. State of the Art in Sound Texture Synthesis. In: *14th International Conference Digital Audio Effects (DAFx)*. Paris, France: Digital Audio Effects (DAFx), pp. 221–231.

Selfridge, R. et al., 2016. Physically Derived Synthesis Model of an Aeolian Tone. In: *141th Audio Engineering Society Convention*, Los Angeles, CA.

Selfridge, R. et al., 2017a. Physically Derived Sound Synthesis Model of a Propeller. In: *ACM Audio Mostly Conference*, Aug. London.

Selfridge, R. et al., 2017b. Physically Derived Synthesis Model of a Cavity Tone. In: *Proceedings of the 20th Digital Audio Effects Conference*. Edinburgh, UK, pp. 5–9.

Selfridge, R. et al., 2017c. Real-Time Physical Model for an Aeolian Harp. In: *International Congress on Sound and Vibration*, Jul. London.

Selfridge, R. et al., 2017d. Real-Time Physical Model for Synthesis of Sword Swing Sounds. In: *International Conference on Sound and Music Computing (SMC)*, Jul. Espoo, Finland.

Selfridge, R. et al., 2017e. Sound Synthesis of Objects Swinging Through Air Using Physical Models. *Applied Sciences*, Nov.

Selfridge, R. et al., 2018a. Creating Real-Time Aeroacoustic Sound Effects Using Physically Informed Models. *Journal of the Audio Engineering Society*, 66, no. 7/8, pp. 594–607, July.

Selfridge, R. et al., 2018b. Physically Derived Synthesis Model of an Edge Tone. In: *Proceedings of the 144th Audio Engineering Society Convention*. Milan, Italy.

Serra, X. and Smith, J., 1990. Spectral Modeling Synthesis: A Sound Analysis/Synthesis System Based on a Deterministic Plus Stochastic Decomposition. *Computer Music Journal*, 14, no. 4, pp. 2–24.

Smith, J.O., 1991. Viewpoints on the History of Digital Synthesis. In: *Proceedings of the International Computer Music Conference*. ICMC, Montreal, Canada.

Turchet, L. et al., 2016. What Do Your Footsteps Sound Like? An Investigation on Interactive Footstep Sounds Adjustment. *Applied Acoustics*, 111, pp. 77–85.

Turner, R.E., 2010. *Statistical Models for Natural Sounds*. PhD thesis, Gatsby Computational Neuroscience Unit, UCL.

Wilkinson, W. et al., 2016. Performable Spectral Synthesis Via Low-Dimensional Modelling and Control Mapping. *DMRN+1: Digital Music Research Network One-Day Workshop*. www.eecs.qmul.ac.uk/~josh/documents/2016/Wilkinson%20DMRN%202016.pdf

Wilkinson, W. et al., 2017. Latent Force Models for Sound: Learning Modal Synthesis Parameters and Excitation Functions from Audio Recordings. In: *Proceedings of the 20th International Conference on Digital Audio Effects*. Edinburgh: DAFx.

Audio Augmented Reality for Interactive Soundwalks, Sound Art and Music Delivery

Dafna Naphtali and Richard Rodkin

14.1.   Introduction

Composers have long sought to utilize public spaces as a means of enhancing audience experience, either by providing a visceral and unique context for performance or by using the environment itself as a collaborator enlisted to influence a piece's outcome. Historically, the desire for this type of engagement with interesting acoustic spaces goes back centuries, if not millennia, to everything from mountain-top yodeling, to sixteenth-century Venetian polychoral music, early church music and antiphonal chants, and in many cases influenced the architecture itself. As such, many historical and contemporary examples of "site-specific" compositions were written for specific acoustic spaces: either natural environments or man-made spaces in interaction with architectural forms or urban design. In some cases, works include the movement of performers or listeners within those spaces, sometimes even turning the spaces themselves into virtual musical instruments.

With advances in technology, the methods for unifying musical experience with place have also become more varied and sophisticated. Now listeners can not only engage with the acoustics of a space, but they can also interact with the location itself, drawing from its inherent symbolism and ambience, thereby heightening the emotional response even further. Now called "location-based music" "location-based sound," or "locative media," the approach embodies many different styles of music and activities, yet it always requires the use of a specific physical space, a method for tracking the listener within that space, and a set of rules for how and when the music is delivered based on the listener's position and/or interactions.

As a subset of the locative media category, audio augmented reality (AAR) is not necessarily new, but the advent of the smartphone and its peripheral technologies have presented an opportunity to express and share these personal and social experiences—even one's life story—in the most engaging and imaginative ways possible, and with far wider audiences. Having collaborated on an AAR installation in New York's Washington Square Park, the authors will reference both the piece—"Walkie Talkie Dream Angles," by Dafna Naphtali—and the system with which it was constructed—Richard Rodkin's U-GRUVE AR—throughout this chapter. The discussion will first examine the available technical components needed to create an AAR system, as well as the decisions and considerations necessary for choosing the appropriate components given certain sets of requirements. The focus will then shift to explore the various compositional techniques and approaches that must be thought through to ensure the most engaging and rewarding experience for the participant.

As a compositional practice, the body of work comprising the locative media category is quite expansive. These projects and experiences— ranging from installations to performances to soundwalks—often operate on multiple levels, as something in between self-guided performances, listening experiences and games. While some are cited throughout the chapter, it is impossible to include them all; a section at the end of the chapter has therefore been provided, which includes a more complete listing of AAR projects and related readings.

14.2. Audio Augmented Reality Defined

It is important to establish a clear definition of "audio augmented reality" early in this discussion, since, as mentioned above, the history and technique for delivering location-based sound art is long and varied. To do this the underlying term "augmented reality" must first be defined.

Augmented reality (AR) has largely been thought of as a visual medium, in which a mobile device—either hand-held or head-mounted—presents an additional, virtual layer of contextual information over top of the user's real-world experience, consisting of either two-dimensional text-and-image display, or three-dimensional imagery that appears to have been inserted directly into his or her surroundings.

AAR employs the same principle as AR, but the overlaid information is presented in audio form. In certain frameworks, AAR strictly refers to non-musical, non-artistic audio information, for example in systems designed to aid visually challenged people, where audio cues can provide wayfinding, traffic and basic point-of-interest information. There is also a branch of AAR focused chiefly on self-guided touring, in which historical information or stories are presented by a narrator as the participant walks through a given area. And while spoken word certainly has a place within artistic performance, narrated tours are not the focus of our discussion.

For this discussion, AAR refers to the real-time delivery of audio artifacts, whether prerecorded, generated or live, in response to the participant's environment, and for the purpose of creating an artistic, performative experience for and with the participant.

Finally, to further clarify, AR/AAR methodology falls into two general types: location-driven and environment-driven. Location-driven systems require a method of detecting the participant's presence, either through a tracking system or through marker/object detection, and then triggering audio events per the composer's rules. An environment-driven system can ostensibly be used anywhere, regardless of specific location, and instead responds to direct stimuli from the participant's surrounding environment (weather, time of day, traffic conditions, environmental sounds, etc.). The discussion that follows will be organized around these two general approaches to AR/AAR.

14.3. Technological Approaches to AAR

14.3.1. Elements of an AAR System

Though not prescribing a specific toolset or software platform upon which to build an AAR system, this discussion focuses on the components and associated factors that need to be considered in the planning and preparation of any viable AAR project. As a basis of comparison, we'll be referencing U-GRUVE AR, as mentioned above, which has been successfully used by the authors for multiple projects over the past several years.

14.3.1.1. System Components

It is useful to review the components comprising any AAR system, as this understanding will enable system designers to clearly traverse the decision tree needed for defining functional requirements. While implementations will vary, all AAR systems consist of: (1) at least one sensor for either tracking the participant or for detecting a specific environmental input, (2) a method for controlling sensor inputs, (3) a method for mapping control events to the composer's rules and conditions and (4) the physical device used by the participant, which houses the sensors, software and playback mechanism.

14.3.1.2. Sensors

AR is a predominantly mobile technology, as its real value lies in its ability to "read" any context and provide the participant with some type of relevant, real-time feedback. The most critical components, therefore, are the sensors used for establishing this context. Sensors can range widely in terms of what they can detect and how, but all are generally used for either tracking the participant's position or for capturing and manipulating environmental elements within the participant's proximity. Tracking systems consist of two subcomponents—the tracking device worn by the person being tracked, and the central tracker used to monitor the tracking device(s).

14.3.1.3. Control Methods

While sensors are used to connect the AAR system with the participant, nothing will happen without a method for invoking the audio elements, either through a system of triggers or by using a data stream to apply continuous control.

Trigger-based controls fire events when specified thresholds are crossed. Location-driven systems using GPS as the primary sensor naturally lend themselves to zone-based trigger methods. Once the zones have been laid out, it is then just a matter of attaching the desired audio artifacts and mapping them to the ruleset defined by the composer.

Camera-based systems use fiducial markers as the trigger method. With this approach, a "match" on an image or pattern can be used to fire an event, in exactly the same way that crossing into a zone would.

For environment-driven systems, triggers are more data-oriented, similar to the GPS example above, but obviously unrelated to the user's location. For example, a thermal sensor could be used to read the temperature around the participant, with discrete temperature values used as trigger points. With a high enough resolution, the sensor could be set up to record ranges of temperature values, which could allow for continuous control over the audio.

As an alternative to triggers, incoming data can be used to apply continuous control to various audio parameters. In a location-driven system, for example, a stream of GPS values could be used to manipulate the resonance on a filter. Similarly, an environment-driven system might also use data such as speed, heading or orientation to affect the sound in some way.

14.3.1.4. Rules and Conditions

Whatever the control method, the heart of the interaction lies with the rules and conditions that are invoked once a control event is registered within the system. This could be as simple as "Play Sound X while the participant stays within Zone A" or something as sophisticated as reading weather data and assigning a randomly selected audio clip. In *Walkie*

Talkie Dream Angles (WTDA) certain sets of sounds were programmed to play depending on time of day, with no sound in a set being selected more than once. These rules are executed via code, and therefore really can run the gamut of complexity, limited only by the composer's imagination and programming skills. That said, a programming or scripting language must be selected to establish and manage the processing between events, rules and audio responses.

14.3.1.5.   The Delivery Mechanism

The device that the participant will use to engage with the system is key to the overall success of the application. It must not only securely house the selected sensor(s), but also must be able to run the software that logs and maps the control events against the rules and then both access and play the requested audio. This could be something as convenient as a smartphone, or something completely custom built, such as a vest embedded with wearable technology.

Having now defined the core set of components needed to construct an AAR system, the project goals and requirements can now be assessed against the range of available solutions.

14.3.2.   Selecting the Right Technology

Before evaluating any AAR technologies, some basic functional requirements must be thought through, as they will determine the participant's core experience. These requirements, of course, should be primarily driven by the artistic intent, though sometimes compromises may be necessary due to cost, feasibility or skill limitations.

First, which mode of AAR will be used—location-driven or environment-driven? Or both? Next, is the chosen location indoors or outdoors? Then, what kind of device will participants use—one that they already own and bring with them, like a smartphone, or one that will be custom built and loaned out? Will the experience be moderated by an attendant, or will participants be able to operate the piece autonomously? Will there be only one composition available, or will there be a variety? There are myriad other aspects to examine, but let's discuss these core questions as they will all influence the construction of the system.

14.3.2.1.   Location-Driven or Environment-Driven? (Or Both?)

As mentioned earlier, AAR experiences utilize either a tracking method for determining the participant's physical location or a set of sensors for collecting information from the participant's immediate surroundings, which are then processed and matched against the composer's ruleset. This is not to say that the two methods are mutually exclusive. It could be quite feasible to implement both a tracking system and an array of sensors for scanning the participant's environment. But the decision needs to be made very early on in the planning process, as it will drive subsequent decisions on both locale and form factor for the delivery device.

14.3.2.2. The Tracking System—Outdoors Versus Indoors

For location-driven AAR experiences, first decide whether the system will be used indoors or outdoors, since the tracking technologies available are largely governed by this factor. In some cases functional limitations will determine the system (GPS, for example, does not work well at all indoors), while in others the project directive itself may drive the decision—if the project is for a museum or another type of indoor installation, then the choice has already been made, and the options are predetermined.

14.3.2.2.1. Outdoor Systems

GPS is the most common form of tracking for outdoor systems. As a near-ubiquitous technology found in most mobile devices and motor vehicles, the costs are relatively low, and there is an abundance of documentation and developer APIs available. There are also very low-cost, micro-module units available, which could be embedded in wearable devices in any number of ways.

Despite its convenience and general accuracy when used for wayfinding, consumer-grade GPS does come with a few caveats. First, positioning is only accurate to ten meters, and while it is wholly possible to achieve better performance in actual practice, artists should plan their pieces according to the ten-meter rule. Secondly, GPS signals can drift from moment to moment, which could result in unpredictable control events. Since these fluctuations aren't handled automatically, this needs to be custom coded by the developer. Finally, GPS signal deteriorates markedly when line-of-sight is broken between the receiver and the satellite. Conditions in urban environments where dense groupings of tall buildings can potentially block satellite signals will therefore need to be further mitigated. Again, software fail-safes will need to be implemented ("If no signal, play sound X," for example), so relevant planning cycles should be allotted for this.

Another option for outdoor tracking is the standard camera-based AR approach for detecting fiducial markers. As this is the era of ARKit and ARCore, the most obvious approach would be to tap into the potential of those systems. The inherent drawback to using a marker-based system, however, is that the markers need to be where they are expected to be at all times. It is highly unlikely that markers placed in a public park will be there the next day. That is not to say there aren't workarounds—projects sponsored by a group that manages the outdoor space may find it possible to construct and position something more durable.

Also consider that the participant must point the camera at the marker for it to be recognized. This means that (1) the participant must always be on the lookout for markers, which may distract from the overall experience, and (2) the participant is forced to always have their device in hand, which may get tiresome or, again, distracting.

An alternative camera-based approach utilizes a form of computer vision in which the camera's inputs are continually scanned and compared to a library of images, and when a match is found, the trigger event is fired. Even more sophisticated computer-vision algorithms can be applied, as with the recently debuted *Play the City*[1] project (nocomputer, 2018). In this case, an automobile was outfitted with a sophisticated rig consisting of cameras and the requisite processor to handle the real-time analysis of potential "targets" while the car was in motion. This is a highly specialized example, but in general, any solution of this nature will require more computational muscle, so it may be prohibitive in cost, programming sophistication and form factor.

Similar to fiducial markers, though a distinctly different technology, are iBeacons. These are small devices that use BlueTooth to detect the presence of a specific app on a smartphone, and deliver programmed content based on the user's proximity to the beacon. iBeacons are mostly used in retail and museum settings, but in certain controlled situations, could be used outdoors, where they would have to be securely mounted out of sight, in a window or kiosk of some sort. While not as flexible as GPS, it is wholly possible to utilize the beacon's three levels of proximity detection (close, mid-range and far) as a means of triggering audio events managed by a mobile app.

In the development of U-GRUVE AR, it was clear from the outset that it would be an outdoor system, as the primary goal was to provide interactive musical backdrops for public spaces such as plazas, parks, gardens and rooftops. Additionally, the goal was to be able to reach as large of an audience as possible, so creating a mobile app for smartphones also made the most sense. The mobile app approach also satisfied our requirement for a self-guided and untimed experience. Given that a GPS sensor is standard equipment in all smartphones, the decision to use GPS for tracking was more or less a foregone conclusion.

14.3.2.2.2. Indoor Systems

For many artists, indoor installations are very appealing because the environment can be controlled and protected, allowing for artistic possibilities that wouldn't otherwise be possible in open, public spaces. While solutions such as iBeacons and fiducial markers will work just as well indoors as they will outdoors, they also bring the same drawbacks, namely, that the participant must actively use their smartphone during the experience.

Alternatively, there are a plethora of indoor positioning technologies that can be used to track individual participants. The most common systems are radio (RFID, BlueTooth), optical (infrared, visible light) and magnetic, among others.[2] It would be far beyond the scope of this discussion to provide an in-depth analysis of each of these technologies, but the key takeaway is that to implement of any of these systems requires custom design, installation and testing. If an installation is touring from site to site, it will need to be fully installed and tested at each new location.

Additionally, most of these technologies require that a physical tracker be either attached to or carried by the participant in some way. In the case of RFID and InfraRed (IR), for example, the user wears a small badge whose signal is read by a central receiver, so this aspect will directly impact the form factor of the delivery device.

14.3.2.3. The Delivery Device

Ultimately, the decisions made about the type of AAR to be used—location-driven or environment-driven—and whether the work will be indoors or outdoors will inherently influence the delivery device that participants will use.

14.3.2.3.1. Handheld

Of course there are other basic factors to consider, in terms of artistic vision and general goals for the work. For example, U-GRUVE AR was designed from the outset to construct a veritable platform that would make the works of many artists, at many locations, available to as many participants as possible. Additionally, the participant would be able to experience a piece without mediation or time limit. Given those requirements, the smartphone, paired with a good set of binaural headphones, was the obvious choice.

There are other options, however. In museum and gallery settings, it's quite common for participants to check-out tablets for self-guided tours, so it wouldn't be unusual to opt for this approach in an indoor setting. Depending on the system, it may also be necessary to design and build a custom device using Arduino or Raspberry Pi.

14.3.2.3.2. Ridable

While the general assumption is that participants will be walking while engaging with an AAR experience, there is precedence for what might be called a "sound drive," for lack of a better term. In one early example of location-driven AAR, composers Jesse Stiles and Melissa St. Pierre outfitted a car with a GPS receiver and a computer to create the "GPS Beatmap,"[3] which was demonstrated on the Bonneville Salt Flats in 2006. To experience the piece the participant would have to drive the car, or be driven, through various zones to trigger the audio, which was transmitted from loudspeakers mounted on the car's roof, as well as through the car's internal speaker system.

The *Play the City* experience mentioned earlier also used a car as the device, with the cameras mounted outside the car, and an AR "viewer" overlaid on top of one of the rear passenger windows. In this case, the participant is always a passenger in the moving vehicle, and can therefore focus solely on the musical experience.

When testing U-GRUVE AR's first version, sound-rides for cyclists were considered a core offering, but in light of concerns over potential accidents and legal issues, this approach was shelved. However, sound artist Kaffe Matthews has found a way to circumvent these issues via her *Sonic Bike* project,[4] which features a self-produced, highly customized bike that is outfitted with speakers, through which GPS-triggered musical events are played—similar to how U-GRUVE works, but played exclusively through the speakers, thereby avoiding the risks incurred by masking environmental sounds with headphones. In addition to using location as a trigger for controlling playback, the output is also influenced by the biker's pedaling speed. The project, which began in 2008, continues to be extended and advanced at the Bicrophonic Research Institute, led by Matthews and Dave Griffiths, and now includes the production of the *Sonic Kayak* and the *Sensory Bike*.[5]

14.3.2.3.3. Wearable

Before the advent of the smartphone, AR applications usually required the construction of some sort of customized wearable setup, utilizing either a vest laden with wiring and sensors and/or a helmet containing a camera or GPS receiver. While technologies have advanced and components have gotten smaller and lighter, wearable systems could still be an attractive solution, especially with the recent trends in the fashion tech industry. Here again, as with custom-built handheld devices, production, distribution and maintenance all need to be considered.

14.3.2.4. *The Software Platform*

Closely linked to the decision on the delivery device, the choice of software platform for pulling everything together—that it is, processing sensor data, setting control methods, defining rulesets and assigning the correlating sound events—is critical, as it will impact factors like cost, extensibility and maintenance. Again, this choice is guided by the overall scope of the project, in that production of a one-off experience may not require the coding rigor or depth of features that is often necessary in a more widely distributed product.

Just as there are integrated development environments (IDEs) such as XCode and Android Studio, there are also complete platforms to consider, such as the game engines used for developing today's vast array of mobile apps. Often these packages, such as the Unity Game Engine and Unreal Engine, come with features for integrating smartphone capabilities, audio processing and scripting logic, as well as "one-click" deployment to various gaming platforms. That feature alone can't be overstated enough, as it potentially spares the developer from having to recode versions of the application for each deployment target.

14.3.3. Audio Generation/Playback

14.3.3.1. Prerecorded Audio Clips

The most straightforward approach to playing audio is to draw from a set of prerecorded audio clips. Using clips ensures an overall audio consistency, barring any parameterized variations in volume or effects handling. While not offering the flexibility of pure synthesis, there are many techniques to employ for varying the sound, such as programmatic playback of clip segments, randomization of clip sequences, and dynamic looping, to name a few.

Regardless of platform, because AR is delivered through mobile media, clip size and resolution should always be kept in mind during the production process. Whatever device format is chosen, a strategy and method for managing the loading in and out of audio files is needed. In some cases it may make sense to preload all of the audio clips, while in others, a more sophisticated plan may be needed, in which clips are loaded dynamically based on the participants specific location or proximity to certain zones. These strategies should also take into account factors such as the participant's connectivity, be it cellular or wi-fi, and average storage capacity on the expected delivery device.

14.3.3.2. Sound Synthesis

Integrating an actual synthesizer into the sonic palette introduces another level of variation and expression, as participants now have control over not just the starting and stopping of a clip, but over the very elements of the sound itself, from the shape of the filter envelope to an oscillator's waveform.

There are a number of options for harnessing software synthesis in mobile contexts. For the iOS platform, AudioKit offers a comprehensive API for harnessing Apple's AudioUnit SDK, allowing for relatively straightforward implementation of synthesizers and audio processors. Within a narrower scope, another option, which is also Android-compatible, is LibPD, which takes the well-known PureData synth engine and embeds within the platform's framework. This package was developed by Peter Brinkmann, and has been documented extensively through his book *Making Musical Apps*.[6]

If the device is custom built, there are many options for integrating both hardware and software synthesizers with microcomputers such as the Arduino or Raspberry Pi. Synths like Fluxamasynth,[7] which consist of a hardware "shield" component that is controlled via a pre-existing software library, offer a relatively easy path to this realm of music making.

Despite these options, the expressive flexibility that comes with native synth engines also carries one significant caveat, which goes more to understanding the audience. While it may seem like a great idea to give participants complete control over the sound, if they aren't necessarily experienced with sound synthesis, care must be taken to make the experience foolproof so that unpredictable outcomes, such as volume loss or resonance spikes, never happen. As soon as the participant feels like she or he has lost control of instrument, she or he will lose interest in the performance as well. Those with more music-making experience may also have more patience for experimenting with new instruments, so it is imperative that the target audience is clearly understood before committing to this type of sound generation.

14.3.3.3. *Live Sampling/Live Sound Processing*

While certainly not limited to environmental-based AAR, sampling sounds around the participant and processing them in real time is a hallmark technique used in non-location-specific experiences. In fact, this was the signature feature of one of the first AAR apps on the market, RJDJ, and its sister app, Dimensions, released for iOS in 2008. Seeking to establish a genre that co-founder Michael Breidenbruecker referred to as "reactive music," RJDJ's premise was to "sense" the sounds around the participant via the smartphone's microphone, and then route the audio through any number of filters and effects for live processing, which were collectively called a "scene." Whether the participant would actively speak into the microphone or a bus would go whizzing by, the audio would be processed per the rules of a given scene, which could be either preset or defined by the user and controlled further using other built-in sensors (camera, accelerometer, GPS).

It should also be noted that RJDJ didn't just passively receive and process audio. Participants could also manipulate sounds by moving the iPhone itself, waving it or dragging it through the air like a wand. In this case, the phone's built-in accelerometer and gyroscope were used as continuous controllers that applied various effects on the incoming audio signal.

While RJDJ has left the scene, there are audio processing alternatives for mobile platforms, such as SuperPowered,[8] which offers a very robust solution for both iOS and Android. In addition to its low-latency processing muscle, SuperPowered also features real-time pitch shifting and time stretching, a capability that was once a luxury on the desktop and is now available in mobile devices, significantly expanding compositional and performance possibilities.

14.3.4. General Considerations

14.3.4.1. Synchronized Versus Non-synchronized Playback

Some musical styles may warrant that tracks play in sync. We're mentioning it here since, depending on which software platform is chosen, this capability is not necessarily a given. Chances are, the developer will need to either write or purchase some code to enable this feature.

14.3.4.2. 3D Spatialization

"3D spatialization" is the term used for giving a sound attributes so that it is perceived as occupying a distinct location within a three-dimensional environment. This is achieved by attaching the sound to an object that occupies a specific location, and then applying various audio parameters that represent that location in relation to the participant's position. Effectively, this means that as the participant gets closer to the object, the associated sound will get louder, and if he or she approaches the object from the left, the sound will be panned toward the right side of the stereo field.

While normally used in video game settings to help the player accurately orient themselves within the game, 3D spatialization can also be effectively applied in much the same manner as well to help focus the participant's attention on specific objects. For example, an audio clip could be "attached" to a statue in a park to emphasize its significance and otherwise bring it to life. Alternatively, one could create the sounds of a virtual street musician and locate the performance in a specific place to create the illusion of presence.

14.3.5.   User Interface

The final component to consider in designing an AAR system is the user interface. This is the "final" component because its format will largely be driven by previous decisions around the delivery device, its form factor and the manner in which it is going to be used. The important thing is that this aspect not be overlooked or underestimated, as it can be crucial to the project's success. If users become frustrated with merely accessing and managing the piece's content, they simply will turn it off and not use it.

There are generally three types of user interfaces to consider, none of which are necessarily mutually exclusive: the graphical user interface (GUI), the hardware interface and the audio user interface (AUI).

14.3.5.1.   Graphical User Interfaces

In the realm of mobile media, the GUI is the most prevalent user interface, though obviously not without the support of a minimal set of hardware buttons [for unlocking the phone, controlling volume and powering on/off]. If the delivery device is a smartphone, then at least one screen will need to be designed for accessing the app. From there, the depth and complexity is entirely dependent on what the app is meant to do.

In the case of U-GRUVE AR, which actually consists of multiple locations, each of which may have multiple pieces available by multiple artists, the GUI is a critical component not just for experiencing a piece, but for navigating to and selecting it. The conventional navigation patterns most commonly associated with music listening apps were used, such as drill-down lists leading to landing screens for each location, with each of those consisting of the location's description and track list. The side-menu pattern was also utilized to present information about the app, the locations and the composers.

The design also included a screen once a piece was launched that would display a graphical representation of the participant over top of the "MixMap" for that piece, similar to the blue dot that appears when using a mobile map application. Within U-GRUVE AR the MixMap is the visual depiction of the various zones to which audio artifacts are assigned (see Figures 14.1 and 14.2). While the MixMap serves several utilitarian purposes during the production process, it is also useful to participants for establishing their orientation within the environment and for helping them see where they might find various sounds.

14.3.5.2.   Hardware Interfaces

Custom devices may utilize a GUI or AUI, though by virtue of being custom built, the standard GUI may be bypassed altogether in lieu of a purely

Figure 14.1 MixMap for *Orela* (2016), by Barbara J. Weber, Lincoln Center Plaza, New York.

hardware interface. This may consist of physical knobs, sliders, toggles—anything, depending on project requirements—and in some cases this may offer an advantage, depending on who the users are. If, for example, the experience is for preschool children, they may engage more readily with a large button than with a flat screen. The key, though, as with any UI, is to maintain a level of simplicity and avoid "cockpit syndrome" in which users become overwhelmed with too many controls.

14.3.5.3. Audio User Interface

Oddly enough, with all of the recent work in what might be called "clinical" AAR—that being the provision of an audio interface for informational and wayfinding purposes (sometimes referred to as "sonification"), the term AUI has not really found its way into the formal lexicon. While a GUI is normally used to help navigate the application's screen-level content, an AUI would be used as an assistive layer for helping users understand boundaries, locations of new sounds, or other in-experience information that's been deemed relevant for the participant to know. Creating a useful

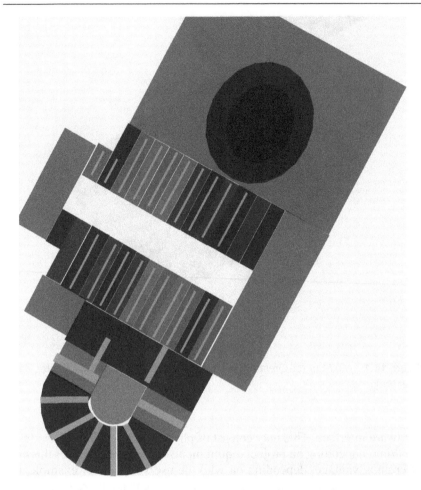

Figure 14.2 MixMap for *Florambula* (2017), by Barbara J. Weber, Conservatory Garden, Central Park, New York.

set of "earcons" could prove very valuable but would also require some degree of usability testing, as interpretations of what a sound means can vary widely, just as with their visual counterparts, icons. Care should also be taken to choose tones that are at once not alarming nor disruptive, yet are still markedly distinct from the actual audio content of the piece.

All of that said, another perfectly valid choice for a UI would be to not create any interface at all, and rather just provide a "start and go" experience in which the participant simply puts on headphones, touches a single button to start the piece, and tucks the device away while listening without distraction of any kind.

14.4. Compositional Considerations

The creative considerations for a composer working with AAR are many and profound. The possible issues, solutions, modes of expressions and content are as multifaceted and vast as the world of music and sound art itself. Successful projects will use interactive or adaptive methods, many drawn from game music composing, sound installations, data sonification and other forms of interactive media. The innumerable ideas about designing interactivity should be considered as independent of the compositional styles used for writing the music.

Using AAR with GPS location affords an opportunity to add several layers of sound and meaning to a composition. Among the possibilities are musical underscoring, sounds giving historical/social context, audio cues and sound elements that imbue the listener with agency to interact directly with and possibly even "play" the music themselves through their movements. Below are outlined ideas and strategies drawn from the authors' own work, interspersed with a discussion of projects by others.

WTDA is a location-driven, electroacoustic soundwalk composed by Naphtali in collaboration with Rodkin using his U-GRUVE AR audio augmented reality platform. Created for NoiseGate, an acoustic ecology festival in New York City in 2016, WTDA celebrates the soundscape of Greenwich Village's historically and sonically rich Washington Square Park, a ten-acre public park in the heart of Greenwich Village.[9] Washington Square Park has been a personal, cultural and political touchstone for many generations of New Yorkers, and a must-see destination for the many tourists who visit each day.

WTDA uses environmental sounds, mostly recorded on-site, and electroacoustically processed compositions created from those sounds to evoke augmented, imagined and/or historical contexts, layering them with percussion and vocal work by the composer. The piece was designed not only to share the personal expression of the composer's experiences in the park (as a lifelong New Yorker) but also to cause the listener to explore and encounter sonically interesting corners of the park in a new way, to underscore the loss of quiet, past and future sounds—a many-pronged approach taken to creating the work, and strategies about site-specific listening, interactivity and attention span.

Although the piece is driven by location, not the environment, it relies on the sonic environment of the park for most of the sound sources. It is assumed that the collages and electroacoustic pieces will be heard, overlaid and aleatorically interwoven with actual live sounds a listener would hear and overhear as they walk through the park. The work is heavily inspired by the work of John Cage, Pauline Oliveros, Hildegard Westerkamp and others who have changed our ideas about the nature of listening, and on a more philosophical/aesthetic level, by the concept of psychogeography[10] and "socio-emotional" concepts about space/sound/emotion and memory.

14.4.1. Planning/Mapping

In the planning for an AAR piece, the first concern will usually be the mapping of the geographical area for the piece into regions or zones. In the case of U-GRUVE, it was easiest to work with an interactive map such as Google Maps[11] to create GPS boundaries ("zones"). Since U-GRUVE AR uses this sort of map directly (called a "MixMap"), drawing out the zones was not only useful for visualizing the sonic structure, but also ensured that the planned arrangement would be translated accurately during the programming phase of the project. The zones may be very large or small, and may be underscored or merely used as triggers to help determine information about how the listener is traversing the area (direction, speed, activity).

The composer will next need to decide what the character is of each zone compositionally, and how the information about a participant's movement is to be used. This is obviously not an ironclad rule, as some composers may do all of the above at the same time on location and others may write the music and then decide how to trigger it later. The relationship between the interactive elements and the audio content itself will likely be different for each composer and each location, as is seen in work by Jesse Stiles, Kaffe Matthews and many other projects listed at the end of this chapter.

For WTDA, Washington Square Park was divided up into multiple overlapping zones/layers, in an approach determined by several factors. Inspired by acoustic ecology "soundwalks" in the tradition of Hildegard Westerkamp, R. Murray Schafer and others, the composer's initial task was to consider the environmental/sonic identity (and actual sounds) in various areas of the park. Each zone related to the physical layout and/or visual attributes of that location, and to the sonic identity that distinguished it from other areas of the park. Other zones were created because they represented inaccessible areas in or near the park whose sounds could be brought out into the public sphere. Still other zones related to emotional/physical memories: of sounds that were either once in the park (in the composer's lifetime or in a more historical context), or sounds that were whimsically imagined could, or should, be there. A unifying factor was that all the zones were viscerally demarcated through some combination of these parameters. It was not simply musical underscoring of the park, but rather a denoting of sounds in a way deeply related to psychogeography and dérive, an attitude toward urban exploration first defined by Guy Debord and Parisian Situationists in 1955. In all of these zones, Naphtali created collages and electronic/

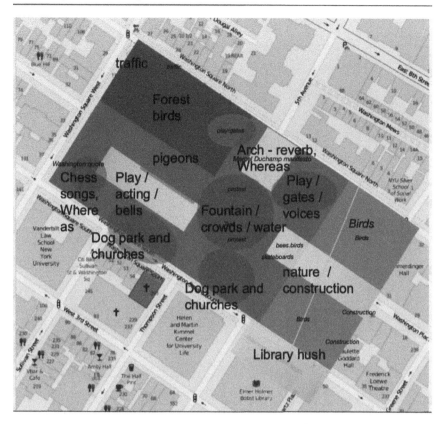

Figure 14.3 MixMap for *Walkie Talkie Dream Angles* (2016), by Dafna Naphtali, Washington Square Park, New York.

electroacoustic pieces based on local recordings and sound effects. Both the acoustic collages and the alternative audio files are triggered from a randomized list. The composer has a long interest in the intersection between birds and technology, and in WTDA one can hear both (in parts of the park where actual construction noise is loud by day, one can still hear birds. At night the virtual construction noises continue to be heard.)

1. *Accounting for participant's perception of time.* Using time as a control factor can be an overlooked way to increase variety and contextualize a piece, and keep it interesting for longer.

    a. *Keep moving.* The tempo and underlying rhythms can be useful for sending cues as to whether to walk or linger in a particular zone. In WTDA, voice and percussion loops, fragments of a wordless

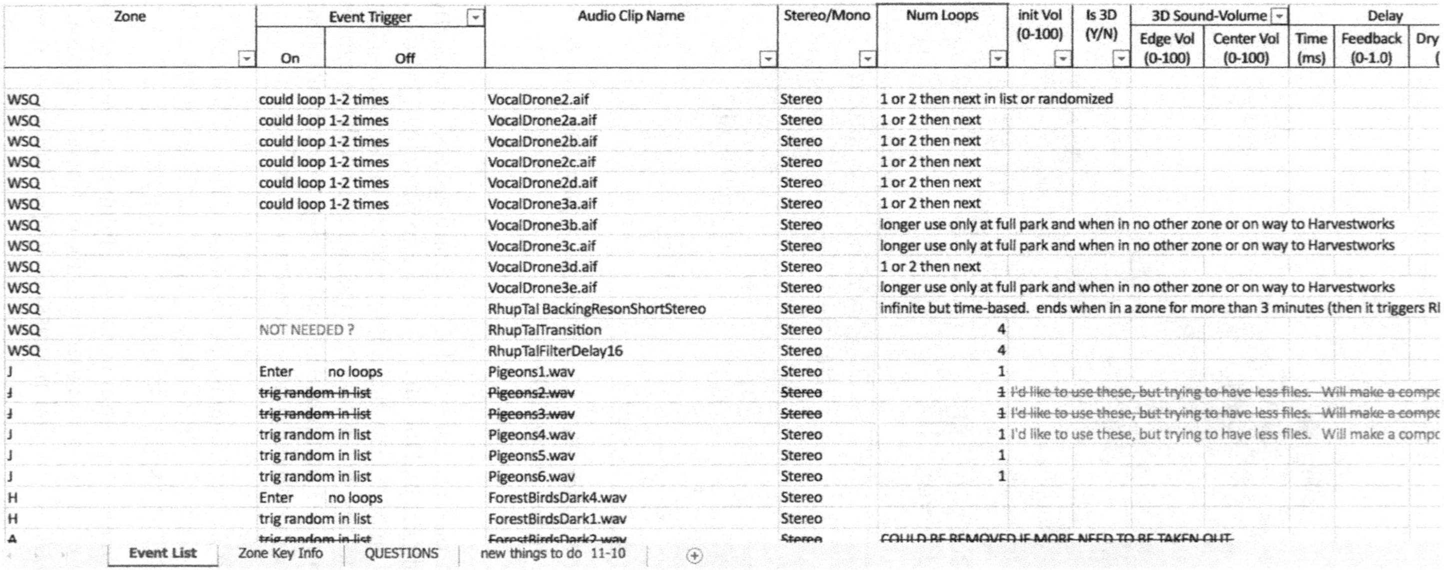

Zone	Event Trigger		Audio Clip Name	Stereo/Mono	Num Loops	init Vol (0-100)	Is 3D (Y/N)	3D Sound-Volume		Delay		
	On	Off						Edge Vol (0-100)	Center Vol (0-100)	Time (ms)	Feedback (0-1.0)	Dry
WSQ	could loop 1-2 times		VocalDrone2.aif	Stereo	1 or 2 then next in list or randomized							
WSQ	could loop 1-2 times		VocalDrone2a.aif	Stereo	1 or 2 then next							
WSQ	could loop 1-2 times		VocalDrone2b.aif	Stereo	1 or 2 then next							
WSQ	could loop 1-2 times		VocalDrone2c.aif	Stereo	1 or 2 then next							
WSQ	could loop 1-2 times		VocalDrone2d.aif	Stereo	1 or 2 then next							
WSQ	could loop 1-2 times		VocalDrone3a.aif	Stereo	1 or 2 then next							
WSQ			VocalDrone3b.aif	Stereo	longer use only at full park and when in no other zone or on way to Harvestworks							
WSQ			VocalDrone3c.aif	Stereo	longer use only at full park and when in no other zone or on way to Harvestworks							
WSQ			VocalDrone3d.aif	Stereo	1 or 2 then next							
WSQ			VocalDrone3e.aif	Stereo	longer use only at full park and when in no other zone or on way to Harvestworks							
WSQ			RhupTal BackingResonShortStereo	Stereo	infinite but time-based. ends when in a zone for more than 3 minutes (then it triggers Rl							
WSQ	NOT NEEDED ?		RhupTalTransition	Stereo	4							
WSQ			RhupTalFilterDelay16	Stereo	4							
J	Enter no loops		Pigeons1.wav	Stereo	1							
J	trig random in list		Pigeons2.wav	Stereo	1	I'd like to use these, but trying to have less files. Will make a compe						
J	trig random in list		Pigeons3.wav	Stereo	1	I'd like to use these, but trying to have less files. Will make a compe						
J	trig random in list		Pigeons4.wav	Stereo	1	I'd like to use these, but trying to have less files. Will make a compe						
J	trig random in list		Pigeons5.wav	Stereo	1							
J	trig random in list		Pigeons6.wav	Stereo	1							
H	Enter no loops		ForestBirdsDark4.wav	Stereo								
H	trig random in list		ForestBirdsDark1.wav	Stereo								
A	trig random in list		ForestBirdsDark2.wav	Stereo	COULD BE REMOVED IF MORE NEED TO BE TAKEN OUT							

Event List | Zone Key Info | QUESTIONS | new things to do 11-10   ⊕

Figure 14.4 Section of the Event List for *Walkie Talkie Dream Angles* (Screenshot).

"song," were used as a way to keep listener-participants moving along, and as red thread throughout the piece. The tempo of the percussion was at a walking pace, and the repetition of various long sung phrases helped to connect the compositionally abstract electroacoustic pieces.

b. *Remember me?* The percussion and voice loops served to remind the listeners, especially those unaccustomed to abstract electroacoustics or soundscape compositions that what they were hearing was music. It was especially needed for audio work that could be mistaken for a similar overheard sound.

c. *How long has this been going on?* Whether using triggered audio files or algorithmic composition, it is good to make decisions regarding what happens if someone is engaging one part of the piece for longer periods of time. In WTDA, the amount of time participants remain within a single zone is measured. If it exceeds three minutes, a successive list of similar but increasingly simpler percussion loops are triggered, until the percussion nearly drops out altogether, along with the voice. When the participant continues their walk (crossing into new zones), the percussion and vocal sounds start up again. However, there is always at least a simple pulse playing, to assure the listener that the soundwalk is still active.

d. *When?* Tracking the time of day, the day of week, or even the time of year is yet another way to bring context or overturn expectations. Imagine a different composition at night, or in winter.

2. *Accounting for and using physical attributes of the environment.* The physical attributes of a particular landscape or architectural space will impact on AAR works. These include not only the *acoustics*, but also *traffic flow* of listeners (pedestrians or cars/bikes), *traffic density* and *what sort of activity is usually present*. All of these attributes can impact on what can be accomplished creatively.

a. *Acoustics*. Washington Arch is a defining visual image of the park. It was decided to work with the acoustics and reverb under the arch by recording some impulse response recordings and creating a convolution reverb with them. As a method for augmented reality, a very heavy dose of this locally sourced reverb was applied to loud "clacks" and noises recorded at the arch, to create electroacoustic pieces based on these sounds. Clips of these reverb-heavy sounds were organized so as to be triggered by any listener anywhere within 50 feet of the arch. Accentuating the reverb greatly expanded the area in which it is experienced. The liberally added reverb functioned as an audio suggestion of "archness," felt by listeners as soon as they approached that zone, in a subtle but reportedly effective way.

b. *As influencing traffic flow.* One should notice where people are walking—the paths or shortcuts ("desire paths") pedestrians take— to see which zones will be visited the most. In Washington Square Park, the central plaza and fountain are the heart of the park. Participants doing the WTDA soundwalk usually come back several times to the center of the park, which is open acoustically, and a giant echo chamber for any louder noises in the entire square. For this central area, a greater range of sounds was used (more audio file choices), to avoid too much repetition of material on the repeat visits. Also taken into account were the acoustics, and the ever-present cacophony of sounds from hawkers, musicians and other street performers, such that when overheard they would assist in creating the needed variety making each moment different from the next.

c. *Centers of activity influencing traffic density.* It is useful to consider where people congregate or make a lot of noise, or where walking might be slow because of traffic density. The fountain is a center of activity in the park. In the summer it's full of children and dogs and their minders cooling off, people yelling and the loud whoosh of water spraying. In the winter, the fountain is off and silent. Recordings made at the fountain were used to create electro-acoustic vignettes. Some were quite realistic (all water and voices), and some very abstract—filtered as to leave only the rhythms and energy. When the water is off (in winter), the audio reminds listeners of the water's absence, even reportedly making some feel as if they might get wet. When the fountain is on in summer, its sound comingles with the composed vignettes.

3. *Use of environmental sounds.* Some decisions should also be made early in the process, about how to interact (or not) with environmental sounds, and about how these sounds might positively impact the composition and contribute to the AAR. All of the following are important to consider:

   a. *Balance.* What will the volume of potential overheard/real sounds be, especially in relation to the recorded/processed/generated sounds?
   - Should the outside sounds be blocked out? Incorporated? Take precedence?
   - What kind of headphones or listening devices will participants be using?
   - Is it the soundwalk solo or group activity? (Will participants interact with each other? How will it impact on the experience of the sounds to be heard?)

- Will the sounds heard need to be synchronized between participants in some way?
- Will participants be asked to use speakers as a way to introduce new sounds to the environment? How will those sounds interact with each other? Consider examples such as Phil Kline's *Unsilent Night* and previous work or interactive pieces with groups walking/bicycling with small speakers such as in Kaffe Matthews's *Sonic Bike* projects, and Lainie Fefferman and Jascha Narveson's *Gaits Soundscape* for Highline Park, NYC.[12]

b. *Mix.* The composer should consider what other aspects of the EQ, mix, effects or audio production might be influenced by his or her choices of balance and sound delivery.

4. *Accounting for natural sounds/animals.* Omnipresent in many outdoor AAR pieces might be sounds of birds, insects and other local animals. In the case of WTDA, a deliberate choice was made to interact with the natural soundscape of Washington Square Park, where the natural sounds are surprisingly dense in some areas and in others disturbingly absent.

   a. *Adding and amplifying.* An "over the top" mix of birds and insect sounds was created for wherever it was felt their presence could be highlighted or given emphasis. If the composer felt these sounds were missing in a particular spot, she put them back (e.g. adding swarms of bees to underscore that there are many fewer bees than just a decade ago). As elsewhere, electroacoustic processed versions were included as well. (Insect sounds are also heard in *Florambula*, Barbara Weber's U-GRUVE piece for Conservatory Garden in New York City.)

   b. *Inventing a fairytale forest.* For WTDA, it was found that the very large and older trees at the northwest corner of Washington Square park (including the 310-year-old Hangman's Elm) suggested an old German fairytale forest scene, especially at sunset. So sounds were added: owls, night birds, both as unprocessed collage and electroacoustic processed sound that maintains a spooky character of the place but is obviously electronic music. This evocative audio is there day and night, but at sunset and at night there are also the resident birds and crickets adding their own sounds to the mix.

   c. *Animal gatherings.* Naphtali also noted that the pigeons congregate by day year-round by a statue in a little plaza in the west part of the park. So, her "flocking" pieces are there too, consisting of clouds of flapping wings and electroacoustic interpretations, to designate that entire area as belonging to those birds.

5. *Man-made sounds.* Urban soundscapes are an opportunity for interaction or exaggeration/commentary.

   a. *Construction sites.* For WTDA it was noticed that on the eastern perimeter of the park there had been continuous construction for several years. So, an electroacoustic construction zone was created from field recordings made on location. Serendipitous interactions ensue between the rhythms in the audio pieces and what a participant hears there when the construction site is active. Unpredictable and fun, these are the clearest and most musically accessible expression of aesthetics of WTDA as related to the aleatoric sound work of John Cage, the listening practices of Pauline Oliveros and others and what the composer hoped would open the participant's ears to new sounds and new ways of hearing.

   b. *Traffic/Street sounds.* Cars driving down fifth avenue must drive around Washington Square park to continue south. Because of this interruption, many cars careen impatiently around the northwest corner of the park to continue their trip downtown as quickly as possible. The juxtaposition of these rushing cars with the tranquility of the park and birds was an excellent opportunity for exaggeration and commentary with electroacoustic pieces based on drag races, raceway sounds and extreme automotive sounds using Doppler effect. At first these added sounds were nearly unnoticeable and needed to be made much louder and more exaggerated because, as discovered, the sounds of the actual cars going by were even more extreme than expected. When WTDA listener/participants realized what was going on, the piece fulfilled its original mission for the acoustic ecology festival—to highlight how much extra noise there is even in a "quiet" corner of this park. Listeners often had not realized how loud and intrusive these sounds were.

6. *What was not recorded.*

   a. *A word about musicians.* Naphtali avoided recording any street musicians in the making of this piece. This was because she is a musician and did not want to take advantage of any of their work, but also because their sounds are always in the park anyway—ever present and serendipitously included in the work, except late at night or in the cold of winter.

   b. *No firetrucks please.* A deliberate choice was made in WTDA to omit any firetruck or emergency response sounds from the piece for two reasons: (1) they are already ubiquitous and unlocatable in a big city and so always part of the general soundscape and (2) in the case of an actual emergency, using them in the walk could present a problem. Civically, the composer feels these sounds should not be used as part of an outdoor piece in an urban environment and that they need to be reserved for their original purpose.

7. *Inaccessible places.* In three places inaccessible sounds were brought outside of their normal environment to be heard and considered by those who are excluded from them. Using accessibility as a determining factor for setting the perimeter of these zones, the physical attributes of the location were included as a means for recognizing and seeking to break down the social/political symbolism of exclusion and boundaries.

   a. *Playgrounds* are only accessible to children and their caretakers (as they should be). They are a cornucopias of rich sounds, when heard up close, but from a distance (from outside) they are at best a rough din of children's voices. Audio pieces were created from these sources, and they were placed in a large area encompassing and extending the reach of two of the playgrounds, particularly accentuating the metal gates which are the literal gateway to this sound world.

   b. *Library.* NYU's Bobst Library is also inaccessible to anyone without privileges there. Libraries are conventionally understood as very quiet spaces, but there are always noises present even there. To highlight both the inaccessibility of the space and its extreme, and supposedly quiet, atmosphere, pieces were created using highly amplified noises from recordings made inside the library. The inaccessible sound world of the interior of the library was brought outside the building onto the sidewalk, overlapping with the park slightly at the southeast corner. This audio can also be heard inside the library for anyone (with access) wanting some distraction from studying.

8. *Walking—travel vectors.* Expanding the idea of a zone as a trigger for an audio event, it is also useful to track the movement between zones, or use the direct mapping of GPS data as trajectories to control some aspect of that audio event or audio generation. This is similar to early computer-vision methods, which divided the field of vision into quadrants and tracked which quadrants were activated to determine direction and speed.

   a. *Direction of travel.* Participants walking north/south from the park to a local gallery heard the sounds in a different sequence[13] simply by activating the zones in a different order. The direction of travel can also be directly tracked using the GPS to deliver completely different experiences, controlling any parameter of the sound, sound generation or mix. Tracking direction of travel could be especially useful for walks that are long but not wide.

   b. *Speed of travel.* For Walkie Talkie Dream Garden (developed by
   the authors and launched in 2018 for the waterfront areas in north
   Brooklyn, NYC, and Hamburg, Germany), there are different experiences for walkers, runners and those who opt to hear the piece
   from the passing ferries out in the middle of the East River. As
   with direction of travel, speed of travel can be used to enhance the
   interactivity of the soundwalk by tracking zone activation, directly
   via the GPS, or simply by careful planning of sizing and placement
   of zones.

   c. *Proximity (to a marker).* Using detailed location information, it is
   possible to point a participant/listener to compose an interactive
   music score dedicated to a specific visual marker or artifact such as
   a sculpture or statue/artwork, in a small zone. Mixed media artist/
   researcher Adrian Hazzard excellently outlined his ideas regarding composing location-based soundtracks to create and resolve
   melodic and harmonic tension within four distinct stages of interaction: approach, arrival, engagement and departure, to create interactive and dramatic scores for works in a sculpture park.[14]

Another example from Barbara Weber's U-GRUVE piece *Orela* (written
for New York's Lincoln Center Plaza) is that she successfully marries the
walking of listeners around the iconic fountain to the cyclical reiteration of
notes in a choral piece. Moving away from the fountain lowers the level of
the drum patterns there. Filtered sounds and darkening timbres of sounds
near the iconic fountain make it seem more intimate and change the mood.
The ultimate effect is that the intensity increases as you get closer to the
center, and the music is underscored by the white noise of the fountain
enhancing the feeling of movement between various corners of the plaza
and in relation to the fountain.

9. *Scene-setting versus underscoring.* With AAR, just as in film music,
   we make choices as to whether we are scene-setting and underscoring, creating literal or metaphorical sounds, to augment or replace the
   original sounds of our scene. Independent of the style of music being
   created, underscoring may be a compositional decision to replace the
   original soundtrack of a location (human or natural sounds) with a
   new one. But how would simply replacing the sounds with new ones
   be different than wearing headphones and listening to tracks of music
   as you walk around? The difference is that in AR the music has been
   consciously written for the place, can be more evocative than descriptive and can be adaptive to your movements. Just because one is walking near a fountain doesn't mean the composer must evoke water, or
   because the route of the walk goes near a statue of an angel that the
   composer must cue up choir sounds. That would be more akin to tone
   painting, and while that may be interesting to kids (maybe), it may not

get to the heart of the psychogeographical experience of the place. It's a state of mind versus state of action. The primary difference between AAR and simply listening to tracks of music is the act of discovery driven by interactivity and indeterminacy.

10. *Historical and cultural context of location* can be a powerful way to engage the listener/participant in a composition, not simply by means of a voiceovered educational/cultural explanation as in a museum audio tour, but by mining hidden meanings/sounds of a particular location. Washington Square Park proved to be a treasure trove of material and inspiration. In some parts of the park, Naphtali was not inspired by the resident sounds, and in some cases she felt that to use them would be "Mickey Mousing" and too literal. She instead chose to connect to these locations in more symbolic, personal ways, mining stories/songs, memory, history and texts, hoping listeners could ponder the essence and themes of the park and location itself.

   a. *The sounds of protest—historical and current.* The center and heart of the park is the plaza, where protests and public gatherings of all kinds, big and small, have been held going back at least 150 years. She used three different protest sound sources as material—two that she recorded on-site, and one from the 1960s. Given no file limitations, she would have included many more protests from different eras, but the three collages of protest sounds have proven to be sufficiently effective at causing listeners to recognize collective protest as a historic function of the central plaza of the park. If a listener passes through when a protest is in progress (which is often enough), they will hear these protest collages as augmenting the experience.

   b. *Hidden stories—Marcel Duchamp.* In 1917, Gertrude Drick, Marcel Duchamp and four other artists snuck through an unguarded door to the spiral stairs inside of Washington Arch, climbed to the top and threw a party with balloons, toy pistols, lanterns and plenty of liquor, even a small fire to keep warm. The "Arch Conspirators" proceeded with pomp, to read aloud their proclamation, declaring independence for the "Free and Independent Republic of Washington Square." In homage in a vocal clip, for the Arch and Chess areas, Naphtali recorded herself singing the manifesto's opening "Whereas, Whereas, Whereas" the one word repeated, ad infinitum, as it was originally read, to create a reminder of the long history of Washington Square as a center of bohemian fun and rule-breaking.[15]

   c. *Chess area—embedded texts, songs of decision and justice.* There have been competitive chess games going on in the southwest corner of the park day and night as long as the composer can remember. Dissatisfied with creating music from clocks and the sounds of chess pieces, she realized this zone should instead tie into a concept of justice and decision making using the symbolism of the chessboard as military and judicial. It was the best spot for the sung text for that zone: the George Washington quotation carved into Washington Arch, "LET US RAISE A STANDARD TO WHICH THE WISE AND HONEST CAN REPAIR." As Marcel Duchamp was also a chess player, the "Whereas, Whereas" audio clip plays here as well.

d. *Church bells*. Continuing on themes of social justice and action, on the south edge of the park across from the historic Judson Church a solemn electroacoustic composition using church bell sounds can be heard.

e. *Shakespearean voices*. In an area that is now a playground and inaccessible to any others than children and their caregivers, Naphtali put Shakespearean voices. These sounds are remembered by older denizens of park from street theater performances that used to be set there in the summer years ago. An older participant on a WTDA soundwalk once reminded the composer that left out of the piece were the voices of drug dealers selling their wares, voices/sounds that were pervasive in the park at that time.

11. *The role of indeterminacy*. The AAR experience can be engaging for a long period of time (or a larger area) by strategically designing the role of indeterminate sounds overheard from the environment and how people interact with the space. WTDA is different every time it is experienced for many reasons, all of them adding to the aleatoric nature of the work.

a. The sonic environment is constantly changing and only partially predictable. The choice was made to mix most of the work to maximize the interference of the outside sounds; the composer therefore prefers if the entire soundwalk is heard with earbuds or open headphones

b. Each participant chooses their own path and, therefore, which zones are activated.

c. Each participant chooses their own pace and, therefore, influences the percussion density (if they linger) and speed of changes of the sounds as they move through zones.

d. The zones in turn contain lists of audio files mostly playing in random order (so two people walking together may still get different versions of the piece). Limits can be set on how many times a file might be repeated.

e. No two soundwalk sessions will be the same, especially given other triggering variables such as time of day, weather, direction of travel (e.g. moving from north to south) and so on.

In the past twenty years, there has been an explosion in the number and types of projects using interactivity, which we encounter in every part of our lives as we work and play and use our smart devices and computers. Ideas about interactivity and indeterminacy for AAR relate directly to all this other work as well as larger-scale work by digital media artists in interactive performance systems and sound installations. Additionally, the literature and thinking on interactivity within the game music world, which has gotten considerably more sophisticated during this time, is vastly useful here.

14.5. Outro. (Conclusion)

AAR is a new medium, allowing the composer to create new site-specific work for many different kinds of spaces and experiences. We have only begun to scratch the surface of what is possible now that personal handheld devices have become so commonplace and GPS tracking have become such universal experiences.

This chapter has outlined many of the aspects and considerations necessary for constructing AAR systems and creating compositions for them. AAR represents the state of the art of interactive, location-based music, and suggests the evolutionary trajectory of our larger music-making experience.

It is a new channel for the public to acquire and enjoy music, as well as a new channel for composers to connect with their audiences. Ultimately, we are working to effect a large-scale shift in the public's view of what music can be, and how it can have emotional impact on our lives in the context of living, while at the same time transforming composers and audiences into collaborators, listeners into performers and environments into instruments.

14.6. Online Resources

Additional online resources related to this chapter can be found online at http://audioar.u-gruve.com/.

14.7. Recommended Projects and Readings

Please visit http:/u-gruve.com/projects-and-readings/ for the complete list.

14.8. Projects

Andreas Zimmermann and Andreas Lorenz
LISTEN: Contextualized Presentation for Audio-Augmented Environ-
    ments (2003)
Virtual/contextual immersive museum guide via custom hardware
http://citeseerx.ist.psu.edu/viewdoc/download;jsessionid=970C577DE
    375C687412D9E70F7F4B2AF?doi=10.1.1.67.3987&rep=rep1&
    type=pdf

The LISTEN project was an early attempt to provide museum visitors
with a contextually driven experience designed to "provide a personalized
immersive augmented environment, an aim which goes beyond the guiding purpose. The visitors of the museum implicitly interact with the system because the audio presentation is adapted to the users' contexts (e.g.
interests, preferences, motion, etc.), providing an intelligent audio-based
environment."

=================

Nigel Helyer and Dr. Daniel Woo
AudioNomad (2004)
Mobile application for Windows Pocket PC

A very early example of an AAR system, in which custom software
is used to plot audio zones and GPS is used for tracking the participant's
position. It also included head-orientation tracking for handling stereophonic panning. One implementation of the technology was for the 2006
presentation of *Syren*, a multichannel audio installation located in Sydney
Harbour (Australia) in which listeners would ride on a boat through an
array of 12 speakers, with the audio changing based on the boat's relative
position to any given speaker.

=================

Janet Cardiff
"Her Long Black Hair" (2005)
Guided Soundwalk with Custom Audio Kit (CD Player/Photographs)
https://phiffer.org/hlbh/
From the site: "Janet Cardiff's *Her Long Black Hair* is a 35-minute journey
    that begins at Central Park South and transforms an everyday stroll in
    the park into an absorbing psychological and physical experience."

"The walk echoes the visual world as well, using photographs to reflect
upon the relationship between images and notions of possession, loss,

history and beauty. Each person receives an audio kit that contains a CD player with headphones as well as a packet of photographs. As Cardiff's voice on the audio soundtrack guides listeners through the park, they are occasionally prompted to pull out and view one of the photographs. These images link the speaker and the listener within their shared physical surroundings of Central Park."

====================

Christina Kubisch
Electrical Walks (2004-present)
Soundwalk with customized wireless, electromagnetic-sensitive headphones
www.christinakubisch.de/en/works/electrical_walks

Kubisch has found a way to transduce the electromagnetic fields within urban environments through customized headphones, allowing listeners to walk through a given city and hear its electromagnetic emissions at various points along the way. Revealing many unexpected sources, each with their own frequency and rhythm, leads to wholly viable indeterminate musical compositions.

====================

Duncan Speakman
Various Works
http://duncanspeakman.net/locative

Speakman's output in the area of location-based music is prolific, with projects going back as far as 2006, in which he's explored and examined a number of formats for delivering locative sound/music experiences.
Here are a few recommended projects:

http://duncanspeakman.net/boundary-songs/
http://duncanspeakman.net/a-song-for-50-hearts/
http://duncanspeakman.net/for-every-step-you-take-i-take-a-thousand/

====================

Bluebrain (Hays Holladay and Ryan Holladay, composition and production)
Bradley Feldman (developer, creator of SSpace, the software used for the app)
Central Park: Listen to the Light (2011)
Mobile App for iOS
https://vimeo.com/channels/539003/29630558
www.nytimes.com/2011/12/08/arts/music/bluebrains-app-central-park-listen-to-the-light.html

Sometimes touted as "the world's first location-aware album," the Holladay brothers created a suite of original music for Central Park that was broken down into its parts and mapped to zones across the park. Using custom software developed by Bradley Feldman, the piece was delivered as an iOS app in 2011. The Holladays have also created similar experiences for the National Mall in Washington, DC, and R Park in Jackson, Wyoming.

======================

Josh Kopeček and Yoann Fauche
Echoes (2013-present)
Soundwalk generation software platform
https://echoes.xyz/

Echoes is a cloud-based platform for creating soundwalks. Though predominantly text-oriented, rich media files can be sound-art/musically oriented, though there is no temporal synchronization featured.

======================

Kaffe Matthews with Dave Griffiths et al.
Sonic Bike (various projects 2008-present, since 2013 as Bicrophonics Research Institute)
Custom fitted bicycles with audio speakers, GPS with custom hardware/ software.
http://sonicbikes.net/sonic-bike/

The *Sonic Bike*'s onboard custom software allows for mapping of sounds onto multiple geographic areas/zones in cities and landscapes, to play content over the speakers "that changes depending on where the cyclist goes and how fast they ride." The composers (Matthews and her collaborators) have created a wide range of community-based experiences and location-driven musical compositions since 2008. Recent extensions of the project include the *Sensory Bike* (*Sonic Bike* as an instrument that can be played) and *Sonic Kayak*.

Notes

1. "The Recording Academy 'plays' New York City through AR leading up. . . ." 26 Jan. 2018, www.thedrum.com/news/2018/01/26/the-recording-academy-plays-new-york-city-through-ar-leading-up-grammys. Accessed 3 Feb. 2018.
2. "Evolution of Indoor Positioning Technologies: A Survey—Hindawi." 21 Feb. 2017, www.hindawi.com/journals/js/2017/2630413/. Accessed 3 Feb. 2018.
3. "GPS Beatmap on Vimeo." 2 Sep. 2009, https://vimeo.com/6402527. Accessed 3 Feb. 2018.

4. "Sonic Bike" http://sonicbikes.net/sonic-bike/. Accessed 15 June, 2018
5. http://sonicbikes.net/about-us/. Access 15 June, 2018
6. "Making Musical Apps—O'Reilly Media." http://shop.oreilly.com/product/063692 0022503.do. Accessed 4 Feb. 2018.
7. "Getting Started with the Fluxamasynth Shield | Modern Device." 9 May. 2013, https://moderndevice.com/documentation/fluxamasynth-quickstart-guide/. Accessed 6 Feb. 2018.
8. "iOS, OSX and Android Audio SDK, Low Latency, Cross Platform, Free." http://superpowered.com/. Accessed 3 Feb. 2018.
9. New York City Department of Parks and Recreation, 2018. Washington Square Park [6 February 2018]. Available from: www.nycgovparks.org/parks/washington-square-park/
10. Defined as "playful, inventive strategies for exploring cities . . . just about anything that takes pedestrians off their predictable paths and jolts them into a new awareness of the urban landscape." *A New Way of Walking*, Joseph Hart, Utne Reader July 2004. www.utne.com/community/a-new-way-of-walking.
11. The Google "My Maps" feature allows for drawing and measurement of boundaries, and easy sharing of maps.
12. Friends of the High Line. The Gaits Soundscape 2017 In: The Highline Blog. 17 December 2017 [11 February 2018]. Available from: www.thehighline.org/activities/the-gaits-soundscape-2017
13. WTDA's first version of the walk for NoiseGate Festival extended into SoHo on two north/south streets.
14. Hazzard, A. 2014. *Principles for composing location based soundtracks* [5 February 2018]. Available from: https://adrianhazzard.com/research/principles-for-composing-location-based-soundtracks/
15. Axelson, E.P., 2007. The Free and Independent Republic of Washington Square (Part II). In: The Daily Plant. 24 January 2007 [viewed 3 February 2018]. Available from: www.nycgovparks.org/parks/washington-square-park/dailyplant/20026

Further Reading

Augoyard, J.F., Mccartney, A., Torgue, H. and Paquette, D., 2006. *Sonic Experience: A Guide to Everyday Sounds*. Montreal: McGill-Queen's University Press.

Filimowicz, M., 2015. The Mobile Augmented Soundscape: Defining an Emerged Genre. *Hz Journal*, no. 20. Viewed Feb 7, 2018 www.hz-journal.org/n20/filimowicz.html

Hart, J., 2004. A New Way of Walking. *Utne*, Ogden Publications, Inc, June 27, 2018. Viewed Feb 7, 2018 www.utne.com/community/a-new-way-of-walking

Hazzard, A., 2014. *Principles for Composing Location Based Soundtracks*. Viewed Feb 7, 2018 https://adrianhazzard.com/research/principles-for-composing-location-based-soundtracks/

Hazzard, A., Benford, S. and Burnett, G., 2014. You'll Never Walk Alone: Composing Location-Based Soundtracks. In: *Proceedings of the International Conference on New Interfaces for Musical Expression. NIME'14*, June 30–Jul 03, 2014. London: Goldsmiths, University of London, pp. 411–414.

Horowitz, S. and Looney, S.R., 2014. *The Essential Guide to Game Audio: The Theory and Practice of Sound for Games*. Burlington, MA: Focal Press.

Koutsomichalis, M., 2011. Site Specific Live Electronic Music: A Sound-Artist's Perspective. In: *Proceedings of the Electroacoustic Music Studies Conference, Sforzando!* June. New York. Viewed Feb 7, 2018 www.ems-network.org/IMG/pdf_EMS11_ Koutsomichalis.pdf

Lafrance, A., Feb 19, 2016. Hearing the Lost Sounds of Antiquity. *The Atlantic.* Viewed Feb 7, 2018 www.theatlantic.com/technology/archive/2016/02/byzantine-angel-wings/470076/

Matthews, K. et al. Feb 12, 2018. *Bicrophonic Research Institute—Makes Music and Audio Landscapes to Be Triggered by You the Cyclist.* Bicrophonic Research Institute. Viewed Feb 7, 2018 sonicbikes.net/

McCartney, Andra, 2017. Soundwalking: creating moving environmental sound narratives. In: Gopinath, S and Stanyek, J, (Eds.), *The Oxford Handbook of Mobile Music Studies,* Volume 2. New York: Oxford University Press, pp. 212–237. https:// soundwalkinginteractions.wordpress.com/2010/09/27/soundwalking-creating-moving-environmental-sound-narratives/

Oliveros, P., 2005. *Deep Listening: A Composer's Sound Practice.* Lincoln, NE: Deep Listening Publications.

Paquette, David and Andra McCartney, 2012. Soundwalking and the Bodily Exploration of Places. *Canadian Journal of Communication,* 37, no. 1, pp. 135–145. www.cjc-online.ca/index.php/journal/article/viewFile/2543/2286

Serafin, S., Erkut, C., Kojs, J., Nilsson, N.C. and Nordahl, R., 2016. Virtual Reality Musical Instruments: State of the Art, Design Principles, and Future Directions. *Computer Music Journal,* 40, no. 3, pp. 22–40. doi:10.1162/comj_a_00372

Urban Squares website. A "Collection and Analysis of Rediscovered Urban Space" with information about psychogeographical practices and "neighborhood portraits." http://urbansquares.com/07psychogeographyNEW.html

Westercamp, H., 2007. Soundwalking. In: Carlyle, A. (Ed.), *Autumn Leaves, Sound and the Environment in Artistic Practice.* Paris: Double Entendre, p. 49. Viewed Feb 7, 2018 www.sfu.ca/~westerka/writings%20page/articles%20pages/soundwalking.html

Soundscape Online Databases State of the Art and Challenges

Miles Thorogood and Philippe Pasquier

15.1. Introduction

Online audio databases provide a crucial service in many research and creative practices such as sound design, sound art and soundscape studies. Enthusiasts and general users are also engaged in retrieval and contribution of audio files to databases. Because of the range of foci, both regarding audio themes and users, there is a range of tools and models that are implemented to service that database. This chapter describes these tools and models for the subject of online audio databases concentrating specifically on environmental sound recordings.

The rise of online community-generated databases, along with the decreasing costs associated with recording equipment and storage, has facilitated an increased ability for creating and sharing sound recordings. These databases provide unifying mechanisms to acquire and disseminate recordings for the groups of users they serve. Furthermore, the open content philosophy underlying many of these projects promotes the free use of the sound recordings generated by individuals. This rich audio dataset provides a vital source of material for researchers and creative practitioners working with environmental sound recordings.

Each community of users has slightly different expectations of working with sound recordings. Thus, a "one size fits all" approach to the database design is often infeasible. For example, databases may provide online community interaction by employing social network technology. Other databases gather free audio recordings or are for sale from commercial production houses. Others answer project-specific goals either online or on a local machine.

In this article, we outline the theoretical basis and community aspects that have emerged as significant contributors to contemporary audio database projects. First, we describe some of the domains that benefit from databases of audio recordings. Specifically, we focus on sound design in the context of video games and film. Along with soundscape studies and music information retrieval, we feel that these would benefit from a rich source of audio material. Second, we turn the reader's attention toward common mechanisms employed by database projects, including taxonomies, ontologies, collaborative tagging and licences. Next, we detail the common database models and demonstrate a number of the database projects that employ these models. Finally, we present our conclusions and point to some specific concerns about the databases seen here.

We hope that the examination of online audio databases here will elucidate the topics of concern for individuals seeking environmental sound recordings for sound design from online databases. The theoretical considerations covered provide a starting point to those wishing to research online audio databases further.

15.2. Use Cases

In this section, we briefly describe game sound, film sound, soundscape studies and music information retrieval as research and practice domains that use environmental sound recordings. In particular, to indicate the practicality of audio databases, we focus on the role sound plays in these areas.

15.2.1. Game Sound and Film Sound

Well-designed video games engage players through the function of interacting systems such as game mechanics Adams (2009), internal economies Adams and Dormans (2012), and uncertainty Salen and Zimmerman (2004). From a high-level analysis, games can be described regarding their mechanics or rules, how the game play changes during the game and the mediated experience of the game made possible by the graphics and sound. As outlined by Hunicke, Leblanc and Zubek (2004), these independent characteristics are central to understanding video games in regard to user experience.

Sound plays an integral role in the game play experience. As Karen Collins (2008) notes in her examination of the history and theory of game audio, sound design in video games is as much a problem of creating an immersive experience for the player as it is a technical challenge. Further, the ability of authors to include sound recordings both enriches possibilities and introduces new problems with resource management and licensing.

Sound design involves the retrieval and segmentation of audio recordings related to game environment concepts, as well as the processing and mixing of those audio segments.

Similar to sound in games, film sound is an essential element of immersing the viewer in the mediated experience. According to sound designer Walter Murch (2005), sound effects function not only to convey information about specific details in the film but also to represent more abstract qualities (e.g. a particular mood).

The "encoded-embodied" debate put forward by Murch is similarly asserted by Michel Chion (2009), who suggests film sound imparts explicit and implicit information to a viewer through the use of characteristic sounds. These particular sounds are created by modifying audio recordings to engender the appropriate semantic and emotional meaning.

Whether with game sound or film sound, a sound designer can accesses audio recording by either recording the material themselves or accessing sounds from database sound effects. Field recording or Foley recording is a process that has an associated time and equipment cost, and sometimes locations may be unreachable for some people. Further, these activities may be moot points if existing audio recordings fulfill the design requirement, such as could be found in sound effects databases like the accessible user-generated audio database Freesound Font et al. (2012).

15.2.2. Music Information Retrieval

The ability of authors to include sound recordings both enriches possibilities and introduces new problems with resource management and licensing. The MIREX community has generated a great deal of research to advance rigorous measures of accuracy and performance for the comparison of different systems and algorithms in the field of music information retrieval Downie et al. (2010).

15.2.3. Soundscape Studies

The combined sounds in an environment provide information that situates listeners. Through experience, we come to associate those sounds with a corresponding environment. These associations can be specific to an individual, for example, the sound of a particular school bell bringing to mind past events from someone's life. The relationships can also be archetypal within specific contexts, for example, the sound of a cow bell denoting a rural environment.

R. Murray Schafer (1977) explored the intersection of individuals and communities with the sound environment in the 1970s and developed soundscape research. From that early work, soundscape studies emerged as an interdisciplinary approach to investigating facets of the world regarding their acoustic environments. For instance, environmental sound raises awareness and understanding of socioeconomic and ecological matters that can inform urban planning or environmental management.

15.2.4. Soundscape Composition

Soundscape composition is a form of sound art that processes and combines environmental sound recordings. According to Canadian composer Barry Truax (1996), "the aim of soundscape composition is to evoke a listener's associations and memories of real or imagined soundscapes." An essential part of this exercise is the search and retrieval of recordings from a database. Such a database can take the form of a simple hierarchical file structure on the composer's computer or more sophisticated databases using state of the art technologies and search algorithms.

Soundscape composition, as well as many other forms of sound art, relies on recordings of environmental sound. For example, *Audio Metaphor* by Thorogood, Pasquier and Eigenfeldt (2012; Thorogood and Pasquier 2013) generates soundscape compositions from text input. The system accesses either an online or local database to retrieve labeled audio recordings. A supervised machine learning algorithm trained with data from human perceptual classification experiments autonomously segments the audio. The semantic and saliency labeled segments are then processed and combined autonomously based on a composition schema.

15.2.5. Soundscape Composition as a Database Usage Context

Correia (2011) outlined a system named AVClash that allows the creation of audiovisual compositions consisting of combinations of sound and animation loops. The AVClash system has been utilized in noise music performance and is available online. The web interface for AVClash presents the user with the affordance of choosing a sound clip from a list that is generated by the top 11 most popular tags pulled from Freesound. AVClash demonstrates online sound databases as an engaging system that reaches a wide audience.

Adapting sound databases to generative soundscape composition has been explored in the literature by Cano et al. (2004), who propose a system to generate multitrack projects for background ambiance from a commercial database of 80 thousand sound effect recordings using natural language queries. Recordings in the database are labeled by concepts using a group of synonyms of the recording name. Information in the natural language query is mined using WordNet to extract concepts that are used to retrieve recordings matching those concepts. The system selects a subset of random sounds and sequences these by placing more extended duration audio as a background loop while adding shorter sounds probabilistically along the timeline.

Eigenfeldt and Pasquier (2011) autonomously generated soundscape composition using methods from audio information retrieval. In their system, artificial agents negotiate between themselves about which audio to play and process from a database of 227 hand selected files curated from the Freesound collection. The agents base their decisions on audio analysis features, and a small set of tags entered by the researchers. Evaluating compositions of the system showed that listeners preferred the machine compositions more than random compositions.

15.3. Searching for Sound

Most people search for audio recordings in online databases using a keyword search through a web browser. A list of resulting recommendations is presented to the user, accompanied by metadata including the file URI, name, format and duration. Many databases include a text-based description and tags about the audio content, and some databases permit geographic data of the recording location. It is from this information that users will base their decisions to proceed further with downloading the recommendations.

Computer programs can also autonomously search and retrieve audio files from online databases. However, scraping web pages for useful information is not always feasible. In this case, the website of some sound databases provides a machine-readable interface through an XML file of entries, or an API to interact with the database. We will, however, refer to any metadata whether human entered, machine generated, accessible in the front end or available through an API as a form of tagging.

15.3.1. Tagging

Tags are a kind of metadata about a resource. In user-generated tagging systems, social interactions of users tag recordings by an agreed upon classification or categorization scheme. A machine generated tagging system, on the other hand, can automatically tag items by a similar classification or categorization approach, and may perform other feature extraction techniques to represent the sound signal.

In his book on outlining the concepts and methods of tagging, Gene Smith (2008) demonstrates a basic three-part model of tagging systems. Smith's model includes the users who create tags and resources, the resource items that users tag and the tags or keywords added by users and assigned to resources. This model applies to any tagging system that has users, resources and tags. The differences, or similarities, in tagging systems are the rules about what kinds of tags are used, who can tag, what resources users are allowed to tag, and if users can add resources. Some tagging systems such as the photo-sharing site Flickr allow users to submit and tag resources and additionally tag other people's resources, whereas sites including SoundCloud (n.d.) and Freesound (n.d.) enable users to upload and tag only their resources. A tagging system applies trade-offs, or what Smith considers "four tension points" that influence design decisions.

- *Personal*—Social: What is the motivation of users for tagging, is it for personal use, or motivated by social factors.
- *Idiosyncratic*—Standard: Tags could be completely unique or standardized.
- *Freedom*—Control: Is there complete freedom given to users, or is some control mechanism in place that influences their tags.
- *Amateur*—Expert: How qualified are the users who are creating the tags. Is the same level of importance given to general users as to expert users?

Decisions made by the database system designer account for how the balance of these tension points is manifested, ultimately affecting the community of users who will be involved with the database, as well as the quality of resources and tags.

15.3.2. *Taxonomy Classification*

In taxonomy classification systems, a designer sets out a limited number of terms or concepts to use for classification, resulting in a single correct classification of each item. A taxonomy establishes a parent-child relationship in a hierarchical structure that can be used to define broader and narrower concepts. In the domain of sound classification, taxonomies have been designed to address the problem of concept representation from different perspectives.

Gaver (1993) develops a taxonomy that covers a full range of environmental sound events based upon the interaction of materials. The hierarchy of simple sonic events includes vibrating objects (impacts, scraping, others), aerodynamic sounds (explosions, continuous) and liquid sounds (dripping, splashing). The taxonomy provides two levels of classification, a high level or broader scope of sound, and the more narrow classification of the sound making event.

Murray Schafer (1977) defines a taxonomy for the soundscape consisting of three levels of classification. The top level contains general grouping classification classes such as Sounds of Nature, Human Sounds, Sounds and Society, Mechanical Sounds and Sounds as Indicators. These classes define the domain for the classification. The next level contains categories of sounds in the domains. At this level, several categories are below the domain in which they are included. Sampling at this level shows that Internal Combustion Engine and Rapid Transit are below Mechanical Sounds, and that Sounds of Birds and Sounds of Fire are below Sounds of Nature and Sounds of the Voice and Sounds of the Body are below Human Sounds. A third level of the taxonomy shows sound sources. Included in the domain of Mechanical Sounds and the category of Internal Combustion Engines are the sound sources Diesel Engine, Tractor and Car.

Another soundscape taxonomy is outlined Brown et al. (2011). Their taxonomy of the acoustic environment for soundscape and noise studies takes into account the variability of environmental contexts. The broader classes of the taxonomy include indoor and outdoor acoustic environments, which contain urban, rural, wilderness and underwater environments. Each of the environments then contains a hierarchical soundscape classification scheme that closely reflects that defined by Schafer. Brown et al. see this taxonomy as a common framework applied to all types of acoustic environments.

These three examples of sound taxonomies demonstrate the design of a hierarchical system for representation of concepts in a domain. The specific and systematic organization of these taxonomies induces a correct, although limited, vocabulary for facilitating rigorous classification. However, the strictness of taxonomies can be problematic in collaborative environments, where users' agreed upon descriptors tend to be fluid, and the agreement of terms is often an open debate.

15.3.3. Ontologies

An ontology refers to the "explicit formal specifications of the terms in the domain and relations among them" Gruber (1993), which is better able to capture the semantic relationships of concepts than a taxonomy. With an ontology, it is common to use a graph structure like semantic networks Sowa (1987), where nodes are concepts, and the connecting edges represent the relationships. A designer uses relationships that ontologies permit to label the specific relationship of concepts. For example, we could say that between a saxophone and instrument there exists an "is a" relationship. Further, that saxophone has a "type of" association with the concept of woodwind, and, finally, woodwind "is a family of" instruments. Once defined, ontologies can be used to reason about the properties of a domain. Using our simple example, we could ask the ontology a question like "what is a saxophone?" Moreover, through a logical inference of the concepts and relationships, it could return to us that a "saxophone is a woodwind instrument." A growing number of ontologies have been designed to represent different aspects of sound for information management systems. Each of these ontologies has focused on formalizing distinct but related sound concepts and vocabularies.

15.3.4. *Collaborative Tagging*

Whereas taxonomies are hierarchical and exclusive classification systems that provide a strict framework of concepts and their relationships, collaborative tagging is a social categorization process of applying descriptive labels by users to items. Taxonomies and collaborative tagging are both systems of labeling to organize items for searching and navigation by users or machines.

While the function of both taxonomies and collaborative tagging is to organize items, the basis of the structure for these systems is inherently different. Neither is without problems, and each system has strengths that benefit different applications. A major distinction between taxonomies and tagging is that taxonomies are hierarchical and exclusive, while tagging systems are non-hierarchical and inclusive.

Taxonomic classification systems are designed to make the relationships between concepts clear and navigation of a domain less ambiguous by rendering the connections between concepts formal and explicit. A typical example of a taxonomy is the Library of Congress classification system comprising subject classes that contain subclasses; for example, a book on data mining is given a subject designation Q, for science, and the subclass A, for mathematics.

Whereas there is a right way to classify a resource within a taxonomy, collaborative tagging systems remove the hierarchical and explicit constraints imposed by taxonomies. Recently, collaborative tagging systems have grown in popularity with the advent of information sharing on the web. In such systems, users will apply contextual knowledge to items by adding tags. A significant advantage of tagging over taxonomy classification is that they are a casual categorization that makes it is easy for users to add tags, rather than forcing them to navigate a strict vocabulary such as with a taxonomy. There are no formal relationships between concepts in a collaborative tagging system, except for those that may be inferred by a degree of relatedness. The casual nature of collaborative tagging and different motivations of taggers results in dynamic semantic spaces that are prone to inconsistencies but reflect the experiences being shared in the community that organizes the knowledge.

In contrast to taxonomies, in collaborative tagging systems, there is no wrong way to categorize an item, and there is an unlimited way to classify a resource. At the very least, collaborative tagging is sympathetic to the range and inconsistencies of users' behavior and provides a more accurate reflection of the semantic space of the community that organizes the knowledge in a domain Golder and Huberman (2005).

To encourage a wide range of users to be involved in a database project with collaborative tagging, provisions for user behavior are taken into account Marlow et al. (2006). These behaviors become increasingly socially motivated as more people become familiar with online social engagement. In collaborative tagging projects, users participate in a casual, and mostly unrestrictive environment that does not require a considerable amount of time or attention to contribute. Additionally, the flexibility of tagging enables users to apply tags to resources in a way that seems appropriate to them. What's more, the playful nature of tagging fosters self-expression in users who are free to share their opinions and interpretations of content. Finally, there is a sense of community produced when users share information and feel they are contributing (van Dijck 2009).

Users' functional motivations for participating in tagging range from a desire to organize personal information, as with bookmarking websites, to a desire to make a social contribution by adding to shared knowledge. Paul Lamere (2008) summarizes these motivations as a duality consisting of personal and community incentives:

- Memory and context items are tagged to assist personal retrieval of an item or group of items.
- Task organization items are tagged to assist in the organization of music discovery tasks and facilitate the listening experience.
- Social signaling items are tagged as a way for an individual to express their tastes to others.
- Social contribution items are tagged to add to the group's knowledge, for the benefit of the wider audience.
- Play and competition with online communities encourages contribution.
- Opinion expressed items are tagged to convey an individual's opinion about an item.

However, social tagging is not without its issues Golder and Huberman (2006). Firstly, social tagging can be slow to gain popularity as they need users to afford the time and energy to generate content and interest. Then, when a more substantial number of people are contributing, the pool of tags is liable to become diluted leading to sparse tag collections. Some of the common problems associated with sparsity in the literature are misspellings, spelling variants polysemy and synonyms.

Additionally, a small number of individuals generate a set number of tags from a population, thereby only representing a section of culture. Furthermore, there is perceived pressure to conform to a small set of tags that were administered by an even smaller number of perceived expert individuals. However, Halpin, Robu and Shepherd (2007) note that the stabilizing of tag distributions might be a consensus around categorization of information driven by tagging behaviors.

15.3.5. Problems With a Sound Vocabulary

Taxonomies and collaborative tagging systems rely on semantic descriptors and perceptual descriptors for marking content with descriptive terms. For audio recordings, semantic descriptors are textual representations that refer to the source of the sound, such as a bird call, and the context, such as the place of the recording. Perceptual descriptors describe the perceptual qualities independent of the source, including timbre, brightness or loudness, described from auditory perception. Cano et al. (2004) note that audio filing and logging is a labor-intensive and error-prone task. Furthermore, languages are imprecise and informal, and sounds are multimodal and multicultural, and there is no agreement on how to describe them. It is because of these issues that the labeling and placement of a sound in a category is imprecise and an imperfect art.

People's use of words and semantics used to describe sound concepts create problems for classification and labeling of audio recordings. This issue is especially apparent in the variety of research domains that are concerned with sound, such as urban planning, sound design and acoustic ecology, which share similar concerns but employ different vocabularies.

Bringing attention to the differences in a language and semantic categorization of environmental sound, Dubois et al. (2006) present the argument that any such language operates differently between cultural groups. However, within these groups, there is a common language for describing an environmental sound. In urban environments, individuals give meaning to sound based on their everyday experiences, assessing their experiences collectively through language. This shared language forms a collective representation falling into a finite set of semantic categories. Addressing the problem of sound descriptors across domains further, Pedersen and Zacharov (2008) outline techniques from sensory evaluation to find descriptive attributes of sound. Sounds in this descriptive mode may be characterized by physical, perceptual attributes, such as "distant" or "loud," as well as subjective attributes, like "pleasing" or "aggressive." Pedersen and Zacharov argue for using sensory evaluation for the development of a standard set of descriptive attributes. Toward this argument they show that the language for characterizing sound is imprecise and involves words that may have different meanings from person to person. Additionally, some attributes of sound do not have descriptor words, and some descriptors are not one dimensional. The ambiguity of sound descriptors poses challenges for the designer and user of audio databases because the fundamental communicative aspect of these databases relies on the active transmission of concepts for the management and dissemination of audio resources.

15.3.6. Copyright Copyleft Licensing

When copying and using audio recordings in production, it is essential to recognize the licence applied to the document. The licence states under what freedoms the document may be used. Online sound databases may employ one or more licences, which are agreed upon by the author of the audio recording. Online sound databases provide notification of the licence by communicating the name of a licence accompanying the particular recording, or a website in cases where a single licence covers all documents.

The main licences attached to audio recordings are copyright, Creative Commons and public domain. Copyright is a licence giving the creator of an original work exclusive rights to it. Exclusive rights assigned to the holder of a copyright licence restrict others from copying or using the work to create derivative works or display the work publicly unless given express permission by the holder of the copyright licence. Copyright is the most restrictive licence for sampling, changing and redistributing works. Fair use is one doctrine of copyright law that facilitates the use of a copyrighted work without permission, under certain conditions. A summary of the primary fair-use conditions given by Harvard University (n.d.a) lists:

- The purpose and character of the use, including whether the use is of a commercial nature or is for nonprofit educational purposes;
- The nature of the copyrighted work;
- The amount and substantiality of the portion used in relation to the copyrighted work as a whole; and
- The effect of the use upon the potential market for or value of the copyrighted work.

Lawrence Lessig (2004) gives a potent argument for the need for a different model of licensing that encourages creativity. In 2001, Lessig, along with James Boyle, Hal Abelson and Eric Eldred founded the Creative Commons Organization (n.d.), which releases a family of licences that allow authors to select the freedoms they want to grant. The six Creative Commons licences are permutations of the four attributes, no-derivatives, non-commercial, attribution and share-alike. No-derivatives means that the document cannot be modified from its original form and must be left whole. The non-commercial clause states that the document or its derivatives may not be used for commercially. With attribution, the person using the recording must credit the author of the original audio. The credit to the author will be included in the distribution of the produced material (e.g. in the credits of a film or on a website where distributed). Share-alike means that the licence that is attached to the document is the same used for the new creation using the original document. While all six Creative Commons licences require attribution to the author of the document, only two stipulate that no derivatives of the document be created. Given that sound design may manifest derivatives of audio recordings, two-thirds of Creative Commons licences make readily available audio recordings that have those licences attached to them.

The final licensing that we will discuss is public domain. One may use a document for any purpose when in the public domain. For example, anyone can reproduce and modify a document without permission by the author for non-commercial or commercial use. A document may enter public domain status if the copyright licence attached to it has expired or the intellectual property rights have been forfeited or are otherwise inapplicable. A resource for searching public domain material is Public Domain Search (University n.d.b), which attempts to restrict web searches to documents that are in the public domain. By far, works in the public domain are the most accessible and provide the fewest limitations for use in creative practice, such as soundscape composition.

15.4.   Sound Databases With Collaborative Tagging

As more and more people have gotten involved with the content creation and the hardware technology for recording has become more accessible, the acquisition and distribution of audio recordings through sound databases has accelerated in the last years. A small number of databases, including Freesound (n.d.) and SoundCloud (n.d.), have benefited from people's desire to share recordings in an open community, which has provided a growing number and variety of audio recordings online, as well as a forum for the appreciation of the recorded material.

People's incentives for adding to user-generated databases has much to do with the progress of social factors attributed to more time spent online. Bringing these phenomena into question, Roma et al. (2012) investigate the motivation of people to share audio recordings in Freesound, one of the most active community-based audio recording repositories. The breadth of content in the Freesound database covers all categories of sound and encourages sharing through the Creative Commons licence. Users upload sound accompanied by GPS coordinates, relevant keyword and location tags.

Browsing Freesound's repository is facilitated through a tag cloud, an interactive map and a keyword search. There are over 120 thousand uploaded audio clips that vary in their content (e.g. synthesized tones, field recordings), and quality of entry (accuracy of commentary, recording quality, etc.), the progress of which is outlined by Akkermans et al. (2011). The Freesound search engine indexes tags and comments entered by users. An online web interface allows users to browse the database by searching for keywords present in tags and commentary. An API is made available that facilitates using Freesound to access the commentary mentioned above as well as tags and a range of audio analysis features for audio files.

The investigation of motivations for users of Freesound put forth by Roma et al. presents a formal analysis of the elements that compose audio clip sharing systems through a variety of network models, including the download network, semantic network, shared interest network and forum network. Properties are obtained to characterize the networks, including, density, efficiency, clustering coefficient, small world coefficient and modularity. The analysis takes into account novelty and relevance measures derived from the community engagement with the audio clip. The results of the analysis show that creating opportunities for connecting users positively impacts the quality and novelty of sounds.

Sharing the same user-generated model employed by Freesound is SoundCloud, designed as an open platform for sharing sounds on the web. The authors of the project wanted to facilitate sharing sounds on the web in a similar populist fashion as online video, image and text. As an open database, it hosts a range of different types of recordings, from house music to sound effects and field recordings. Contributing users tag and make comments on recordings. There are more than five hundred recordings tagged as "field recording" or "environment." A keyword search returns recommendations from track names, and user-entered tags. From the web interface, results can be viewed as a waveform and downloaded as an MP3 format of the recording. A public API is made available to interface with the SoundCloud database, which facilitates searching, downloading and uploading recordings. Additionally, social features may be accessed such as commenting, sharing, following and liking, which results in a social curation of the audio files.

Both Freesound and SoundCloud share similarities in their mandate to build a platform for users to share sounds. Social factors drive the motivation for users. The conditions of open systems that these sound databases supply results in a diverse range of topics and qualities of contributed audio recordings. Providing access to the database through an API furnishes users with the means to work with the database programmatically. Freesound has by far the most considerable number of audio recordings applicable for soundscape composition, which has afforded research projects the means to leverage the database of audio recordings for soundscape composition.

Soundscape compositions for virtual worlds can use geographic location data to render sounds into simulated environments based on location.

Finney and Janer (2010) demonstrate a system for generating soundscapes using the Freesound database for virtual environments. Audio recordings are handpicked from the Freesound database and classified as either background or foreground based upon semantic identifiers, which assumes classes based on keywords, such as birds or wind belonging to background sound. The composition of the soundscape is made using background and foreground layers, which are placed on a Google map and spatialized depending on the user's position and rotation. The definition of background and foreground by user-entered keywords is prone to produce false positives as a recording of birds could have qualities of being foreground when classified as background.

Janer et al. (2011) outlines a system for augmented reality audio by creating soundscapes to augment the acoustic information in a physical location. Janer's strategy is to place virtual sound objects in geographical areas that allow people to explore the soundscape generated through a Google Map interface, which produces an ambiance that spatially covers a given location. The authoring interface of the system allows a user to design a soundscape from a concept name or tags based on Gaver's (1993) sound taxonomy, and position it within a GPS location. A sound concept is a sound containing the higher-level concept name, the sound taxonomy parameters and location. The Freesound database is used to retrieve audio recordings, and a machine learning algorithm rates the audio recommendations based on the sound taxonomy classes. The system augmented the sound environment in an open-world game, allowing multiple users to access the soundscape from different positions in the Second Life environment.

15.5. Sound Mapping

A small number of online database projects use interactive maps with audio recordings set at specific coordinates. Online sound maps such as Radio Aporee (n.d.), Montreal Sound Map (n.d.), Sounds Like Staten Island (n.d.), Sound Transit (Holzer n.d.) or Soundcities (Stanza n.d.) provide a platform for user-generated audio recordings similar to Freesound or SoundCloud, but with the explicit aim of gathering field recordings from geographic areas. The Montreal Sound Map is an initiative from Concordia University to create an archival database of sound recordings from all over Montreal. The Montreal Sound Map is an ongoing and continually evolving project with the goal of acquiring new sounds of the city. An interactive map interface lets users navigate the recordings based on location on the map, and a drop-down list shows the different tags that have been attributed to individual recordings. These tags include recent uploads, date, locations, soundscape categories and contributors. Clicking on a tag either expands the list or pops up an entry on the map interface. The pop-up contains the recording download link, which is in FLAC format, a location, time, duration and description of the recording. There are over 250 recordings of which over one hundred are binaural recordings.

In the same way that Montreal Sound Map focuses on recordings from the sound environment from a specific locale, Sounds Like Staten Island is a soundscape project that revolves around Staten Island. There are soundscape related initiatives such as listening exercises and sound activities, as well as a Google Map and field recording database with around 150 recordings in MP3 format. Searching the database can be done by a keyword search can be done with all the words, exact phrase, none of the words, or at least one of the words in a query. Users categorize recordings into groups such as outdoors, people, history, streets, towns and author. Word clouds for each of the categories visualize the frequency words within categories. An icon in the browser starts and stops a stream of the recordings in the database.

Montreal Sound Map and Sounds Like Staten Island are just a couple of the examples from the growing number of sound map projects that link a place to its sonic representations. A smaller number of online sound maps facilitate the addition of location-based audio recordings from a larger contextual frame. Radio Aporee is a series of open and public, global sound map projects with a focus on urban, rural and natural field recordings. The intention is to have people contribute and listen to recordings singularly or in mixes. The main project is the Maps project with over ten thousand recordings, where a map interface allows a user to navigate and listen to recordings of an area. A user downloads a recording in MP3 format using a pop-up text box. Along with the recording, a short description provides more information on the audio length and recording location. Listening to the recording on a map can also be done in a random sequence, by all visible recordings, in a geo-mixing mode that plays sounds like a journey and in a streaming mode that continually plays selections of recordings from the database. A mobile app is available for download that operates as a mobile radio player for Aporee. Radio Aporee has several mechanics for searching the database of recordings. A places tab allows the most recently added entries of the last 24 hours, week and month; it offers an additional option to view the recording places that have been most popular over the past month. The web interface returns an unordered list of locations or results from a keyword search. An advanced search provides a keyword search with negation and search by map coordinates. The advanced search results indicate the recording names, artist name and location IDs, and when selected a user can see a waveform representation of the audio and present the recording in map mode or download an MP3 version of the recording. A new sound feed link produces a flat XML file of the most recent recordings added to the Radio Apogee.

Another online sound map, Sound Transit, aims to create an online community for field recording artists through a web interface. The website allows an artist to upload a sound file with a description and location information. A search permits a user to retrieve sound files based on the keywords, artist name and location entered by the artists. Audio file recommendations are made containing a description, length and URI of the recording. Recordings are in MP3 format. Sound Transit also permits a user to select a departure and arrival location with a selectable number of stopovers. The transit route is then translated into a continuous sound file with a linear fade between sound files. A flat XML file of the uploaded sounds and last transits is available for two RSS feeds that can be used in computational systems.

Similarly, Soundcities is an online database of field recordings from cities around the globe. The database exists as a repository for users to contribute and browse various city sounds, and for remixing and creative practice. Users can search for sounds using city names, or by one of several categories from the GUI. Sound categories include ambient, atmosphere, beat, birds, boredom, churches, industrial, mechanical, music, noisy, people, rhythm, sirens, speaking, traffic, travel, voices and weather. A Google map also provides a mechanism for exploring sounds by navigating the map. A flat XML file makes available the database entries, which contain the file URI, title, category, description, date recorded, quality in kbps, length, size, GPS coordinates and city name. All recordings are in MP3 format, at a variety of qualities, as specified by an element in the XML file. The database administrator curates the databases to preserve the quality and relevance of the recordings to the location.

15.5.1. *Sound Databases Without Collaborative Tagging*

Sound databases made possible by the user-generated model such as Freesound, or location technologies such as sound maps, capitalize on crowdsourcing to explore new ways of engaging with databases of audio recordings. Alternatively, sound databases that exploit conventional models of audio libraries tend not to have an open contribution system. The trade-off is a more structured, high-quality set of recordings at the expense of quality control from social factors. Typically, a user accesses these databases with a subscription and searches for recordings using keywords or tags.

As an example of this type of database, SoundJax (n.d.) is semiautonomous sound search engine that indexes and renders waveforms and tags and catalogs audio files from other websites. At the time of this writing, there have been 18,380 sounds cataloged. A keyword search is provided by the web interface, which matches any words in a query, including partial matches. The interface also provides a tag cloud showing the most popular searches. The user receives recommendations as a list of pages, with five results per page. Each result has the file name, the number of channels, format, length, bit rate/sampling frequency, the URI to the source and a URI to the cataloged audio data in MP3 format. There is no API or XML file. The licence is dependent on the source. Quality of the recordings is dependent on the cataloged source.

SoundDogs (n.d.) is another commercial repository of sound effects and audio recording for use in production. It contains recordings from many sound effects suppliers. There is a full breadth of content to cater to commercial productions such as movies and advertising. Without paying, users can listen to or download previews in MP3 format, and a smaller number of recordings are downloadable for free. Searching the free downloads involves selecting categories and drilling down into subcategories. An advanced search allows the user to search by exact match, all words or any words, as well as to filter by a supplier. Search results are returned as a table of file names, brief description and recording length. There is no API for interfacing with the database, and a computer program would need to scrape the well-formed HTML page. Quality of recordings is commercial grade, but from the non-paid interface are 32kbps MP3 are only made available.

A few stock sound databases, such as pdSounds Mobius (2009) and Audio Micro Audio Micro (n.d.), make available assets that, although not helping the web experience, allow for machine access to the database of audio recordings. pdSounds provides a forum for sound appreciation of the recordings submitted by users. The primary themes that pdSounds addresses are field recordings and environmental sound recordings with just shy of one thousand recordings in total. Searching the database is facilitated by a keyword search, and a tag cloud also provides access to recordings. Search results contain the file name, recording length, format, number of times downloaded and user-entered description of the recorded material. Recordings are in MP3 format at a bit rate submitted by the contributing user. RSS feeds for individual tags, users and latest sounds link to a flat XML file.

Whereas recordings from pdSounds may be used freely in the public domain, Audio Micro is a commercial collection of music and sound effects aimed at the media industry. This database hosts over 265 thousand sound effects, of which around two thousand are free. Recordings are grouped into categories and divided again further by a substantial number of subcategories. A user searches Audio Micro by a keyword search in the web interface, with recommendations returned as a list of pages.

Each result contains the file name, length and description of the recording content. An API is available for registered partners. The API is based on RESTful queries and has a Python programming interface that allows a user to browse the category tree, find subcategories, search through categories, get the list of tracks and track details, preview the MP3 track and track the original file URL.

15.5.2. World Soundscape Project Tape Library: A Database With a Taxonomy

The WSPTL contains five unique collections of soundscape recordings and a subject index. These collections were captured between 1972 and 2010 and captured across Canada and Europe. Researchers at Simon Fraser University, Canada, have digitized the earlier collections from the original analog tape format. As a detailed study of the acoustic environment on the cusp of post-industrial society, the early WSPTL provides a significant contribution as the only intentional collection of soundscape recordings of that era. In response to the call of the World Soundscape Project for the documentation and preservation of meaningful acoustic environments, a group of young composers and students including Peter Huse, Bruce Davis and Howard Broomfield created in 1972 the first installation of the Tape Collection recording meaningful sites in Vancouver and British Columbia. In 1973, members of that first group set about recording soundscapes from all provinces across Canada. Furthering this remarkable endeavor, R. Murray Schafer led the original group involved with the Vancouver and British Columbia Collection in 1975 to record European soundscapes, now contained in the Europe Collection, which led to the fundamental soundscape analysis Five Village Soundscapes. Later collections, from 1991 to1995 (Vancouver DAT Collection) and 2010 (Vancouver 2010 Collection), revisit those sites in Vancouver where recordings were made in the early collections and provide an excellent source for comparative analysis of changes in the soundscape in those locales. Vancouver recordings from each decade are located on an interactive Google map. Regarding tagging, each recording is assigned a subject classification and is accompanied by expert commentary.

15.6. Conclusion

As the number of audio recordings available from online sound databases grows, sound designers are the beneficiaries of the increasing availability and quality of these audio recordings. Although the list of sound databases examined in this chapter is not exhaustive, it highlights the different approaches to the state of the art of sound databases, their mechanisms and focus for soundscape composition. Similarly, the related work referenced by this chapter demonstrates the research and projects that have used sound databases as the foundation of their work.

References

Adams, E., 2009. *Fundamentals of Game Design*, 2nd edition. Thousand Oaks, CA: New Riders Publishing, chapter 10.

Adams, E. and Dormans, J., 2012. *Game Mechanics: Advanced Game Design*, 1st edition. Thousand Oaks, CA: New Riders Publishing.

Akkermans, V., Font, F., Funollet, J., de Jong, B., Roma, G., Togias, S. and Serra, X., 2011. Freesound 2: An Improved Platform for Sharing Audio Clips. In: *International Society for Music Information Retrieval Conference (ISMIR 2011)*. Late-Breaking Demo Session.

Audio Micro, n.d. Viewed Apr 12, 2012 www.audiomicro.com/

Brown, A., Kang, J. and Gjestland, T., 2011. Towards Standardization in Soundscape Preference Assessment. *Applied Acoustics*, 72, no. 6, pp. 387–392.

Cano, P., Fabig, L., Gouyon, F. and Loscos, A., 2004. Semi-Automatic Ambiance Generation. In: *Proceedings of 7th International Conference on Digital Audio Effects*, pp. 1–4.

Cano, P., Koppenberger, M., Herrera, P., Celma, O'. and Tarasov, V., 2004. Sound Effect Taxonomy Management in Production Environments. In: *25th AES International Conference*. Metadata for Audio.

Chion, M., 2009. *Film, a Sound Art, Film and Culture*. New York: Columbia University Press.

Collins, K., 2008. *Game Sound: An Introduction to the History, Theory, and Practice of Video Game Music and Sound Design*. Cambridge, MA: The MIT Press.

Correia, N.N., 2011. AV Clash, Online Audiovisual Project: A Case Study of Evaluation in New Media Art. In: *Proceedings of the 8th International Conference on Advances in Computer Entertainment Technology*. ACM, article no. 44, pp. 44:1–44:8.

Downie, S., Ehmann, A., Bay, M. and Jones, C., 2010. The Music Information Retrieval Evaluation Exchange: Some Observations and Insights. In: *Advances in Music Information Retrieval*, Vol. 274. Berlin and Heidelberg: Springer, pp. 93–115.

Dubois, D., Guastavino, C. and Raimbault, M., 2006. A Cognitive Approach to Urban Soundscapes: Using Verbal Data to Access Everyday Life Auditory Categories. *Acta Acustica United with Acustica*, 92, no. 6, pp. 865–874.

Eigenfeldt, A. and Pasquier, P., 2011. Negotiated Content: Generative Soundscape Composition by Autonomous Musical Agents in Coming Together: Freesound. In: *Proceedings of the Second International Conference on Computational Creativity*, pp. 27–32.

Finney, N. and Janer, J., 2010. Soundscape Generation for Virtual Environments Using Community-Provided Audio Databases. In: *W3C Workshop: Augmented Reality on the Web*, June 15. Barcelona, Spain.

Font, F., Roma, G., Herrera, P. and Serra, X., 2012. Characterization of the Freesound Online Community. In: *Third International Workshop on Cognitive Information Processing*.

Freesound, n.d. Viewed Apr 12, 2012 www.freesound.org/

Gaver, W.W., 1993. What in the World Do We Hear? an Ecological Approach to Auditory Event Perception. *Ecological Psychology*, 5, pp. 1–29.

Golder, S.A. and Huberman, B.A., 2005. The Structure of Collaborative Tagging Systems. *Journal of Information Science*, 32, no. 2, pp. 198–208.

Golder, S.A. and Huberman, B.A., 2006. Usage Patterns of Collaborative Tagging Systems. *Journal of Information Science*, 32, no. 2, pp. 198–208.

Gruber, T.R., 1993. A Translation Approach to Portable Ontology Specifications. *Knowledge Acquisition*, 5, no. 2, pp. 199–220.

Halpin, H., Robu, V. and Shepherd, H., 2007. The Complex Dynamics of Collaborative Tagging. In: *Proceedings of the 16th International Conference on World Wide Web*. ACM, pp. 211–220.

Harvard University, n.d.a. Copyright and Fair Use. Viewed Sept 17, 2012 www.ogc. harvard.edu/copyright docs/copyright and fair use.php

Harvard University, n.d.b. Copyright and Fair Use. Viewed Sept 17, 2012 www.appro pedia.org/Appropedia%27s

Holzer, D., n.d. Sound Transit. Viewed Apr 12, 2012 http://turbulence.org/soundtransit/ index.html

Hunicke, R., Leblanc, M. and Zubek, R., 2004. Mda: A formal Approach to Game Design and Game Research. In: *Proceedings of the Challenges in Games AI Workshop, Nineteenth National Conference of Artificial Intelligence*, pp. 1–5.

Janer, J., Kersten, S., Schirosa, M. and Roma, G., 2011. An Online Platform for Interactive Soundscapes with User-Contributed Audio Content. In: *Audio Engineering Society Conference: 41st International Conference: Audio for Games*.

Lamere, P., 2008. Social Tagging and Music Information Retrieval. *Journal of New Music Research*, 37, no. 2, pp. 101–114.

Lessig, L., 2004. The Creative Commons. *The Montana Law Review*, 65, pp. 1–13.

Marlow, C., Naaman, M., Boyd, D. and Davis, M., 2006. Ht06, Tagging Paper, Taxonomy, Flickr, Academic Article, to Read. In: *Proceedings of the 17th Conference on Hypertext and hypermedia*. HYPERTEXT '06. ACM, pp. 31–40.

Mobius, S., 2009. pdSounds. Viewed Apr 12, 2012 www.pdsounds.org/

Montreal Sound Map, n.d. Viewed Apr 12, 2012 www.montrealsoundmap.com/?lang=en

Murch, W., 2005. Dense Clarity Clear Density. *Transom Review*, 5, no. 1, pp. 7–23.

Organization, C.C., n.d. Creative Commons. Viewed Apr 12, 2012; visited on Sept 17, 2012 http://creativecommons.org/licenses/

Pedersen, T.H. and Zacharov, N., 2008. How Many Psycho-Acoustic Attributes Are Needed. *Journal of Acoustic Society of America*, 123, no. 5, p. 3163.

Radio, A., n.d. Viewed Apr 12, 2012 http://aporee.org/maps/

Roma, G., Herrera, P., Zanin, M., Toral, S.L., Font, F. and Serra, X., 2012. Small World Networks and Creativity in Audio Clip Sharing. *International Journal of Social Network Mining*, 1, no. 1, pp. 112–127.

Salen, K. and Zimmerman, E., 2004. *Rules of Play: Game Design Fundamentals*. Cambridge, MA: The MIT Press.

Schafer, R.M., 1977. *The Soundscape: Our Sonic Environment and the Tuning of the World*. Rochester, VT: Destiny Books.

Smith, G., 2008. *Tagging: People-Powered Metadata for the Social Web*. New Riders.

SoundCloud, n.d. Viewed Apr 12, 2012 www.soundcloud.com/

SoundDogs, n.d. Viewed Apr 12, 2012 http://sounddogs.com

SoundJax, n.d. Viewed Apr 12, 2012 www.soundjax.com/

Sounds Like Staten Island, n.d. Viewed Apr 12, 2012 http://soundslikestatenisland.com/ audio database

Sowa, J.F., 1987. Semantic Networks. In: *Encyclopedia of Artificial Intelligence*. New York: Wiley.

Stanza, n.d. Sound Cities. Viewed Apr 12, 2012 www.soundcities.com

Thorogood, M. and Pasquier, P., 2013. Computationally Generated Soundscapes with Audio Metaphor. In: *Proceedings of the 4th International Conference on Computational Creativity*, pp. 1–7.

Thorogood, M., Pasquier, P. and Eigenfeldt, A., 2012. Audio Metaphor: Audio Information Retrieval for Soundscape Composition. In: *Proceedings of the 6th Sound and Music Computing Conference*, pp. 372–378.

Truax, B., 1996. Soundscape, Acoustic Communication and Environmental Sound Composition. *Contemporary Music Review*, 15, no. 1–2, pp. 49–65.

Truax, B., 2015. World Soundscape Project Tape Library. Viewed Jan 12, 2015 www.sfu.ca/sonic-studio/srs/index2.html

van Dijck, J., 2009. Users Like You? Theorizing Agency in User-Generated Content. *Media Culture Society*, 31, no. 1, pp. 41–58.

Index

Note: numbers in *italics* indicate figures, and page numbers in **bold** indicate tables on the corresponding pages.

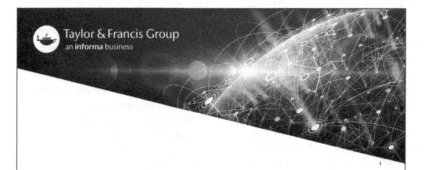